AF851237

SOLUTIONS AND APPLICATIONS OF SCATTERING, PROPAGATION, RADIATION AND EMISSION OF ELECTROMAGNETIC WAVES

Edited by **Ahmed Kishk**

Solutions and Applications of Scattering, Propagation, Radiation and Emission of Electromagnetic Waves
http://dx.doi.org/10.5772/3369
Edited by Ahmed Kishk

Contributors

Gregorio José Molina-Cuberos, María J. Núñez, Angel J. García-Collado, Jose Margineda, Ernesto Martín, Irismar Da Paz, Maria Carolina Nemes, Jose Geraldo Peixoto de Faria, Marcos D. R. Sampaio, Luis Cabral, José G. G. O. Junior, Saulo Moreira, Gennadiy Vorobyov, Aleksandr Rybalko, Ivan Barsuk, Larissa Vietzorreck, Ada Vittoria Bosisio, Yuri Alvarez Lopez, Fernando Las-Heras, Cebrián García González, Elena Vinogradova, Lyudmyla Velychko, Yuriy Sirenko, Jaime Laviada-Martinez, Marcos Rodríguez-Pino, Ismael Barba, Ana Grande, Ana Cristina López Cabeceira, Álvaro Gómez Gómez, José Represa, José Antonio Pereda, Hector Torres-Silva, Farhang Alijani, Nikolai Shul'Ga, Sergii Trofymenko, Lutz Angermann, Vasyl Yatsyk

Published by InTech

Janeza Trdine 9, 51000 Rijeka, Croatia

Copyright © 2012 InTech

All chapters are Open Access distributed under the Creative Commons Attribution 3.0 license, which allows users to download, copy and build upon published articles even for commercial purposes, as long as the author and publisher are properly credited, which ensures maximum dissemination and a wider impact of our publications. After this work has been published by InTech, authors have the right to republish it, in whole or part, in any publication of which they are the author, and to make other personal use of the work. Any republication, referencing or personal use of the work must explicitly identify the original source.

Notice

Statements and opinions expressed in the chapters are these of the individual contributors and not necessarily those of the editors or publisher. No responsibility is accepted for the accuracy of information contained in the published chapters. The publisher assumes no responsibility for any damage or injury to persons or property arising out of the use of any materials, instructions, methods or ideas contained in the book.

Publishing Process Manager Dragana Manestar

Technical Editor InTech DTP team

Cover InTech Design team

First published November, 2012

Printed in Croatia

A free online edition of this book is available at www.intechopen.com

Additional hard copies can be obtained from orders@intechopen.com

Solutions and Applications of Scattering, Propagation, Radiation and Emission of Electromagnetic Waves, Edited by Ahmed Kishk

p. cm.

ISBN 978-953-51-0838-2

We are IntechOpen,
the world's leading publisher of
Open Access books
Built by scientists, for scientists

4,000+
Open access books available

116,000+
International authors and editors

120M+
Downloads

151
Countries delivered to

Our authors are among the
Top 1%
most cited scientists

12.2%
Contributors from top 500 universities

WEB OF SCIENCE™

Selection of our books indexed in the Book Citation Index
in Web of Science™ Core Collection (BKCI)

Interested in publishing with us?
Contact book.department@intechopen.com

Numbers displayed above are based on latest data collected.
For more information visit www.intechopen.com

Meet the editor

Dr. Kishk is a Professor at Concordia University, Canada and Canada Research Chair in Advanced Antenna Systems. He is an Editor of Antennas & Propagation Magazine. His research interest are in the areas of electromagnetics, dielectric resonator antennas, printed antennas, RFID antennas for readers and tags, Multi-function antennas, microwave circuits, EBG, artificial magnetic conductors, soft/hard surfaces, phased array antennas, and computer aided design for antennas. He has published over 230-refereed Journal articles and 380 conference papers. He is a coauthor of four books and several book chapters and editor of one book. He offered several short courses in international conferences. He is a recipient of several awards. Two papers received outstanding paper awards from the Applied Computational Electromagnetic Society Journal. In addition, he received the 2004 Microwave Theory and Techniques Society Microwave Prize. Kishk is a Fellow of IEEE, Fellow of Electromagnetic Academy, and Fellow of Applied Computational Electromagnetic Society.

Contents

Preface

This book consists of four sections that are constructed from twelve chapters presented by several international specialized authors with physics and engineering background in different aspects of Electromagnetic waves. The reader may jump directly to the chapter of his interest as each chapter is almost self-contained. Wide range of different topics related to analytical as well as numerical solutions of problems related to scattering, propagation, radiation, and emission in different medium are discussed. Design of several devices and their measurements aspects are introduced. Topics related to microwave region as well as Terahertz and quasi-optical region are considered. Bi-isotropic metamaterial in optical region is investigated. Interesting numerical methods in frequency domain and time domain for scattering, radiation, forward as well as reverse problems and microwave imaging are summarized. Therefore, the book will satisfy different tastes for engineers interested for example in microwave engineering, antennas, and numerical methods.

Chapter 1 presents practical methods to characterize the electrical properties of chiral media. Detailed review of the propagation of electromagnetic waves in unbound chiral media is revised and theoretical relations between constitutive parameters and the scattering parameters are developed. A review for the different theoretical solutions, techniques and experimental setups developed in order to characterize chiral media is presented. Special attention to the retrieval algorithms is given.

Chapter 2 presents a short review on the chiral propagation of electron waves in monolayer graphene and optical simulation with optical field in the negative-zero-positive index metamaterial (NZPIM) and its close connection. The study verifies that the chiral field near the Dirac point is a diffusive wave.

Chapter 3 is a review and modifications to several numerical techniques in both frequency domain and time domain to deal with *Bi-isotropic* materials. The extension of the traditional methods is necessary to deal with the constitutive parameters of the bi-isotropic materials, as well as their inherently dispersive behavior. The extension is necessary especially when dealing with time-domain methods, in which the frequency dispersion must be translated to time-domain. Finite Differences Time Domain technique (FDTD), Transmission Line Matrix (TLM) and Multiresolution in Time Domain (MRTD) have been modified to include the frequency dispersion. In the frequency domain, an extended hybrid technique, based upon the Coupled Mode Method (CMM), which is originally formulated for closed structures and the mode matching method (MMM), are modified to deal with open structures, like free propagation inside a bi-isotropic medium. The MMM is included to characterize discontinuities in the power propagation direction.

Chapter 4 presents the key aspects of the analysis of an electromagnetic problem with characteristic basis functions (CBFs) that are a subtype of macrobasis functions. Thus, the generation of the CBFs and their inclusion the conventional method of moments is explained. In addition, several enhancements of the method such as the multilevel generation of CBFs are detailed. The most important applications, where the method becomes very useful, are covered. In addition, the recursive formulation for generating nested CBFs is shown in order to improve the compression of the reduced matrix to show that the reduced size of the system of equations enables the partial solution of electrically large geometries.

Chapter 5 presents the necessary theoretical background to the numerical solution of initial boundary value problems as applied to two-dimensionally periodic structures. Correct and efficient truncation of the computational space in the problems describing spatial-temporal electromagnetic wave transformation in two-dimensionally periodic structures is presented. Some important characteristics and properties of transient and steady-state fields in rectangular parts of the Floquet channel are discussed. The method of transformation operators is described to show that the computational resources can be optimized when calculating a multi-layered periodic structure or a structure on a thick substrate.

Chapter 6 presents a summary of the source reconstruction method (SRM)), an integral equation-based inverse technique, and its applications on antenna diagnostics and the recent advances on imaging applications. Implementation and technical details, as well as a compilation of SRM applications for antenna diagnostics and characterization, inverse scattering and imaging applications are presented.

Chapter 7 focuses on the waveguide directional couplers and gives finalized designing formulas and tables. A full generalized field theory for designing based on multi-hole coupling structure is introduced. A recursive procedure is provided to adjust the required directivity, coupling and frequency or waveguide type to provide the number of holes, individual diameters, and their locations besides coupler length, length of matched load and other sizes of structure for synthesis. Different methods such as binomial, Chebyshev are used to achieve the wide bandwidth response.

Chapter 8 presents an experimental setup and procedure for measuring electrodynamic characteristics of periodic metal and metal–dielectric structures, which can be used for practical devices operating in millimeter, sub millimeter, and terahertz wavelength ranges. A directional coupler quasioptical power dividers formed by grids of metal bars and a dielectric waveguide are considered. Peculiarities of electromagnetic characteristics of surface waves diffraction on the periodic heterogeneities are realized to create the sources for radiation on the Smith- Purcell effect.

In chapter 9 the transfer theory is applied to investigate propagation within an indoor environment to evaluate the performance of an indoor communication system. For simplicity, a two-dimensional environment is considered and the assumption that the propagation environment can be modelled as a layered parallel plane medium. This is a first approximation of an office indoor environment made of a succession of tables where each table is modelled as random medium containing randomly placed scatterers that are treated as infinite circular cylinder so as to simplify the solution of the radiative transfer equation.

Chapter 10 studies electromagnetic field space-time evolution and its spectral-angular characteristics in the processes of high-energy electron instantaneous scattering and its traverse of thin metallic plate. It involves the scrupulous analysis of evolution of packets of free electromagnetic waves by which the investigated field can be represented. Attention is drawn to the fact that the state of electron with nonequilibrium field can manifest itself not only in the case of ultra-high particle energies but also for electron energies of several tens Mev in the millimeter region of radiated waves. The problem of relativistic electron's transition radiation on thin metallic ideally conducting plate is considered onward. The space-time evolution of the electromagnetic field, which arises before and after passing of the electron of the metallic plate, is discussed. The main attention here is drawn to the effects in backward transition radiation in this process. The wave packets of the field reflected from the plate in this case are the packets of free electromagnetic waves, which gradually transform into transition radiation field.

Chapter 11 investigates the effects of weak fields at multiple frequencies on the scattering and generation of waves by an isotropic, non-magnetic, linearly polarized (E-polarisation), layered, cubically polarisable, dielectric structure. In the domain of resonance frequencies, wave packets are consisting of both strong electromagnetic fields at the excitation frequency of the nonlinear structure are considered leading to the generation of waves, and of weak fields at the multiple frequencies, which do not lead to the generation of harmonics but influence on the process of scattering and generation of waves by the nonlinear structure. The electromagnetic waves for a nonlinear layer with a cubic polarisability of the medium can be described by an infinite system of nonlinear boundary-value problems in the frequency domain. The effect of weak fields at multiple frequencies on the scattering and generation of waves is investigated numerically. The results indicate a possibility of designing a frequency multiplier and nonlinear dielectrics with controllable permittivity.

Chapter 12 presents theoretical aspects of the Gouy phase for classical electromagnetic waves. It shows how the phase can be measured from measuring quadratures, which are related by the determinant of a covariance matrix just in the way it is done in usual quantum mechanics. The geometrical construction is explicated and the relationship with matter waves is established.

Dr. Ahmed Kishk
University of Mississippi
USA

Section 1

Waves in Chiral Media

Chapter 1

Electromagnetic Characterization of Chiral Media

J. Margineda, G.J. Molina-Cuberos, M.J. Núñez,
A.J. García-Collado and E. Martín

Additional information is available at the end of the chapter

http://dx.doi.org/10.5772/51539

1. Introduction

The electromagnetic characterization of materials is a fundamental problem in many research areas of Electromagnetism. The goal is to determine the electrical permittivity and magnetic permeability, which make up the usual constitutive relations: $\vec{D}=\varepsilon_0\varepsilon_r\vec{E}$, $\vec{B}=\mu_0\mu_r\vec{H}$, where ε_r and μ_r are, for isotropic and homogeneous materials, two complex frequency-dependent parameters.

During recent decades, a great variety of novel and complex materials have been designed with promising practical applications. It has been found, however, that in many cases the relations between electric and magnetic fields cannot be described by standard constitutive equations. Among these new materials we may mention, for example, chiral, nonreciprocal, gyrotropic and negative refractive index media. Bi-isotropic media are the most general linear, homogeneous and isotropic materials, and they respond to electromagnetic excitation according to the following relations:

$$\begin{aligned} \vec{D} &= \varepsilon_0\varepsilon_r\,\vec{E}+\left(\chi - j\kappa\right)\sqrt{\varepsilon_0\mu_0}\,\vec{H} \\ \vec{B} &= \mu_0\mu_r\,\vec{H}+\left(\chi + j\kappa\right)\sqrt{\varepsilon_0\mu_0}\,\vec{E} \end{aligned} \tag{1}$$

where κ is the chirality, a dimensionless parameter for describing the handedness of the material, and χ is the Tellegen parameter, a dimensionless quantity for the degree of inherent nonreciprocity in the medium [1]. The reciprocal chiral medium, also known as Pasteur medium, is a class of bi-isotropic medium characterized by $\chi=0$. The electromagnetic behavior of these materials reflects two effects: electromagnetic rotatory dispersion, which causes a

INTECH
open science | open minds

© 2012 Margineda et al.; licensee InTech. This is an open access article distributed under the terms of the Creative Commons Attribution License (http://creativecommons.org/licenses/by/3.0), which permits unrestricted use, distribution, and reproduction in any medium, provided the original work is properly cited.

rotation of the polarization direction for a linearly polarized wave, and circular dichroism (a change in the polarization from linear to elliptical) due to the different absorption coefficients of right- and left-handed circularly polarized waves.

Electromagnetic activity at optical frequencies was first observed in the XIX century, [2, 3]. Although nature presents many examples of chiral media at optical frequencies, there is, to the best of our knowledge, no natural medium able to produce electromagnetic activity at microwave frequencies. At these frequencies, the first man-made chirals were manufactured by embedding a random distribution of conducting helices with the same handedness into a host medium. The first experimental studies were developed using basic free-wave systems with the aim of verifying electromagnetic activity at microwave frequencies. Lindman [4] and later Tinoco and Freeman [5] found a critical wavelength for copper helix at $\lambda \cong 2L$, where L is the length of the wire in the helix; this constitutes a rule of thumb to select the frequency range in which to search for interesting effects. By far, the most used structure for chiral manufacture has been the metallic helix [6-10], although other elements such as metallic cranks have also been used [11]. Manufacturing techniques have evolved from the random inclusion of metallic particles to the presently used alternatives, which are based on periodic distribution of planar or quasi-planar chiral particles printed on a circuit board or even integrated circuits. The interested reader can find a review in [12].

The ordered distribution of resonant structures in a periodic lattice enhances the electromagnetic effects and is able to produce larger electromagnetic activity and circular dichroism. These enhanced chirals are often known as chiral metamaterials. It has been found that a material with very high electromagnetic activity can also possess a negative refractive index [13]. During the last decade, a large variety of resonant structures have been designed and analyzed in order to produce high values of optical activity, circular dichroism and negative refractive index; for example, see [14-15] and references therein.

The presence of the chirality parameter in the constitutive equations prevents the use of standard characterization procedures. Furthermore, the existence of resonant structures, which finally produce strong variations in the characteristic parameters, makes the retrieval algorithms more complicated. In this paper, the main techniques for chiral characterization are reviewed. The second section is devoted to free-wave methods, which are the most widely used systems for chiral characterization. The guided methods presented in the third section, usually provide more accuracy than the free-wave methods, although the search for electromagnetic rotatory dispersion limits the guided systems to those of axial symmetry. In section four, resonator techniques are considered. Finally, the accuracy and feasibility of each technique is discussed.

2. Electromagnetic fields in chiral media

The constitutive relations (1) for reciprocal media ($\chi = 0$), are:

$$\begin{aligned} \vec{D} &= \varepsilon_0 \varepsilon_r \vec{E} - j\kappa / c \vec{H} \\ \vec{B} &= \mu_0 \mu_r \vec{H} + j\kappa / c \vec{E} \end{aligned} \tag{2}$$

where $c=(\varepsilon_0\mu_0)^{-1/2}$. The rotational Maxwell equations present a coupling between the electric and magnetic fields. In frequency domain and for the sourceless case, the curl of the electric and magnetic fields are:

$$\begin{aligned} \nabla \times \vec{E} &= -j\omega \left(\frac{j\kappa}{c} \vec{E} + \mu \vec{H} \right) \\ \nabla \times \vec{H} &= j\omega \left(\varepsilon \vec{E} - \frac{j\kappa}{c} \vec{H} \right) \end{aligned} \tag{3}$$

It is possible to find a linear combination of the E and H fields, called normal modes and written $\{\vec{E}^n, \vec{H}^n\}_{n=1,2}$, that satisfy independent rotational Maxwell equations so that:

$$\left\{ \begin{aligned} \nabla \times \vec{E}^n &= -j\omega\mu^n \vec{H}^n \\ \nabla \times \vec{H}^n &= j\omega\varepsilon^n \vec{E}^n \end{aligned} \right\}_{n=1,2} \tag{4}$$

These combinations of electromagnetic fields propagate through the chiral material as through an isotropic non-chiral medium, with equivalent constitutive parameters: $\{\varepsilon^n, \mu^n\}_{n=1,2}$.

The normal modes satisfying eq. (4) are:

$$\begin{aligned} \vec{E}^1 \equiv \vec{E}_+ &= \tfrac{1}{2}(\vec{E} - j\eta\vec{H}), \quad \vec{H}^1 \equiv \vec{H}_+ = \tfrac{j}{\eta_+}\vec{E}_+ \\ \vec{E}^2 \equiv \vec{E}_- &= \tfrac{1}{2}(\vec{E} + j\eta\vec{H}), \quad \vec{H}^2 \equiv \vec{H}_- = \tfrac{-j}{\eta_-}\vec{E}_- \end{aligned} \tag{5}$$

the medium parameters being:

$$\begin{aligned} \varepsilon_\pm &= \varepsilon(1 \pm \kappa_r) \\ \mu_\pm &= \mu(1 \pm \kappa_r) \end{aligned} \tag{6}$$

where $\kappa_r = \kappa / \sqrt{\varepsilon_r \mu_r}$ is the relative chirality, and:

$$\eta = \eta_{\pm} = \sqrt{\mu_{\pm} / \varepsilon_{\pm}} = \sqrt{\mu / \varepsilon} \tag{7}$$

2.1. Plane wave propagation in chiral media

A propagating wave can always be expressed as a linear combination of the propagation of normal modes. The propagation constant for each mode $\{\vec{k}_{+}, \vec{k}_{-}\}$ could be, in principle, different:

$$\begin{aligned} \vec{E}_{\pm}(\vec{r}) &= \vec{E}_{\pm} exp\left(-j\vec{k}_{\pm} \bullet \vec{r}\right) \\ \vec{H}_{\pm}(\vec{r}) &= \vec{H}_{\pm} exp\left(-j\vec{k}_{\pm} \bullet \vec{r}\right) \end{aligned} \tag{8}$$

If the wave is propagating in the $\hat{u}$ direction, the modes propagation constants are:

$$\vec{k}_{\pm} = \hat{u} k_{\pm} \tag{9}$$

where $k_{\pm} = \omega\sqrt{\varepsilon_{\pm}\mu_{\pm}} = \omega\sqrt{\varepsilon\mu}(1 \pm \kappa_r) = \omega\, n\,(1 \pm \kappa_r) / c$. The refraction indices for each mode are

$$n_{\pm} = n(1 \pm \kappa_r) \tag{10}$$

It can be deduced, from eq. (4, 5), that the vectors $\hat{u}$, $\vec{E}_{+}$ and $\vec{H}_{+}$ are mutually orthogonal. And the same happens to $\hat{u}$, $\vec{E}_{-}$ and $\vec{H}_{-}$. Also, according to eq. (5) the propagation modes are circularly polarized electromagnetic waves to the right (RCP) $\{\vec{E}_{+}, \vec{H}_{+}\}$ and to the left (LCP) $\{\vec{E}_{-}, \vec{H}_{-}\}$.

3. Free-wave

3.1. Experimental systems

For regular materials where ε and μ are the only parameters to be retrieved, the experimental determination of the complex scattering coefficients S_{11} and S_{21} are sufficient to solve the problem. But there remains an extra parameter to be determined, the chirality κ. To do this, we are forced to experimentally measure one complex or two real additional quantities. This is usually achieved in terms of the reflection coefficient and the two transmission coefficients corresponding to the parallel and perpendicular direction to the incident wave. Sun et al. [16] measured the complex reflection coefficients of short-circuit (backed with metal) and open-circuit (not backed with metal) samples. For transmission, they measured the rotation angle and the axial ratio.

The usual free-wave experimental setup consists of a pair of transmitting and receiving antennas, a sample holder, a vector network analyzer (to determine magnitude and phase of the scattering parameters) and a calibration kit. In order to minimize the diffraction effects at the edges of the sample it is necessary to decrease the size of the beam or place the sample close to the antennas. But if the antenna is close to the sample, the conditions of plane wave are not accomplished and the far-field retrieval algorithms cannot be applied. Moreover, the effects of multiple reflections between the antenna and the sample are enhanced, which introduces extra errors into the measurements. The size of the beam can be reduced by placing focusing devices, such us spot focusing lens in the antennas [17] or an ellipsoidal reflector mirror after the transmitting antenna [18]. Behind the focus, the wave is not plane, but takes the configuration of a Gaussian beam. In this case, the plane wave assumption does not introduce serious errors in the measurements [19].

Figure 1 shows a schematic diagram of the experimental system developed by Varadan et al. [6, 17, 20]. It makes use of a lens in each antenna to focus the wave; the sample is located in the common focal plane. The transmitting and receiving antennas are mounted on a carriage and the distance between them can be adjusted for calibration. This setup was initially developed by Ghodgaonkar [21] to measure permittivity and permeability, and later adapted to determine rotation angle, ellipticity and power absorption. Making use of this system, Guire et al. [20] measured the normal incidence reflection of linearly polarized waves of metal-backed chiral composite samples at microwave frequencies. Ro et al. [6], determined the ellipticity, rotation angle and power absorption for samples containing helix and observed that, at resonance, the angle of rotation rapidly increases or decreases or even changes its sign, a phenomenon also known as the Cotton effect. They also reported that this phenomenon is observed in the frequency region where maximum electromagnetic power absorption occurs.

Figures 2 and 3 show a picture and a schematic diagram, respectively, of the free-wave experimental system developed by the authors. The setup is based on a previous one for permittivity and permeability measurements [22], and adapted to measure electromagnetic activity [10]. This system has been used to characterize chiral materials based on random distribution of elements and periodical lattices [10-11, 23-25].

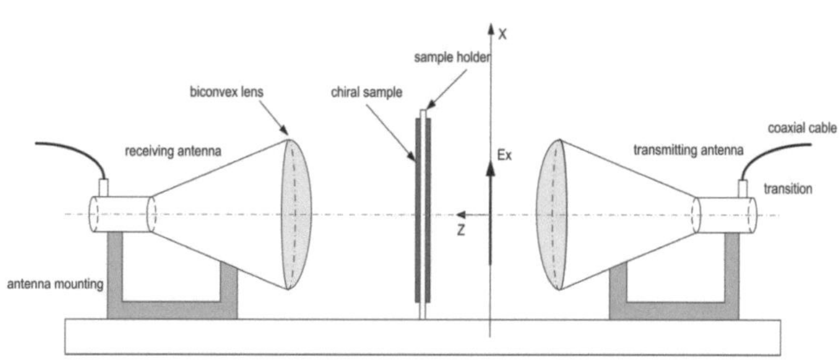

Figure 1. Setup for free-wave measurement, modified from [6, 17, 21].

The incident beam is focused by an ellipsoidal concave mirror, with the transmitting antenna placed at one of the mirror foci and the sample at the other. The position of the receiving antenna is not crucial in this configuration. Figure 2 shows a diagram where the focusing mirror and sample holder are not shown for clarity. The incident wave is linearly polarized in the x-axis direction. The rotation on the polarization angle, θ, is defined as the difference between the polarization direction of the incident wave and the direction of the major axis of the transmitted elliptically polarized wave.

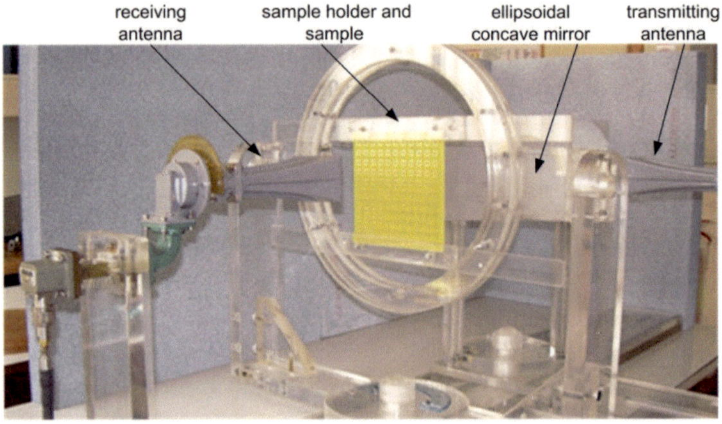

Figure 2. Photo of the free-wave setup in the X-band developed by the authors.

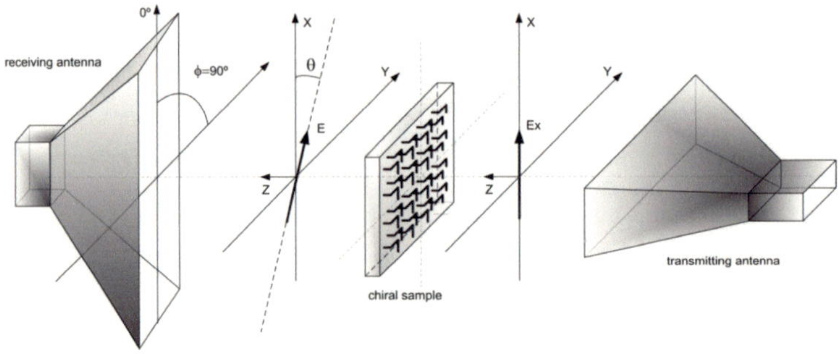

Figure 3. Schematic diagram of figure (2), where the receiving antenna with a rotation angle of 90° to measure T_{CR}. The focusing mirror and sample holder are not shown for clarity.

The experimental system presented by Wang et al. [14] to characterize chiral metamaterials with high electromagnetic activity makes use of a pair of standard gain horn antennas, which are located close to the sample, without any focusing device. This configuration leads to multiple reflections and border diffractions, which affect the experimental reflection and transmission coefficients.

The scattering parameters are measured using a vector network analyzer. First, a two port calibration should be made, for which purpose the TRL (Transmission, Reflection, Line) technique is one of the most frequently used. The basic TRL calibration process is a two port calibration widely used in non-coaxial environments, such as waveguide and free-wave systems[1]. TRL uses a 12-term error model with three calibration standards: THRU, REFLECT and LOAD. For the THRU step, the test ports are connected and the four S-parameters are measured in both directions. For the REFLECT step, a highly reflective device is connected to each port and the reflection coefficients are measured. In the LINE step, a short transmission line is inserted and, again, the four S-parameters are measured.

The reference planes of the calibration can be located at the sample surface [17] or elsewhere, for example, at the two waveguide flanges where the antennas are connected [22]. In the first case, Ghodgaonkar et al. [17, 21] implemented the TRL calibration for free-wave systems: The THRU standard was configured by keeping the distance between the two antennas equal to twice the focal distance, without the sample. The LINE standard was achieved by separating the focal plane of transmitting and receiving antennas by a distance equal to a quarter of the free space wavelength at the center of the band. The REFLECT standards for both ports were obtained by placing a metal plate in the focal planes of the transmitting and receiving antenna, respectively.

The second option is the one used by the authors. In this case, it is possible to use standards for calibration in rectangular waveguide: a short-circuit is located at the antenna flanges for the REFLECT standard and a rectangular guide of $\lambda/4$ length for the LINE standard. A second calibration is necessary to shift the reference planes to the sample surface. To do that, the reflection coefficient, with a metal plate in sample position, and the transmission coefficient without the sample were used as reference measurements. Then, the ratio between the measurement with and without sample provides the magnitude of the scattering parameters of the sample. The phase of the reflection coefficient must be corrected by a π factor introduced by metal plate. For the phase of the transmission coefficient, it is necessary to subtract a quantity equals to$k_0 L$, where k_0 is the vacuum propagation constant and L is the sample width.

At first sight, the second TRL calibration option seems more complicated; however, it has the advantage of using LINE and THRU standards in rectangular waveguide, which are more accurate and easier to use than standards obtained by moving the antennas.

A time domain (TD) transform can be used to filter out mismatches from the antennas, edge diffraction effects and unwanted reflections from antennas, sample, mirrors or elsewhere. The network analyzer makes measurements in the frequency domain, and the time domain response is obtained by applying an inverse Fourier transform. The unwanted signals, which have longer paths than the direct signal, are separated from the main signal in the TD response and then can be removed by gating the signal. In order to obtain good filtering, it is necessary to clearly identify the sample response in the time domain and to choose the most suitable filter parameters [26]. The Fourier transform of the gated time-domain response provides the filtered signal. As will be shown later, this gating process greatly improves the experimental results.

1 Agilent Technologies Inc. Applying Error Correction to Network Analyzer Measurements, Application Note 2002, 1287-3, 5965-7709E.

3.2. Retrieval algorithm

The standard retrieval procedure for a normal medium relates the transmission and reflection coefficients with the electromagnetic properties of the medium. As mentioned above, for a chiral material a new experimental quantity is necessary. Here reflection coefficient, *R*, and the transmission coefficients parallel (co-polar, T_{CO}) and perpendicular (cross-polar, T_{CR}) to the polarization direction of the incident wave are the experimental data.

Consider a chiral slab of thickness *L*, which is illuminated by a normally incident linearly polarized plane wave $E_0\hat{x}$, Figure 4.

The incident wave may be decomposed as:

$$E_0\hat{x} = \frac{E_0}{2}(\hat{x} - j\hat{y}) + \frac{E_0}{2}(\hat{x} + j\hat{y}) \tag{11}$$

The reflected wave in *z=0* is $RE_0\hat{x}$, the co-polar and cross-polar transmitted fields in *z=L* are $T_{CO}E_0\hat{x}$ and $T_{CR}E_0\hat{y}$, respectively.

The reflected and transmitted electric fields are,

$$\vec{E}_r = \frac{1}{2}R_R E_0(\hat{x} - j\hat{y}) + \frac{1}{2}R_L\ E_0(\hat{x} + j\hat{y}), \tag{12}$$

$$\vec{E}_t = \frac{1}{2}T_R E_0(\hat{x} - j\hat{y}) + \frac{1}{2}T_L\ E_0(\hat{x} + j\hat{y}), \tag{13}$$

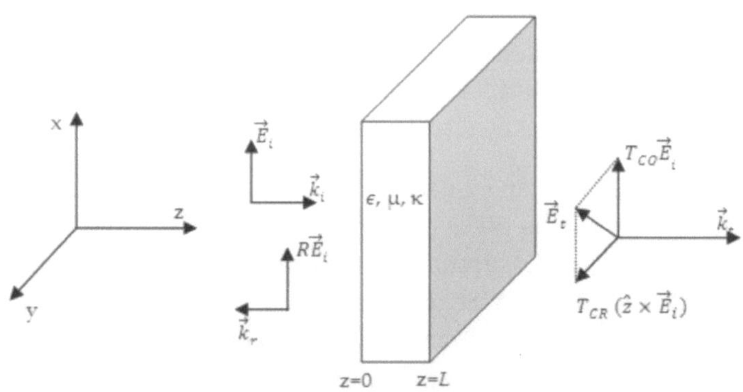

Figure 4. Schematic diagram of the experimental transmission and reflection coefficients and field of a chiral slab, where $\vec{E}_i$, $\vec{E}_r$ and $\vec{E}_t$ are the incident, reflected and transmitted fields, respectively.

where the subscript *R* (*L*) refers to right-hand (left-hand) circular polarization and subscript r to the reflected wave. The magnetic field can be expressed in a similar way in terms of wave impedance.

Applying the condition of continuity of tangential electric and magnetic fields at *z=0* and *z=L*, the reflection coefficients for RCP and LCP waves can be expressed as:

$$R_R = R_L = -\frac{j}{2}\frac{\sin(nk_0L)(1-\eta_r^2)}{2\eta_r\cos(nk_0L) + j(\eta_r^2+1)\sin(nk_0L)} \tag{14}$$

The experimental reflection coefficient is $R = R_R + R_L$. The transmission coefficients for the RCP and LCP waves are

$$T_R = \frac{2\eta_r e^{-j\kappa k_0 L}}{2\eta_r\cos(nk_0L) + j(\eta_r^2+1)\sin(nk_0L)} \tag{15}$$

$$T_L = \frac{2\eta_r e^{+j\kappa k_0 L}}{2\eta_r\cos(nk_0L) + j(\eta_r^2+1)\sin(nk_0L)} \tag{16}$$

These are related with the experimental transmission coefficients T_{CO} and T_{CR} by:

$$T_{CO} = \frac{1}{2}(T_R + T_L) \tag{17}$$

$$T_{CR} = \frac{-j}{2}(T_R - T_L) \tag{18}$$

It is useful to consider the following additional parameters:

$$T^2 \equiv T_R T_L \tag{19}$$

$$\Gamma \equiv -\frac{(T^2 - R^2 - 1) \pm \sqrt{(T^2 - R^2 - 1)^2 - 4R^2}}{2R} = \frac{\eta_r^2 - 1}{(\eta_r + 1)^2}, \tag{20}$$

$$P^2 = e^{-2jnk_0L} = \frac{R - \Gamma}{R\Gamma^2 - \Gamma} \tag{21}$$

We are now ready to compute the electromagnetic parameters. The relative impedance η_r is calculated from equation (20), where the correct sign in square root is selected according to $\mathrm{Re}(\eta) > 0$.

The refractive index is obtained from equation (21) and is:

$$n = j\frac{c}{4\pi fL}\ln|P^2| - \mathrm{Phase}(P^2)\frac{c}{4\pi fL} - \frac{c}{2fL}m \tag{22}$$

where *m* is an integer related to the branch index of the logarithm function. Only the imaginary part of *n* is unequivocally determined, but ambiguity arises for the real part because an infinite number of solutions exist for different choices of m. The value of m can be determined by applying Kramers-Kronig relations to the wavenumber [28] or by expanding equation (22) in a Taylor series [29].

From nand η_r permittivity and permeability can be obtained:

$$\varepsilon_r = \frac{n}{\eta_r} \tag{23}$$

$$\mu_r = n\,\eta_r \tag{24}$$

The chirality parameter is obtained from T_L andT:

$$\frac{T_L}{T} = e^{\,j(\kappa k_0 L\,)}\ \kappa = \frac{-j}{k_0 L} ln\left|\frac{T_L}{T}\right| + \frac{1}{k_0 L} Phase\left(\frac{T_L}{T}\right) + \frac{2\pi}{k_0 L} p \tag{25}$$

where $p$is an integer that can be determined by assuming null chirality far from resonance, continuity conditions, and that both real and imaginary parts resonate at the same frequency.

The retrieval procedure may fail and problems arise when the thickness L of the effective slab is not estimated accurately [30], when S_{11} and S_{21} are very small in magnitude [31] or when the first boundary of the effective homogeneous slab is not well determined [29]

The rotation angle θand the ellipticity φof the transmitted wave can be calculated by:

$$\theta = \frac{1}{2} arg\left(\frac{T_R}{T_L}\right), \tag{26}$$

$$\varphi = \frac{1}{2} tan^{-1}\left(\frac{|T_R|^2 - |T_L|^2}{|T_R|^2 + |T_L|^2}\right) \tag{27}$$

3.3. Results

As already mentioned, periodical structures provide certain advantages over random distributions: for example, they are easier to manufacture and are more homogeneous and they present extreme values for the rotation angle and chirality parameter. Figure 5 shows the reflection and transmission coefficients, and the retrieved parameters ε_r, μ_r, andκ_r, for a typical chiral slab composed of a periodical lattice of conductive cranks printed on a circuit board substrate. The dimensions of the crank are similar to those proposed by [23, 24]. Experimental measurements without and with TD filtering are shown in the first and second columns, respectively. The third column shows the results obtained through simulations based on the finite-difference time-domain method (FDTD). It can be seen that the TD filtering eliminates ripples in the measured transmission and reflection coefficients, which are usually associated with multiple reflections or residual post calibration errors.

A clear resonance at around 11 GHz is present in all the plots, both measured and simulated. Below the resonance, the retrieved parameters correspond to the blank FR4 board, which does not present electromagnetic activity (ε_r~4.2, μ_r~1.0). As the frequency approaches the resonance, all the parameters increase, with a maximum chirality in the order ofκ~2.5. At resonance, the chirality changes the sign and its imaginary part peaks. The frequency dependence of permittivity and permeability observed in the plot are the same as those in metamaterials with a negative refractive index [14, 15, 24, 25]. The chirality parameter depends on frequency, which follows a Condon model for a homogeneous chiral medium [32].

There is good agreement between the retrieved parameters obtained from experiments with TD filtering and simulated data. It is important to highlight that the correct selection of TD parameters is crucial for obtaining good results in the retrieval process. The values of the time width (span) of the gating and the window shape affect the final results and, in this respect, the use of experimental and numerical data improves the calculations. Two propagation modes with two different velocities inside the chiral slab can be observed in the signal received by the receiving antenna when TD is applied, specially when the TCR coefficient is measured. Two Gaussian curves arrive at two different times, which implies a time span greater than that used for standard dielectric material.

4. Waveguides

4.1. Experimental systems

Waveguide techniques for characterizing chiral media are based on the same principles as free-wave techniques, and it is necessary to measure three complex quantities. The experimental setup, therefore, must allow measurements in, at least, two different polarizations for the transmission wave. All the experimental systems reported in the literature [7-9, 33-35] use circular waveguides because in rectangular waveguides the effects related to rotatory dispersion cannot be observed. Lubkowski [36] proposed a square waveguide for measurements with SRR (Split Ring Resonators). There, the orthogonal modes TE_{10} and TE_{01} are degenerated and, therefore, allow propagation for co-polar and cross-polar polarization. This could be a good alternative for measurement involving chiral materials but, to the best of our knowledge, such a system has not been implemented experimentally.

Figures 6 and 7 show a picture and a schematic diagram, respectively, of the experimental setup using a circular waveguide developed by the authors. It is similar to that used by Brewitt-Taylor [7] et al. and Liu et al. [34]. The circular measurement cell containing the chiral sample (2 in Figures 6, 7) is fed with a linearly polarized wave through a rectangular-to-circular waveguide transition (1 in Figures 6, 7). The transmission wave passes to the rectangular waveguide through another transition which can rotate around the longitudinal axis (3 in Figures 6, 7), enabling measurements in different polarizations. The waveguide transition contains a section of resistive film to absorb any field cross-polarized from the fields being measured, which prevents them from being reflected from the waveguide transition back on to the sample.

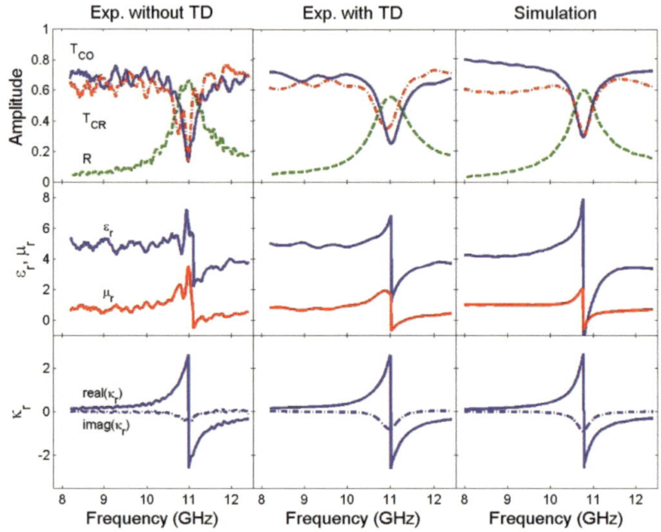

Figure 5. Transmission and reflection amplitudes (top panel), retrieved values for permittivity and permeability (middle panel) and chirality (bottom panel). Experimental results without time domain filtering are shown in the left column, with time domain filtering in the middle and simulation on the right.

S-parameters are measured using a network analyzer and the TRL technique is used for calibration as described for free-wave measurements. The REFLECT standard is achieved by shorting with a metal plate the flanges where the measurement cell is connected, which defines the reference planes. The TRHU standard is achieved simply by connecting directly these flanges without the measurement cell. Finally, the LINE standard is a section of the circular waveguide quarter wavelength at midband.

Hollinger et al. [33] measured some effects of chiral materials including rotatory dispersion and circular dichroism, but they did not study the inversion procedure. Liu et al. [9, 34] used a slotted line instead of the network analyzer to measure the reflection coefficients of short-circuited and open-circuited samples, completing the experimental measurements with the rotation angle and axial ratio. Reinert et al. [8, 35] used coaxial-to-circular waveguide transition to excite the fundamental mode in circular waveguide TE_{11}, avoiding rectangular-to-circular transitions.

Two aspects that can produce important errors must be taken into account in waveguide measurements: the presence of local density variations in the sample, which causes fluctuations to the measurements, depending on sample orientation, and the existence of gaps between the sample and the waveguide walls. In order to improve the accuracy of the results, the samples are rotated and the measured scattering parameters for several positions are averaged in an attempt to overcome the sample variations and inhomogeneities [8, 9, 33, 34]. The influence of the gap depends on sample manufacturing and can be overcome, for instance, by locating the chiral particles in non-conducting spheres [8], since the multiple air gaps between the spheres are part of the sample, or by using an elastic material as dielectric host for distributing the chiral particles [12].

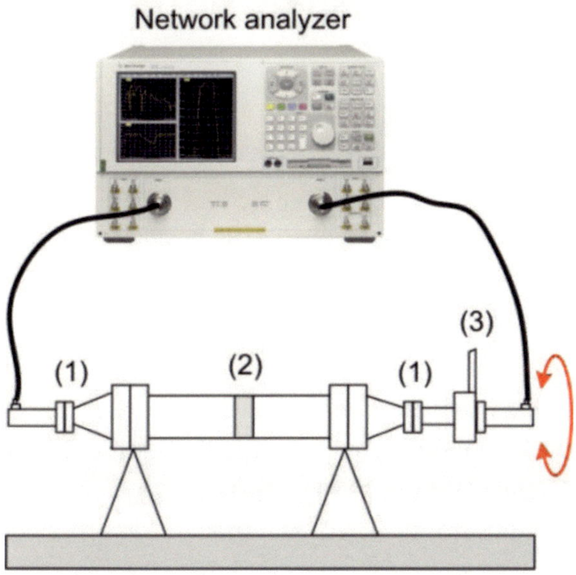

Figure 6. Schematic diagram of the X-band circular waveguide experimental system developed by the authors.

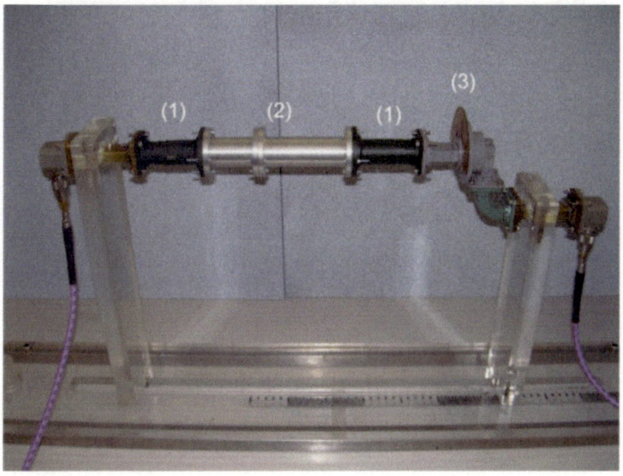

Figure 7. Photo of the experimental diagram of Figure 6.

4.2. Modes in circular waveguides

The propagation of electromagnetic fields in circular waveguides filled with chiral media, also known as chirowaveguides, can be better understood by using the normal modes formulation. Assuming, as usual, solutions of the type $e^{-\gamma z}$, the longitudinal component of the electric field satisfies the Helmholtz equation [37],

$$\left(\nabla_t^2 + K_\pm^2\right)E_{\pm z} = 0 \tag{28}$$

where

$$K_\pm = \left(k_\pm^2 - \gamma^2\right)^{1/2} \tag{29}$$

Once $E_{\pm z}$ are known, the transversal components can be obtained by:

$$E_t(\vec{r}) = \frac{1}{K_+^2}\left(-\gamma\nabla_t - k_+\hat{z}\times\nabla_t\right)E_{+z} + \frac{1}{K_-^2}\left(-\gamma\nabla_t - k_-\hat{z}\times\nabla_t\right)E_{-z} \tag{30}$$

Similar expressions for the magnetic components can be deduced. Assuming an azimuthal dependence as $e^{-jm\varphi}$, the longitudinal component of the electric field is a linear combination of the longitudinal components of the normal modes:

$$E_z(\vec{r}) = \sum_{m=-\infty}^{\infty}\sum_{l=1}^{\infty}\left\{A_{m,l,+}J_m(K_{l,+}\rho) + A_{m,l,-}J_m(K_{l,-}\rho)\right\}e^{-jm\varphi}e^{-\gamma_l z} \tag{31}$$

where classification of the normal modes is described by $\{m, l\}$ indices. $A_{m,l,\pm}$ is the amplitude of the mode, J_m is the m order Bessel function and J'_m its derivate. If the boundary is ideally conducting, the following eigenvalue equation is obtained:

$$\frac{m\gamma_l}{a}\left(\frac{1}{K_{l,+}^2} - \frac{1}{K_{l,-}^2}\right) + \frac{k_+}{K_{l,+}^2}\frac{J'_m(K_{l,+}a)}{J_m(K_{l,+}a)} + \frac{k_-}{K_{l,-}^2}\frac{J'_m(K_{l,-}a)}{J_m(K_{l,-}a)} = 0 \tag{32}$$

where a is the waveguide radius. For a pair of solutions satisfying $\gamma_{l'} = -\gamma_l$ and $m' = -m$, the parameters K_+ K_- are coincident. The wave number is, in general, a complex quantity $\gamma_l = \alpha_l + j\beta_l$.

Figure 8 shows the dispersion diagram of a circular waveguide for X band (a=1.19cm) filled with a chiral material ($\varepsilon_r = 3.0$, $\mu_r = 1.0$, $\kappa_r = 0.8$) for $m = \pm 1$. This diagram is, clearly, much more complex than the one corresponding to a non-chiral waveguide. A detailed discussion on the modal expansion can be found in [37].

4.3. Retrieval algorithm

When several modes propagate in the chirowaveguide it is not possible to obtain inverse equations. Only if just one mode propagates can a simple retrieval algorithm be developed. However, this is only valid for low values of the chirality parameter.

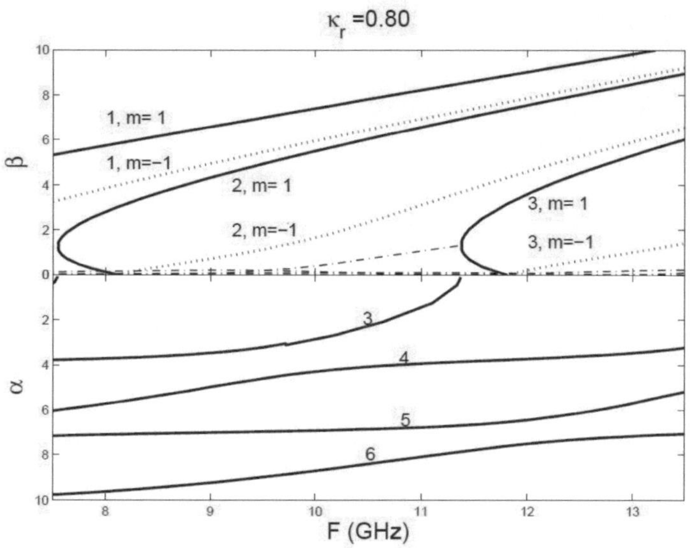

Figure 8. Dispersion diagram of chirowaveguide (1.19 cm radius, $\varepsilon_r = 3.0$, $\mu_r = 1.0, \kappa_r = 0.8$), for $m = 1$(solid line), $m = -1$ (dotted line). The propagation constant at frequencies below the cut-off is shown by dash-dot lines.

Liu et al. [9, 34] assumed that only the fundamental mode, $HE_{\pm 11}$ ($m = \pm 1, j = 1$), propagates in the sample, and this is approximately treated in the same way as the TE_{11} mode, the fundamental mode in the empty waveguide. In this case, the procedure for obtaining the inversion equations is similar to the one used in the free-wave technique. The above mentioned authors used the following experimental quantities: $\Gamma_1$andΓ_2, the complex reflection coefficients of the open circuit sample (sample in transmission situation) and the short circuit sample (sample backed with metal plate) in the circular waveguide, respectively, $\theta$is the rotation angle and R_A is the axial ratio, for the electromagnetic waves travelling through chiral material in the circular waveguide.

The characteristic parameters of the material are calculated from the measured quantities by:

$$Z_0' = \frac{\omega \mu_0}{\left[\omega^2 \varepsilon_0 \mu_0 - k_c^2\right]^{1/2}} \tag{33}$$

$$Z_\omega = Z_0' \left(\frac{\Gamma_1 \Gamma_2 + \Gamma_1 + \Gamma_2 + 1}{\Gamma_1 \Gamma_2 - 3\Gamma_1 + \Gamma_2 + 1} \right)^{1/2} \tag{34}$$

$$A = \frac{1}{2d} arccos \left(\frac{\Gamma_1 \Gamma_2^2 + \Gamma_2^2 - 2\Gamma_1 \Gamma_2 - \Gamma_1 + 1}{2(\Gamma_1 - \Gamma_2)} \right) \tag{35}$$

$$B = \frac{1}{2d}\left(2\theta - j\tanh^{-1}\frac{2R_A}{1 + R_A^2}\right) \tag{36}$$

$$\mu = \frac{Z_\omega A}{\omega} \tag{37}$$

$$\kappa = \frac{B\left(A^2 + k_c^2\right)^{1/2}}{a\mu\omega} \tag{38}$$

$$\varepsilon = \frac{\left(A^2 - B^2\right)\mu\kappa^2}{B^2} \tag{39}$$

where d is the thickness of the chiral sample and k_c the cut-off constant for the fundamental mode TE_{11} in the empty waveguide.

These equations are relatively easy to use but are only valid for media with small values of the characteristic parameters. For example, Figure 8 shows two propagating modes in the X-Band for a chirality relative constant κ_r=0.8, and, so, the previous retrieval algorithm would not be useful in that case. Other authors consider more modes. For instance, Busse et al. [8, 35] used five modes but could not find inversion equations to calculate the permittivity, permeability and chirality of the medium. Instead, they processed the direct problem numerically, in an iterative way, to fit measured quantities: the reflection coefficient R, the co-polar transmission coefficient T_{CO}, and the cross-polar transmission coefficient T_{CR}. Using the expression for waveguide modes in the empty and filled regions and starting with approximate values for permittivity, permeability and chirality, the material properties are obtained by iterative fitting of the calculated S parameters to the experimental ones.

Brewitt-Taylor et al. [7] followed a similar procedure to calculate the characteristic parameters. More specifically, they used four modes and eight different points to perform the modal analysis in the interfaces. In this case, mode matching was performed at the two interfaces simultaneously. The number of iterations required varied from 5 to 30, although the authors admit that high chirality samples could require more iterations.

Both Reinert et al. and Brewitt-Taylor et al. established the number of modes empirically, this number increasing with the values of the characteristic parameters. For instance, the samples used by Brewitt-Taylor et al. presented maximum values of 4 for the permittivity, 0.5 for the chirality and around 1 for the permeability, while the corresponding rotation angle was always less than 20º. Samples manufactured by the authors present values of up to 7 for permittivity, and 2 for chirality and permeability (see Figure 5), finding rotation angles in the range of 90º (Figure 9). The number of modes which should be used is, probably, more than five, increasing the computation requirements.

http://dx.doi.org/10.5772/51539

5. Resonators

Some authors propose using resonator techniques to characterize chiral media [38, 39]. Such techniques are usually the most suitable method for obtaining high precision measurements. However, there are several reasons why they are difficult to use with chiral media. In particular, resonators work at discrete frequencies this is not useful to deal with dispersive media; only two real quantities (resonant frequency and quality factor) can be measured and, as mentioned above, three complex quantities have to be measured to extract permittivity, permeability and chirality. Measurements using different resonant modes and samples in several positions is a possibility, but, in this case, the experimental procedure becomes very complicated.

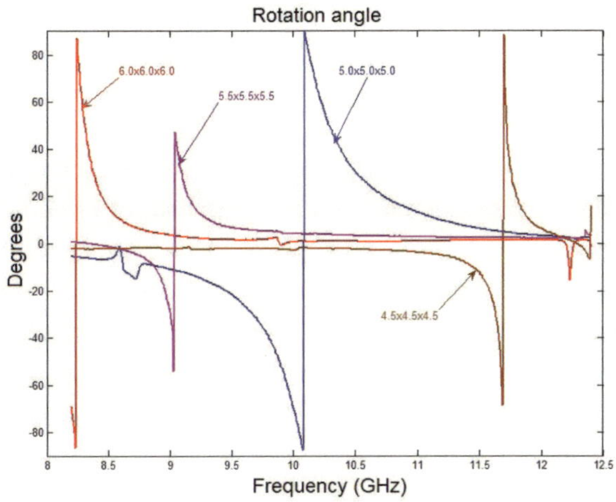

Figure 9. Rotation angle produced by several samples composed of four cranks in foam [12]. Labels indicate the size (in mm) of the cranks.

Tretyakov et al. [38, 39] proposed a perturbation technique, although this gives rise to a new problem. It is well known that perturbation techniques are only valid for small samples. But small samples of chiral media will have a small number of inclusions, perhaps only one, and so it makes no sense to treate them as a homogeneous material.

In our opinion, resonant techniques are not suitable for characterizing chiral media. In fact, although Tretyakov et al. studied the proposed technique theoretically, none was implemented experimentally.

6. Discussion and conclusions

In this work, microwave techniques for characterizing chiral media have been revised, and free-wave, waveguide and resonator techniques have been analyzed. Since chiral materials are characterized by three complex parameters, permittivity, permeability and chirality, at least, three experimental quantities have to be measured. All the techniques considered present some difficulties, while free-wave techniques are, in our opinion, the most suitable: i) They do not need any special component. ii) Quasi-plane wave propagation can be assumed in the sample, so that inversion equations for extracting characteristic parameters can be developed without difficulty. Usually, the measured quantities are the reflection coefficient, the co-polar and cross-polar transmission coefficients.

Free-wave methods present an inherent problem due to the finite size of the sample, which should be large enough to avoid diffraction at the borders. Several techniques found in the literature overcome this problem by using focusing systems, lenses or ellipsoidal mirrors. They concentrate the propagated wave at a tiny spot, where the sample is situated and so reasonably sized samples can be used. Many authors use time domain transformations to remove unwanted reflections from the experimental measurements.

Waveguide measurements may be considered a possible alternative. The measuring principle is similar to that of free-wave techniques, but they present the advantage of avoiding many kinds of external interference and use only small samples, although good machining is needed. However, they present one serious problem: usually several modes are propagated in the sample and inversion equations cannot be developed. All the techniques found in the literature use a circular waveguide because axial symmetry is necessary for measurements in different polarizations (only waves with polarization parallel to the short side can be propagated in rectangular waveguide). Unfortunately many circular waveguide components are not commercially available. A common solution is to use a rectangular waveguide and to feed the measurement cell, made in the circular waveguide, by means of two rectangular to circular waveguide transitions. The transition contains a resistive film to absorb any fields cross-polarized from those being measured. Square waveguides could, in principle, be another alternative for chiral characterization, although we have not found any theoretical nor experimental study in the current literature.

Finally, it should be emphasized that resonator techniques are not suitable for characterizing chiral media. Only one research group has proposed a resonator technique in a theoretical study without experimental implementation. In fact, resonators work at a fixed frequency and are therefore not useful for measurements in dispersive media.

Acknowledgements

This work was supported by the Dirección General de Investigación (TEC 2010-21496-C03-02) of the Spanish Ministry of Education and Science and by Fundación Séneca (11844/PI/09) Región de Murcia (Spain).

Author details

J. Margineda[1*], G.J. Molina-Cuberos[1], M.J. Núñez[1], A.J. García-Collado[1,2*] and E. Martín[1]

1 Universidad de Murcia, Murcia, Spain

2 Universidad Católica San Antonio, Murcia, Spain

References

[1] Lindell, I. V., Sihvola, A. H., Tretyakov, S. A., & Vitanen, A. J. (1994). Electro-magnetic Waves in Chiral Media, Boston, USA: Artech House.

[2] Arago, D. F. (1811). Mémoire sur une Modification Remarquable Qu'éprouvent les Rayons Lumineux dans leur Passage a Travers Certains Corps Diaphanes, et sur Quelques Autres Nouveaux Phénoménes D'optique*Mémoires de la Classe des Sciences Mathématiques et Physiques de l'Institut Impérial de France*, 1-93.

[3] Biot, B. (1812). Mémoire sur un Nouveau Genre D`Oscillation que les Molécules de la Lumiére Éprouvent en Traversant Certains Cristaux. *Mémoires de la Classe des Sciences Mathématiques et Physiques de l'Institut Impérial de France*, 1.

[4] Lindman, K. F. (1914). Om en Genom ett Isotropt System av Spiralformiga Resonatorer Alstrad Rotationspolarisation av de Elektromagnetiska Vagorna. Öfversigt af Finska Vetenskaps-Societetens förhandlingar, A. *Matematik och naturvetenskaper* LVII (3) 1-32.

[5] Tinoco, I., & Freeman, M. P. (1957). The Optical Activity of Oriented Copper Helices. I. Experimental, *J Physical Chemistry*, 61-1196.

[6] Ro, R., Varadan, V. V., & Varadan, V. K. (1992). Electromagnetic Activity and Absorption in Microwave Composites, *Proc. Inst. Elect. Eng.* H, 139(5), 441-448.

[7] Brewitt-Taylor, C. R., Lederer, P. G., Smith, F. C., & Haq, S. (1999). Measurement and Prediction of Helix-Loaded Chiral Composites, *IEEE Transactions on Antennas and Propagation*, 47(4), 692-700.

[8] Busse, G., Reinert, J., & Jacob, A. F. (1999). Waveguide Characterization of Chiral Material: Experiments. *IEEE Transactions on Microwave Theory and Techniques*, 47(3), 297-301.

[9] Sun, G., Yao, K., & Liu, Z. (1999). Influence of Helix Parameters on the Reflectance of Microwave by Chiral Composites, *IEEE Transactions on Electromagnetic Compatibility*, 411-350.

[10] Gómez, A., Lakhtakia, A., Margineda, J., Molina-Cuberos, G. J., Núñez, M. J., Ipiña, S. A., & Vegas, A. (2008). Full-Wave Hybrid Technique for 3-D Isotropic-Chiral-Ma-

terial Discontinuities in Rectangular Waveguides: Theory and Experiment, *IEEE Transactions on Microwave Theory and Techniques*, 56(12), 2815-2824.

[11] Molina-Cuberos, G. J., García-Collado, A. J., Margineda, J., Núñez, M., & Martín, E. (2009). Electromagnetic Activity of Chiral Media Based on Crank Inclusions, *IEEE Microwave and Wireless Components Letters*, 19(5), 278-280.

[12] Barba, I., Cabeceira, A. C. L., García-Collado, A. J., Molina-Cubero, G. J., Margineda, J., & Represa, J. (2011). Quasi-planar Chiral Materials for Microwave Frequencies. In: Kishk A. (ed.) *Electromagnetic Waves Propagation in Complex Matter, Intech*, 978-9-53307-445-0, http://www.intechopen.com/books/electromagnetic-waves-propagation-in-complex-matter.

[13] Pendry, J. B. (2004). A Chiral Route to Negative Refraction, *Science* 306 1353--1355.

[14] Wang, B., Zhou, J., Koschny, T., Kafesaki, M., & Soukoulis, C. M. (2009). Chiral Metamaterials: Simulations and Experiments, *J. Opt. A: Pure Appli. Opt.* 11: 114003.

[15] Mackay, T. G., & Lakhtakia, A. (2012). Negatively Refracting Chiral Metamaterials: *A Review, SPIE Reviews* 1: 018003.

[16] Sun, G., Yao, K., Liu, Z., & Huang, Q. (1998). A Study on Measuring the Electromagnetic Parameters of Chiral Materials. *J. Phys. D: Appl. Phys.*, 31-2109.

[17] Chen, L. F., Ong, C. K., Neo, C. P., Varadan, V. V., & Varadan, V. K. (2004). Microwave Electronics Measurement and Materials Characterization, *Chinchester, England* John Wiley & Sons.

[18] Lynch, A. C., & Simkin, D. (1990). Measurement of Permeability and Permittivity Of Ferrites. *Measurement Science Technology*, 1-1162.

[19] Rickard, Petersson., & Smith, G. S. (2002). An Estimate of the Error Caused by the Plane-Wave Approximation in Free-Space Dielectric Measurement Systems, *IEEE Transactions on Antennas Propagation*, 50(6), 878-887.

[20] Guire, T., Varadan, V. K., & Varadan, V. V. (1990). Influence of Chirality on the Response of EM Waves by Planar Dielectric Slabs, *IEEE Transactions on Electromagnetic Compability*, 32(4), 300-303.

[21] Ghodgaonkar, D. K., Varadan, V. V., & Varadan, V. K. (1989). A Free-Space Method for Measurement of Dielectric Constants and Loss Tangent at Microwave Frequencies. *IEEE Transactions on Instrumentation and Measurement*, 38-789.

[22] Muñoz, J., Rojo, M., Parreño, A., & Margineda, J. (1998). Automatic Measurement of Permittivity and Permeability at Microwave Frequencies Using Normal and Oblique Free-Wave Incidence with Focused Beam, *IEEE Transactions on Instrumentation and Measurements*, 47(4), 886-892.

[23] García-Collado, A. J., Molina-Cuberos, G. J., Margineda, J., Núñez, M. J., & Martín, E. (2010). Isotropic and Homogeneous Behavior of Chiral Media Based on Periodical Inclusions of Cranks, *IEEE Microwaves and Wireless Components Letters*, 20(3), 176-177.

[24] Molina-Cuberos, G. J., García-Collado, A. J., Barba, I., & Margineda, J. (2011). Chiral Metamaterials with Negative Refractive Index Composed by an Eight-Cranks Molecule, *IEEE Antennas and Wireless Propagation Letters*, 10-1488.

[25] García-Collado, A. J., Molina-Cuberos, G. J., Núñez, M. J., & Margineda, J. (2012). Negative Refraction of Chiral Metamaterial Based on Four Crank Resonator, *Journal of Electromagnetic Waves and Applications*, 26-986.

[26] Agilent Technologies Inc. (2012). Time Domain Analysis Using a Network Analyzer, *Application Note* 1287-12, 5989-5723.

[27] Bohren, C. F. (1974). Light Scattering by an Optically Active Sphere. *Chemical Physics Letters*, 29-458.

[28] Varadan, V. V., & Ro, R. (2007). Unique Retrieval of Complex Permittivity and Permeability of Dispersive Materials from Reflection and Transmitted Fields By Enforcing Causality, *IEEE Transactions on Microwave Theory and Techniques*, 55(10), 2224-2230.

[29] Chen, X., Grzegorczyk, T. M., Wu-I, B., Pacheco, J., & Kong, J. A. (2004). Robust Method to Retrieve the Constitutive Effective Parameters of Metamaterials. *Physical Review* E.; 70 016608.

[30] Smith, D. R., Schultz, S., Markoš, P., & Soukoulis, C. M. (2002). Determination of Effective Permittivity and Permeability of Metamaterials from Reflection and Transmission Coefficients. *Physical Review B*. 65 195104.

[31] Ziolkowski, R. W. (2003). Design, Fabrication, and Testing of Double Negative Metamaterials. *IEEE Transactions on Antennas and Propagation*, 7-1516.

[32] Condon, E. U. (1937). Theories of Optical Rotatory Power, *Review Modern Physics*, 9-432.

[33] Hollinger, R., Varadan, V. V., & Varadan, V. K. (1991). Eigenmodes in a Circular Waveguide Containing an Isotropic Chiral Material, *Radio Science*, 26(5), 1335-1344.

[34] Liu, Z., Sun, G., Huang, Q., & Yao, K. (1999). A Circular Waveguide Method for Measuring the Electromagnetic Parameters of Chiral Materials at Microwave Frequencies. *Measurement Science Technolology*, 10-374.

[35] Reinert, J., Busse, G., & Jacob, A. F. (1999). Waveguide Characterization of Chiral Material: Theory. *IEEE Trans. Microwave Theory and Techniques*, 47(3), 290-296.

[36] Lubkowski, G. (2009). Simulation of Electromagnetic Fields in Double Negative Metamaterials, Technishen Universität Darmstadt, *PhD Dissertation*.

[37] Barybin, A. A. (1998). Modal Expansions and Orthogonal Complements in the Theory of Complex Media Waveguide Excitation by External Sources for Isotropic, Anisotropic, and Bianisotropic Media, *Progress in Electromagnetic Research, PIER*, 19, 241-300.

[38] Tretyakov, S. A., & Viitanen, A. J. (1992). Perturbation Theory for a Cavity Resonator with a Biisotropic Sample: Application to Measurement Techniques, *Microwave and Opt. Tech. Lett.*, 5(4), 174-177.

[39] Tretyakov, S. A., & Viitanen, A. J. (1995). Waveguide and Resonator Perturbation Techniques Measuring Chirality and Nonreciprocity Parameters by Biisotropic Materials, *IEEE Transactions on Microwave Theory and Techniques*, 43(1), 222-225.

Chapter 2

Chiral Waves in Graphene Medium and Optical Simulation with Metamaterial

H. Torres-Silva

Additional information is available at the end of the chapter

http://dx.doi.org/10.5772/51328

1. Introduction

In modern age Engineers have paved the way for a new generation of faster, more powerful cell phones, computers and other electronics by developing a practical technique to replace silicon with carbon on large surface. The capability of silicon, the material at the heart of computer chips has been harnessed beyond its limits by engineers and carbon has come up as an integrating replacement for the same. The material called"Graphene" which is a single layer of atoms arranged in honeycomb lattice could let electronics to process information and produce radio transmission 10 times better than silicon based devices.

For theorists, such a system is also of great interest because it provides a physical realization of two-dimensional field theories with quantum anomalies. Indeed, the continuum limit of the effective theory describing the electronic transport in graphene is that of two-dimensional massless Dirac fermions. The reported and predicted phenomena include the Klein paradox (the perfect transmission of relativistic particles through high and wide potential barriers), the anomalous quantum Hall effect induced by Berry phases and its corresponding modified Landau levels and the experimental observation of a minimal conductivity.

From the point of view of its electronic properties, graphene is a two-dimensional zero-gap semiconductor with the cone energy spectrum, and its low-energy quasiparticles are formally described by the Dirac-like Hamiltonian [1, 2].

$$H_0 = -i\hbar v_F \sigma \cdot \nabla$$

Where $v_F \approx 10^6 ms^{-1}$is the Fermi velocity and $\sigma = (\sigma_x, \sigma_y)$ are the Pauli matrices. The fact that charge carriers are described by the Dirac-like equation, rather than the usual Schrödinger equation, can be seen as a consequence of graphene's crystal structure, which consists of two equivalent carbon sublattices [1, 2]. Quantum mechanical hopping between the sublattices

INTECH open science | open minds

© 2012 Torres-Silva; licensee InTech. This is an open access article distributed under the terms of the Creative Commons Attribution License (http://creativecommons.org/licenses/by/3.0), which permits unrestricted use, distribution, and reproduction in any medium, provided the original work is properly cited.

leads to the formation of two cosine-like energy bands, and their intersection near the edges of the Brillouin zone yields the conical energy spectrum. As a result, quasiparticles in graphene exhibit the linear dispersion relation $E_G = E = \hbar k v_F$, as if they were massless relativistic particles with momentum k (for example, photons) but the role of the speed of light is played here by the Fermi velocity $v_F \approx c/300$. Owing to the linear spectrum, it is expected that graphene's quasiparticles will behave differently from those in conventional metals and semiconductors where the energy spectrum can be approximated by a parabolic (free-electron-like) dispersion relation.

From a crystallographic point of view, the graphene is a triangular Bravais lattice with a diamond-shaped unit tile consisting of two sites so one gets the honeycomb structure. The very unique feature of the graphene band structure is that the two lowest-energy bands, known as the valence and the conduction bands, touch at two isolated points located at the corners of the Brillouin zone. In the immediate vicinity of these degeneracy points, known as the Dirac points, the band structure is a cone. In natural graphene samples, there is exactly one electron per site, and thus, at zero temperature, all levels in the valence band are filled (a situation known as half-filling). As a result, the energy of the last occupied level precisely slices the band structure at the Dirac points. The low-energy excitations of this system are then described by the massless two-dimensional Weyl-Dirac equation and their energy dispersion relation $\omega = v_F k$ is that of relativistic massless fermions with particle-hole symmetry. In graphene these massless fermions propagate with a velocity v_F. The maths is simple but the principles are deep. We will review the formulation of graphene's massless Dirac Hamiltonian, under the chiral electromagnetism approach, like a metamaterial media, hopefully demystifying the material's unusual chiral, relativistic, effective theory. The novel result here is that in our theory we do not make $c \to v_F$ [1, 2, 27-35], but we obtain v_F as $v_F = c(1 - k_0 T)$ if $k_0 T > 0$ or $v_F = c(1 + k_0 T)$ if $k_0 T < 0$. These results are derived of the Chiral Electrodynamics with T as the chiral parameter and $k_0 = \omega / c$ when the electric wave E is quasi parallel to the magnetic wave H.

With this approach we can review different phenomena such: The Dirac point with a double-cone structure for optical fields, an optical analogy with Dirac fermions in graphene, can be realized in optically homogenous metamaterials. The condition for the realization of Dirac point in optical systems is the varying of refractive index from negative to zero and then to positive.

Also we give a support to the similitude of the band structure of a macroscopic photonic crystal with the electronic band structure of grapheme, which is experimentally much more difficult to access, allows the experimental study of various relativistic phenomena. With our analytical and numerical analysis we hope to verify that, similar behaviors exist to electrons in graphene treated as mass-less particle, ie, electron wave propagation.

This chapter presents a short review on the chiral propagation of electron waves in monolayer graphene and optical simulation with optical field in the negative-zero-positive index metamaterial NZPIM and its close connection. Section II presents an enhanced vector diagram of Maxwell's equations for chiral media with quasi parallel electromagnetic fields,

$E \parallel H$. Chiral waves in graphene acting as metamaterial media are discussed in section III. In section IV, two component equations and tunneling rate of Dirac electron in graphene are derived. Section V is about *Zitterbewegung* of optical pulses near the Dirac point inside a negative-zero-positive index metamaterial, and shows that the chiral field near the Dirac point becomes a diffusive wave The last sections describe the theoretical description of minimal conductivity in graphene under chiral approach and absorption of light by quasi 2D Dirac fermions.

2. A vector diagram of Maxwell's equations for chiral media with $E \parallel H$

The idea of representing Maxwell's time-harmonic equations in homogeneous isotropic media by vector diagram as put forward by Wilton [3] and by S. Uckun [4] deserves consideration. All the common relations between field and potential quantities implied by Maxwell's equations can be represented by a diagram. It is started that the diagram not only illustrates Maxwell's equations, but also many of the methods for constructing diagram are based on the formal similarity between many theorems of vector calculus and those of vector algebra.

An isotropic chiral medium is a macroscopically continuous medium composed of equivalent chiral objects that are uniformly distributed and randomly oriented. A chiral object is a three-dimensional body that cannot be brought into agreement with its mirror image by translation and rotation. An object of this sort has the property of handedness and must be either left-handed or right-handed. An object that is not chiral is said to be achiral, and thus all objects are either chiral o achiral. Due to their novel properties and wide applications in microwave and radar engineering, chiral media has been undergoing extensive research during the last years. That is why this study aims to cover chiral medium for the representation of Maxwell's equations in vector diagram form. In a chiral media a cross coupling between electric and magnetic filed exists. Thus, the vector diagram has vectors along all three coordinate axes whereas the vector diagram presented by Wilton [3] for achiral media has vectors only in one plane with H vector normal to it.

2.1. Vector diagram construction

Assumingexp($i\omega t$) time dependence, Maxwell's time-harmonic equations [5, 6] for isotropic, homogeneous, linear media are:

$$\nabla \times \boldsymbol{E} = -j\omega \boldsymbol{B} \tag{1}$$

$$\nabla \times \boldsymbol{H} = j\omega \boldsymbol{D} + \boldsymbol{J} \tag{2}$$

$$\nabla \cdot \boldsymbol{B} = 0 \tag{3}$$

$$\nabla \cdot \boldsymbol{D} = \rho \tag{4}$$

Chirality is introduced into the theory by defining the following constitutive relations to describe the isotropic chiral medium [5]

$$\boldsymbol{D} = \varepsilon \boldsymbol{E} + j\omega\varepsilon T \boldsymbol{B} \tag{5}$$

$$\boldsymbol{H} = j\omega\varepsilon T \boldsymbol{E} + \frac{1 - k_o^2 T^2}{\mu} \boldsymbol{B} \tag{6}$$

Where the chirality admittance $-\omega\varepsilon T$ indicates the degree of chirality of the medium, and the ε and μ are permittivity and permeability of the chiral medium, respectively. Since **D** and **E** are polar vectors and **B** and **H** are axial vectors, it follows that ε and μ are true scalars and $-\omega\varepsilon T$ is a pseudoscalar. This means that when the axes of a right-handed Cartesian coordinate system are reversed to form a left-handed Cartesian coordinate system, $-\omega\varepsilon T$ changes in sign whereas ε and μ remain unchanged.

For a graphical representation of the above relationships, following Wilton's procedure [3], let us assume vector-differential operator, ∇ is an ordinary vector and treat the divergence and curl operations in equations (1) to (4) as ordinary scalar (dot) and vector (cross) products, respectively. Equation (3) implies that ∇ is perpendicular to **B** and the vector $\nabla \times \boldsymbol{B}$ must be perpendicular to both ∇ and **B**.

As shown in Figure 1, three transverse coordinate axes are chosen as ∇, from Equation (1) $\boldsymbol{B} = -\nabla \times \boldsymbol{E} / (j\omega)$ and, $\nabla \times \boldsymbol{B} / (j\omega\varepsilon\mu') = \nabla \times \nabla \times \boldsymbol{E} / k^2$ where $k^2 = \omega^2 \varepsilon\mu'$.

Since $\nabla \cdot \boldsymbol{B} = 0$ always, this conditions will hold identically if **B** is expressed as the curl of a vector potential **A** since the divergence of the curl of a vector is identically zero, thus

$$\boldsymbol{B} = \nabla \times \boldsymbol{A} \tag{7}$$

and **A** must be perpendicular to both ∇ and **B** and lie in ∇ and $\nabla \times \boldsymbol{B}$ plane. However, **A** is not unique since only its components perpendicular to ∇ contribute to the cross product. Therefore, $\nabla \cdot \boldsymbol{A}$, the component of **A** parallel to ∇, must be specified. The curl equation for **E**, as in Equation (1), and Equation (7) give $\nabla \times (\boldsymbol{E} + j\omega\boldsymbol{A}) = 0$ where the quantity in parentheses should be parallel to ∇ and the curl of the gradient of a scalar function ϕ is identically zero; so the above equation is $\boldsymbol{E} + j\omega\boldsymbol{A} = -\nabla\phi$ or

$$\boldsymbol{E} = -j\omega\boldsymbol{A} - \nabla\phi \tag{8}$$

That is shown in Figure 1.

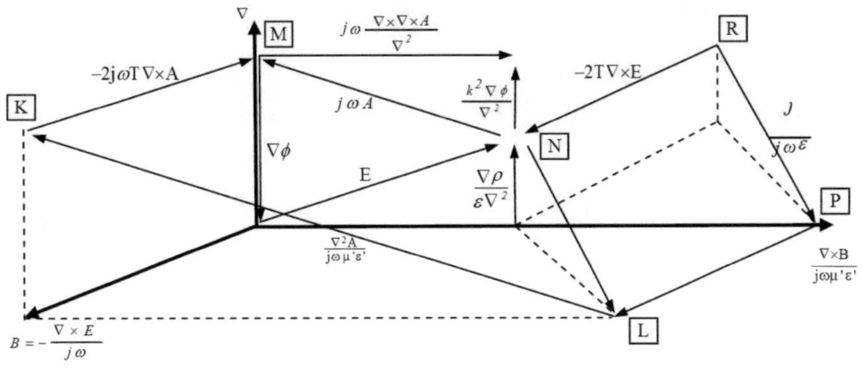

Figure 1. Diagram of the full Maxwell System for a chiral media with Lorentz gauge $\frac{\mu\varepsilon}{1-k_o^2T^2}=\mu'\varepsilon'$.

Following the Uckun's approach [5], we substitute Equation (7) into Equation (6) having

$$H=\frac{1-k_o^2T^2}{\mu}\nabla\times A+j\omega\varepsilon TE \tag{9}$$

Substituting Equation (9) and Equation (5) into Equation (2) gives
$\nabla\times\nabla\times A+j\frac{\omega\mu\varepsilon T}{1-k_o^2T^2}\nabla\times E=j\frac{\omega\mu\varepsilon}{1-k_o^2T^2}E-\frac{\omega^2\mu\varepsilon T}{1-k_o^2T^2}B+\frac{\mu}{1-k_o^2T^2}J$

$\nabla\times\nabla\times A+j\frac{\omega\mu\varepsilon T}{1-k_o^2T^2}\nabla\times E=j\frac{\omega\mu\varepsilon}{1-k_o^2T^2}E-\frac{\omega^2\mu\varepsilon T}{1-k_o^2T^2}B+\frac{\mu}{1-k_o^2T^2}J$, playing the value of $\nabla\times E$ from Equation (1) into the above equation $\nabla\times\nabla\times A+2\frac{\mu\omega^2\varepsilon T}{1-k_o^2T^2}B=j\frac{\omega\mu\varepsilon}{1-k_o^2T^2}E+\frac{\mu}{1-k_o^2T^2}J$ and using the vector identity $\nabla\times\nabla\times A=\nabla(\nabla\cdot A)-\nabla^2A$ enables us to write the above equation as
$\nabla^2A+2\frac{\omega^2\mu\varepsilon T}{1-k_o^2T^2}\nabla\times A=\nabla(\nabla\cdot A)-j\frac{\omega\mu\varepsilon}{1-k_o^2T^2}E-\frac{\mu}{1-k_o^2T^2}J$

$\nabla^2A+2\frac{\omega^2\mu\varepsilon T}{1-k_o^2T^2}\nabla\times A=\nabla(\nabla\cdot A)-j\frac{\omega\mu\varepsilon}{1-k_o^2T^2}E-\frac{\mu}{1-k_o^2T^2}J$, so using Equation (8), we have

$$\nabla^2A+\frac{k_0^2}{1-k_0^2T^2}A+2\frac{\omega^2\mu\varepsilon T}{1-k_o^2T^2}(\nabla\times A)=\nabla(\nabla\cdot A-j\frac{\omega^2\mu\varepsilon T\phi}{1-k_o^2T^2})-\frac{\mu}{1-k_o^2T^2}J \tag{10}$$

Here $\nabla\cdot A$is arbitrary, so in order to specify $\nabla\cdot A$, for unique A, we may choose

$$\nabla\cdot A=j\frac{\omega\mu\varepsilon\phi}{1-k_o^2T^2} \tag{11}$$

And eliminate the term in parentheses in Equation (10). The choice in Equation (11) can be known as a chiral Lorentz gauge. Then Equation (10) will be simplified to

$$\nabla^2 \boldsymbol{A} + \frac{k_0^2}{1-k_0^2 T^2}\boldsymbol{A} + 2\frac{\omega^2 \mu \varepsilon T}{1-k_o^2 T^2}(\nabla \times \boldsymbol{A}) = -\frac{\mu}{1-k_o^2 T^2}\boldsymbol{J} \tag{12}$$

Divide both sides of Equation (12) by $(j\frac{\omega\mu\varepsilon}{1-k_o^2 T^2})$ and reorganize it to get

$$(1-k_o^2 T^2)\frac{\nabla^2 \boldsymbol{A}}{j\omega\mu\varepsilon} = j\omega \boldsymbol{A} + 2j\omega T(\nabla \times \boldsymbol{A}) - \frac{1}{j\omega\varepsilon}\boldsymbol{J} \tag{13}$$

(Figure 1, shows this vectorial equation).

The difference between our approach and the Uckun's procedure [4], is that we take the chiral media characterized by $\boldsymbol{D}=\varepsilon(\boldsymbol{E}+T\nabla\times\boldsymbol{E})$ and $\boldsymbol{B}=\mu(\boldsymbol{H}+T\nabla\times\boldsymbol{H})$. In this form we can obtain the condition of quasi spatial parallel condition between $\boldsymbol{B}$ and $\boldsymbol{E}$ where the main equation is like a Beltrami equation which is important for the numerical simulation of graphene systems.

Placing the value of $\boldsymbol{B}$ from Equation (1) into Equation (5) $\boldsymbol{D}=\varepsilon(\boldsymbol{E}+T\nabla\times\boldsymbol{E})$ will be obtained. Placing the value of $\boldsymbol{H}$, from Equation (6), and D into Equation (2) will give

$$\nabla \times (j\omega\varepsilon T\boldsymbol{E} + \frac{1-k_o^2 T^2}{\mu}\boldsymbol{B}) = j\omega(\varepsilon\boldsymbol{E} + \varepsilon T\nabla\times\boldsymbol{E}) + \boldsymbol{J}$$

by rearranging this equation we obtain

$$(1-k_o^2 T^2)\frac{\nabla \times \boldsymbol{B}}{j\omega\mu\varepsilon} = \boldsymbol{E} + 2T\nabla\times\boldsymbol{E} + \frac{\boldsymbol{J}}{j\omega\varepsilon} \tag{14}$$

will be obtained as shown in Figure 1. In this figure we put $\mu \to \mu/(1-k_0^2 T^2)$.

Taking divergence of Equation (5) and using Equations (3) and (4) in it

$$\rho = \varepsilon\nabla\cdot\boldsymbol{E} \tag{15}$$

will be derived. To find the projection of $\boldsymbol{E}$ onto ∇, from Equation(15) $\nabla\cdot\boldsymbol{E}=\rho/\varepsilon$, take the gradient of both sides and divide by scalar value ∇^2 to normalize ∇ to a unit vector. So

$$\frac{\nabla(\nabla\cdot\boldsymbol{E})}{\nabla^2} = \frac{\nabla\rho}{\varepsilon\nabla^2} \tag{16}$$

Similarly, getting gradient of both sides of Equation(11), using the vector identify $\nabla\times\nabla\times\boldsymbol{A}+\nabla^2\boldsymbol{A}$ for $\nabla(\nabla\cdot\boldsymbol{A})$ and normalizing by ∇^2 we have:

$$jw\frac{\nabla\times\nabla\times\boldsymbol{A}}{\nabla^2} + jw\boldsymbol{A} = k^2\frac{\nabla\phi}{\nabla^2} \tag{17}$$

http://dx.doi.org/10.5772/51328

So it will be obtained as parallel component of A to ∇ coordinate.

By using Equations (8), (13), (14), (16), and (17) the vector diagram of Lorenz gauge can be completed as shown in Figure 1, where all Maxwell's relations and potential quantities appear.

Now let us examine derivation of some relations from the diagram. For example, it is seen that the component of $\boldsymbol{E}$ and $\boldsymbol{J}/(j\omega\varepsilon)$ parallel to ∇ must be equal and opposite. By taking the divergence of Equation (14) and using $\nabla \cdot \boldsymbol{E} = \rho/\varepsilon$ can be shown that

$$\nabla \cdot \boldsymbol{E} = -\frac{1}{j\omega\varepsilon}\nabla \cdot \boldsymbol{J} = \frac{\rho}{\varepsilon} \tag{18}$$

Taking the gradient in both sides of Equation (18) and dividing it by scalar value ∇^2will give the same value as Equation (16) with opposite sign. From the right side of Equation (18) it is seen that

$$\rho = -\frac{1}{jw}\nabla \cdot \boldsymbol{J} \tag{19}$$

This is the known continuity equation. Since the divergence of the curl of any vector is identically zero, the divergence of Equation (2) yields.$0 = j\omega\nabla \cdot \boldsymbol{D} + \nabla \cdot \boldsymbol{J}$. Using Equation (4) convert this immediately into continuity equation as, expected. Again, as seen in Figure 1, $2T\nabla \times \boldsymbol{E}$and $-j\omega 2T\nabla \times A$ are equal and opposite vectors. From Equation(8), taking curl of both side and using the vector identity $\nabla \times \nabla\phi = 0$ will show that

$$2T\nabla \times \boldsymbol{E} = -j\omega 2T\nabla \times \boldsymbol{A} \tag{20}$$

as expected. By using the vector calculus a few possible equations from the vector diagram can be written as follows

$$2T\nabla \times \boldsymbol{E} + \frac{\boldsymbol{J}}{j\omega\varepsilon} - (1-k_o^2T^2)\frac{\nabla \times \boldsymbol{B}}{j\omega\mu\varepsilon} - \nabla\phi - j\omega\boldsymbol{A} = 0 \tag{21}$$

$$j\omega\frac{\nabla \times \nabla \times \boldsymbol{A}}{\nabla^2} - k^2\frac{\nabla\phi}{\nabla^2} - \boldsymbol{E} - \nabla\phi = 0 \tag{22}$$

$$j\omega\frac{\nabla \times \nabla \times \boldsymbol{A}}{\nabla^2} - k^2\frac{\nabla\phi}{\nabla^2} + \frac{\boldsymbol{J}}{j\omega\varepsilon} + (1-k_o^2T^2)\frac{\nabla^2\boldsymbol{A}}{j\omega\mu\varepsilon} - 2j\omega T\nabla \times \boldsymbol{A} = 0 \tag{23}$$

$$(1-k_o^2T^2)\frac{\nabla^2\boldsymbol{A}}{j\omega\mu\varepsilon} - 2j\omega\beta\nabla \times \boldsymbol{A} + \nabla\phi + (1-k_o^2T^2)\frac{\nabla \times \boldsymbol{B}}{j\omega\mu\varepsilon} - 2T\nabla \times \boldsymbol{E} = 0 \tag{24}$$

For example, adding Equations (13) and (14) side by side and using Equation (8) will give Equation (24) which shows the correctness of the equation derived from the diagram 1. Instead of Lorenz gauge we can choose Coulomb's gauge.

$$\nabla \cdot \boldsymbol{A}_c = 0 \tag{25}$$

In Equation (10) so that it will take the form

$$\nabla^2 \boldsymbol{A}_c + k^2 \boldsymbol{A}_c - 2\frac{\omega^2 \mu\varepsilon T}{1-k_o^2 T^2}(\nabla \times \boldsymbol{A}_c) = j\frac{\omega\mu\varepsilon}{1-k_o^2 T^2}\nabla\phi - \frac{\mu}{1-k_o^2 T^2}\boldsymbol{J}$$

where the subscript "c" is used it indicate Coulomb's gauge. Using Equation (8) and (20) in the above equation

$$-(1-k_o^2 T^2)\frac{\nabla^2 \boldsymbol{A}_c}{j\omega\mu\varepsilon} - 2T\nabla \times \boldsymbol{E} = \boldsymbol{E} + \frac{\boldsymbol{J}}{j\omega\varepsilon} \tag{26}$$

will be obtained. Placing the values of Equations (5) and (6) into Equation (2) will give

$$\nabla \times (j\omega\varepsilon T\boldsymbol{E} + \frac{(1-k_o^2 T^2)}{\mu}\boldsymbol{B}) = j\omega(\varepsilon\boldsymbol{E} + j\omega\varepsilon T\boldsymbol{B}) + \boldsymbol{J}$$

and value of $\nabla \times \boldsymbol{E}$ from Equation (1) will give

$$(1-k_o^2 T^2)\frac{\nabla \times \boldsymbol{B}}{j\omega\mu\varepsilon} + 2j\omega T\boldsymbol{B} = \boldsymbol{E} + \frac{\boldsymbol{J}}{j\omega\varepsilon} \tag{31}$$

.

Combining these equations with Equation (26) and using Equation (7) we have

$$(1-k_o^2 T^2)\frac{\nabla \times \boldsymbol{B}}{j\omega\mu\varepsilon} = -(1-k_o^2 T^2)\frac{\nabla^2 \boldsymbol{A}_c}{j\omega\mu\varepsilon} \tag{27}$$

By using the same coordinates axes∇, $\boldsymbol{B}$and $(1-k_o^2 T^2)\nabla \times \boldsymbol{B}/(j\omega\mu\varepsilon)$ and Equation (1), (8), (16), (26) and (27) for the Coulomb gauge. It is clear from Equation (25) that the component of the vector A parallel to∇ is equal to zero.

As seen in Figure 1, Lorenz gauge are the best choice because these make $\boldsymbol{A}$ either parallel or perpendicular to any of the other vectors and simplify its relationship to those vectors. In Figure 1a, if the chirality factor$T$, goes to zero, point$K$, L and R approach point$M$, P, and N respectively, in which case the diagram will be the same as in Reference [3] for linear, homogeneous, isotropic achiral medium. If $(1-k_o^2 T^2) \to 0$ then E is quasi parallel to B, and parallel to$\boldsymbol{A}$so all vectors remain in an only plane. (see Figure 2).

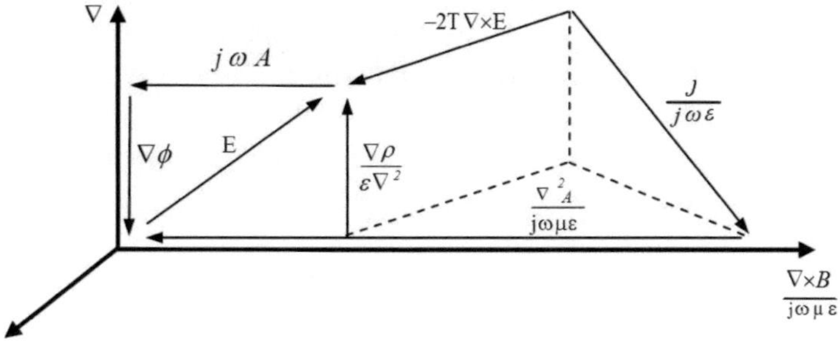

Figure 2. Two dimensional condition when $(1-k_o^2T^2)\rightarrow 0$ for a graphene system, the thickness is about one carbon atom

This Beltrami condition is useful to numerical calculations in graphene. We apply this approach to a two dimensional chiral graphene slab. This result cannot be obtained with the Uckun's approach [4]. In terms of chiral magnetic potential $\boldsymbol{A}_c$ with $\boldsymbol{J}=0$, and $\nabla\rho=0$ we have the wave equation

$$\nabla^2\boldsymbol{A}_c+\frac{k_0^2}{1-k_0^2T^2}\boldsymbol{A}_c+2\frac{\omega^2\mu\varepsilon T}{1-k_o^2T^2}(\nabla\times\boldsymbol{A}_c)=0 \tag{28}$$

As $k=\sqrt{k_x^2+k_y^2+k_z^2}$, if $k_x=k\sin\theta$, $k_y=0$, and $k_z=k\cos\theta$, we have the matrix:

$$\begin{pmatrix} -k^2(1-k_0^2T^2)+k_0^2 & -2jk_0^2kT\cos\theta & 0 \\ 2jk_0^2kT\cos\theta & -k^2(1-k_0^2T^2)+k_0^2 & -2jk_0^2kT\sin\theta \\ 0 & 2jk_0^2kT\sin\theta & -k^2(1-k_0^2T^2)+k_0^2 \end{pmatrix}\begin{pmatrix} A_x \\ A_y \\ A_z \end{pmatrix}=0 \tag{29}$$

The dispersion relation of the transversal wave is $(-k^2(1-k_0^2T^2)+k_0^2)^2-4k_0^4k^2T^2(\sin^2\theta+\cos^2\theta)=0 \Rightarrow k=k_\pm=\pm k_0/(1\pm k_0T)$.

That is

$$k_\pm=\pm k_0\big/(1\pm k_0T)=\pm\tfrac{\omega}{c}(1\pm k_0T)^{-1} \tag{30}$$

So we have $k_\pm=\pm\frac{\omega}{c}(1\pm k_0T)^{-1}=\omega\big/v_F$ with $v_F=c(1\pm k_0T)$ if we consider a bi dimensional graphene system.

The novel result here is that in our chiral theory we do not make $c\rightarrow v_F$ but we obtain v_F as $v_F=c(1-k_0T)$ if $k_0T>0$ or $v_F=c(1+k_0T)$ if $k_0T<0$. These results are derived of the Chiral Electrodynamics with T as the chiral parameter and$k_0=\omega / c$ [5, 6, 7]. This situation corresponds to a model of chiral electron where the E field is almost parallel to the$\boldsymbol{H}$ field, the movement of electrons is helical in the valence band of the membrane graphene. Semi classically, for a hydrogen atom, this condition was shown by Huang [36] for E wave perpendicular to $\boldsymbol{H}$ wave with radiation and by Torres-Silva for E parallel to$\boldsymbol{H}$, without radiation. See refs ([5, 37, 38]). Here, we extend this result to the electron bound to carbon atom.

In the next section we study the situation when the refractive index is negative.

3. Chiral waves in graphene acting as metamaterial media

Metamaterials are composite materials in which both permittivity and permeability possess negative values at some frequencies has recently gained considerable attention [see e.g., [8-12]. This idea was originally initiated by Veselago in 1967, who theoretically studied plane wave propagation in a material whose permittivity and permeability were assumed to be simultaneously negative [11]. Recently Shelby, Smith, and Schultz constructed such a composite medium for the microwave regime, and experimentally showed the presence of anomalous refraction in this medium [10]. Previous theoretical study of electromagnetic wave interaction with omega media using the circuit-model approach had also revealed the possibility of having negative permittivity and permeability in omega media for certain range of frequencies [9].

The anomalous refraction at the boundary of such a medium with a conventional medium, and the fact that for a time-harmonic monochromatic plane wave the direction of the Poynting vector is antiparallel with the direction of phase velocity, can lead to exciting features that can be advantageous in design of novel devices and components. For instance, as a potential application of this material, compact cavity resonators in which a combination of a slab of conventional material and a slab of metamaterial with negative permittivity and permeability are possible. The problems of radiation, scattering, and guidance of electromagnetic waves in metamaterials with negative permittivity and permeability, and in media in which the combined paired layers of such media together with the conventional media are present, can possess very interesting features leading to various ideas for future potential applications such as phase conjugators, unconventional guided-wave structures, compact thin cavities, thin absorbing layers, high-impedance surfaces, to name a few. In this section, we will first present a brief overview of electromagnetic properties of the media with negative permittivity and permeability, and we will then discuss some ideas for potential applications of these materials.

Such a medium is therefore termed left-handed medium [12]. In addition to this "left-handed" characteristic, there are a number of other dramatically different propagation characteristics stemming from a simultaneous change of the signs of ε andμ, including reversal of both the Doppler shift and the Cerenkov radiation, anomalous refraction, and even reversal of radiation pressure to radiation tension. This configuration exhibit also chirality and a rotation of the polarization so the analysis of metamaterial presented by several authors provides a good but not exact characterization of the metamaterial [9]. The evidence of chirality behavior suggests that if it is included in the conditions to obtain a metamaterial behavior of a medium futher progress will be obtained. In this short paper, we propose to investigate the conditions to obtain a metamaterial having simultaneously negative ε and negative μ and very low eddy current loss. As a initial point, we consider a media where the electric polarization depends not only on the electric field$\boldsymbol{E}$, and the magnetization depends not only on the magnetic field$\boldsymbol{H}$, and we may have, for example, constitutive relations given by the Born-Federov formalism [12].

$$\boldsymbol{D}(\vec{r},\,\omega)=\varepsilon(\omega)(\boldsymbol{E}(\vec{r},\,\omega)+T(\omega)\nabla\times\boldsymbol{E}(\vec{r},\,\omega)) \tag{31}$$

$$\boldsymbol{B}(\vec{r},\,\omega)=\mu(\omega)(\boldsymbol{H}(\vec{r},\,\omega)+T(\omega)\nabla\times\boldsymbol{H}(\vec{r},\,\omega)) \tag{32}$$

The pseudoscalar Trepresents the chirality of the material and it has length units. In the limit$T\to 0$, the constitutive relations (31) and (32) for a standard linear isotropic lossless dielectric with permittivity ε and permeability μ are recovered.

According to Maxwell's equations, electromagnetic waves propagating in a homogeneous dielectric magnetic material are either positive or negative transverse circularly polarized waves, and can be expressed as

$$E^{\pm}(\vec{r},\,t)=\hat{E}_{0}{}^{(\pm)}e^{-j(k_{\pm}z-\omega_0 t)} \tag{33}$$

$$\boldsymbol{H}^{\pm}(\vec{r},\,t)=\hat{H}_{0}{}^{(\pm)}e^{-j(k_{\pm}z-\omega_0 t)} \tag{34}$$

where$E_0^{\pm}=E_0(\hat{x}\pm i\hat{y})$, and$\nabla\times\vec{E}^{\pm}(\vec{r},\,t)=\mp k_{\pm}\vec{E}^{\pm}$, $k_{\pm}\geq 0$is the chiral wave number.

If the phase velocity and energy flow are in the same directions, and from Maxwell's equation, one can see that the electric$E$ and magnetic field and $\boldsymbol{H}$ and the wave vector $\vec{k}$ will form a right-handed triplet of vectors. This is the usual case for right-handed materials. In contrast, if the phase velocity and energy flow are in opposite directions, and$E$, $\boldsymbol{H}$, and $\vec{k}$ will form a left-handed triplet of vectors. This is just the peculiar case for left handed materials where the effective permittivity $\varepsilon(1+T\nabla\times)$ and the effective permeability $\mu(1+T\nabla\times)$ are simultaneously negative. So, for incident waves of a given frequencyω, we can determine whether wave propagation in the composite is right handed or left handed through the relative sign changes of$k$.

To advance in our propose we considerer other more popular representation to describe a chiral medium, [12] as

$$\boldsymbol{D}=\varepsilon_P E+(\chi+i\kappa)\boldsymbol{H}, \tag{35}$$

$$\boldsymbol{B}=\mu_P\boldsymbol{H}+(\chi-i\kappa)E, \tag{36}$$

in which electromagnetic coupling terms are added to the basic terms. Bi-isotropy or bianisotropy is used for calling such constitutive equations, according to the parameters to be scalars or tensors. If $\kappa=0$ and$\chi\neq 0$, it is the Tellegen medium; if $\kappa\neq 0$and$\chi=0$, as the requirement of reciprocity, it is the Pasteur medium:

$$\boldsymbol{D}=\varepsilon_p E+i\kappa \boldsymbol{H}, \tag{37}$$

$$\boldsymbol{B}=\mu_p \boldsymbol{H}+i\kappa E, \tag{38}$$

There is a long dispute on strong chiral medium since it was introduced theoretically. Traditional electromagnetic conclusions have limited us to understand strong chirality, i.e. $k_{+}T\geq 1$ or$\mu_p\varepsilon_p\kappa^2$, until we see the fact that artificial Veselago's medium [9] was successfully realized in certain frequency bands [3]. For the case, $E \parallel \boldsymbol{H}$, we have to ask the following question: can strong chiral medium exist?

In Ref. [14], the reason for traditional restriction of chirality parameters was concluded

as: 1) The wave vector of one eigenwave will be negative; 2) The requirement of a positive definite matrix to keep positive energy:

$$\begin{pmatrix} \varepsilon_p & i\kappa \\ -i\kappa & \mu_p \end{pmatrix} \tag{39}$$

With the exploration of backward-wave medium, we know that negative wave vector, or opposite phase and group velocities, are actually realizable. And there is an unfortunate mathematical error in the second reason: in linear algebra, only if it is real and symmetric, positive definite matrix is equivalent to that all eigenvalues should be positive. The matrix (14) is a complex one, making the analysis on restriction of positive energy meaningless.

Actually, in a strong bi-isotropic medium with constitutive relations as Eqs. (1) and (2), the energy can be drawn as

$$w=w_e+w_m=\frac{\boldsymbol{D}\cdot E}{2}+\frac{\boldsymbol{B}\cdot \boldsymbol{H}}{2}=\varepsilon\frac{|E|^2}{2}+\mu\frac{|\boldsymbol{H}|^2}{2} \tag{40}$$

Mathematically, the amount of energy density propagated is proportional to the magnitude of the Poynting vector$\vec{S}$, where$S=E\times \boldsymbol{H}$. With the condition$E \parallel \boldsymbol{H}$, $S(r,\,t)=0$so we find right circularly polarized wave or left circularly polarized wave.

The concept of parallel fields is important in the theoretical formulation of: Space electromagnetism and vacuum, the classical and quantum gravitational fields, the study of elementary particles, operator and Dirac matrices, fields and chiral electrodynamics [15, 16]. If we put $E=\alpha B$ and solving the Maxwell equations with the Born Fedorov relations we find $\alpha=i/\sqrt{\mu\varepsilon/4}$ $E=i\eta \boldsymbol{H}=i\sqrt{\frac{\mu/2}{\varepsilon/2}}\boldsymbol{H}$ so $\varepsilon(\omega)$, $\mu(\omega)$ and T are transformed as:

$$\varepsilon_{\|}=\frac{\varepsilon}{2}=\varepsilon_p\left(1-\frac{\kappa^2}{\mu_p\varepsilon_p}\right) \tag{41}$$

$$\mu_{\|}=\frac{\mu}{2}=\mu_p\left(1-\frac{\kappa^2}{\mu_p\varepsilon_p}\right) \tag{42}$$

$$T=\frac{\kappa}{\omega(\mu_p\varepsilon_p-\kappa^2)} \tag{43}$$

In this case the total density energy is may be

$$w=w_e+w_m=0 \tag{44}$$

In this special case where the energy propagated in one direction is equal to that propagated in the opposite direction, there is no net energy flow in the medium and the sum of the two TEM waves form what is generally known as a standing wave. The condition for a standing wave is that the time average of S vanishes. This can be achieved if S is zero all the time everywhere in the region of space under consideration, i.e., $S(r,\ t)=0$ (see Fig. 4). Examination of Eq. (39) shows that $S=0$ if $E\ \|\ \boldsymbol{H}$. In this last case, a particular solution of Eq. (31,3 2) is when $k_0^2T^2=1$, where we have the condition $E\ \|\ \boldsymbol{H}$, and $E=i\eta\boldsymbol{H}$, so we find the Beltrami force free equation $E+2T\nabla\times E=0$ and the vector Poynting $S(r,\ t)$ vanishes [15-16].

In terms of Eqs. (41-42), if $\mu_p\varepsilon_p<\kappa^2$ for whole frequency range, the energy will still keep positive as long as the permittivity and permeability are positive, under the weak spatial dispersion condition. Therefore, the real reason for traditional strong-chirality limitation is neither negative wave vector nor energy conversation. Next we will point out two other important reasons.

First, with the assumption that $\varepsilon_p>0$, $\mu_p>0$, $\kappa>0$ and $\kappa>\sqrt{\mu_p\varepsilon_p}$, we easily show that $\varepsilon_{\|}$, $\mu_{\|}$ and T turn to negative from the transformation between Pasteur constitutive relations and BF relations shown in Eqs. (41)-(43). This is absolutely unacceptable before people realizing Veselago's medium. Actually, strong chiral medium can be equivalent to Veselago's medium for the right circularly polarized wave [9-16]. The negative ε and μ have shown such a point. Hence the negative sign in the BF model is not strange at all, since we realize effective double-negative with strong chirality parameter instead of simultaneously frequency resonances. For a limiting case, the chiral nihility [10], in which $\mu_p\to 0$ and $\varepsilon_p\to 0$ while $\kappa\neq 0$, the parameters in DBF representation become $\varepsilon_{\|}\to\infty$, $\mu_{\|}\to\infty$ and $T=-1/\omega\kappa$, remaining a finite value after a simple mathematical analysis. There is no evidence that strong chirality cannot exist in this aspect.

Second, it is the effectiveness of linear models. Similar to the case that linear optical and electromagnetic models can no longer deal with very strong optical intensity and electromagnetic field, we introduce nonlinear optics to take into account the higher order terms of polarization.

If the spatial dispersion is strong enough, the higher order coupling terms cannot be neglected as before. People used to mistake strong chirality with strong spatial dispersion, hence adding a limitation to chirality parameter,$\kappa\sqrt{\mu_P \varepsilon_P}$. We believe that this is the most probable reason.

However, the strong spatial dispersion is embodied in the BF model, e.g. the value of T, while the strong chirality is represented by the Pasteur model, e.g. the ratio of $\kappa$to$\sqrt{\mu_P \varepsilon_P}$. That is to say, strong chirality does not necessarily lead to strong spatial dispersion.

Based on Eqs. (41)-(43), we have computed T and $\varepsilon_{||} / \varepsilon_P$or $\mu_{||} / \mu_P$versus$\kappa / \sqrt{\mu_P \varepsilon_P}$, as shown in Figs. 2 and 3. When $\kappa$is very close to$\sqrt{\mu_P \varepsilon_P}$, the value of T is quite large, indicating a strong spatial dispersion. Hence the singular point is the very point of traditional limitation. However, withκcontinuously increasing, the spatial dispersion strength falls down very quickly. Therefore, if $\kappa$is not around$\sqrt{\mu_P \varepsilon_P}$, e.g. $\kappa < 0.8\sqrt{\mu_P \varepsilon_P}or\kappa > 1.2\sqrt{\mu_P \varepsilon_P}$, we need not take nonlinear terms into consideration at all. Hence the strong spatial dispersion and nonlinearity cannot put the upper limitation to chirality parameters either.

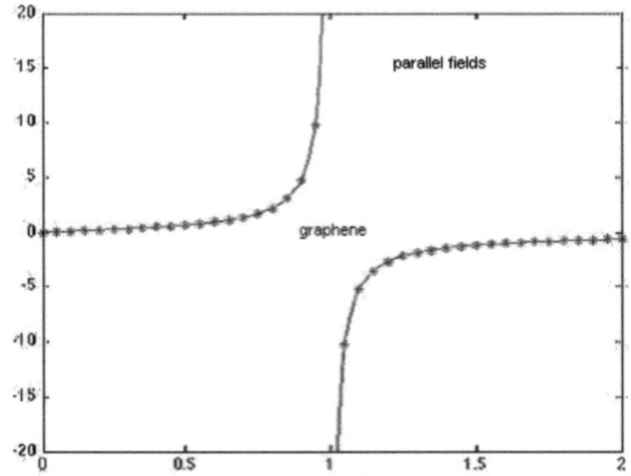

Figure 3. The strength relationship of chirality and spatial dispersion. $\omega T / c$versus $\kappa / \sqrt{\mu_P \varepsilon_P}$ The point of $\kappa / \sqrt{\mu_P \varepsilon_P} = 1$ is singularity, corresponding infinite spatial dispersion coefficient T. When$\kappa / \sqrt{\mu_P \varepsilon_P} 1$, T becomes negative for keeping the positive rotation term coefficients with negative $\mu_{||}$ and$\varepsilon_{||}$. The 2-D grapheme system may be modeled when $\omega T / c \sim 0.9 - 1.1$ and $\kappa / \sqrt{\mu_P \varepsilon_P} \sim 1$

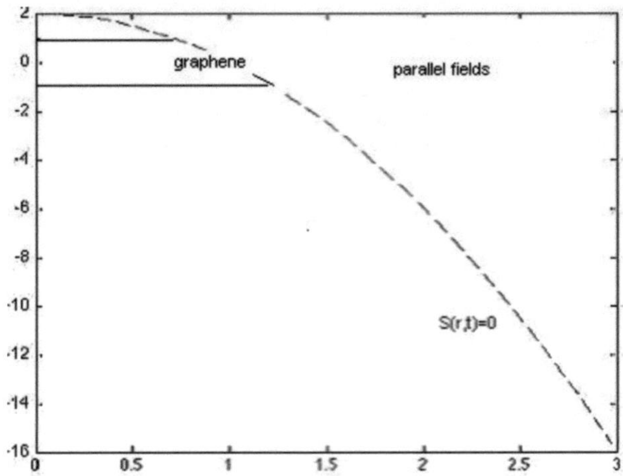

Figure 4. $\varepsilon_{\|} / \varepsilon_p$, $\mu_{\|} / \mu_p$ versus $\kappa / \sqrt{\mu_P \varepsilon_P}$. With chirality strength increases, $\varepsilon_{\|}$ and $\mu_{\|}$ reduces quickly from ε_p and μ_p to $-\infty$. The case 2-D graphene corresponds to $\kappa / \sqrt{\mu_P \varepsilon_P} \sim 1$.

If $\omega | T | / c = 0.9927$ or $\omega | T | / c = 1.0073$, we have that the Fermi velocity is given by $v_F = c(1 \pm k_0 T)$, so it appears a connection between the chiral electrodynamics and α the fine structure constant.

From Figure 4. we see that near of $\varepsilon_{\|} / \varepsilon_p$, $\mu_{\|} / \mu_p \approx 0$ the graphene have a linear behavior so the Beltrami equation is

$$\nabla \times \vec{E}^{\pm}(\vec{r}, t) = \mp k_{\pm} \vec{E}^{\pm} \tag{50}$$

Solving Beltrami's equation with boundary conditions, we can demonstrate the optical *Zitterbewegung* effect by means of electromagnetic pulses propagating through a negative-zero-positive index metamaterial (NZPIM) if we make
$k_{\pm} = \mp k_0 / (1 \pm k_0 T) = \mp \omega / v_F$ with

$$v_F = c(1 \pm k_0 T) \tag{51}$$

Thus in optics, the Beltrami's equation for electromagnetic waves can be reduced to the Helmholtz equation.

In later sections we discuss some effects such as *Zitterbewegung* of optical pulses, diffusion phenomenon and tunneling rate of dirac electron in graphene.

4. Two component equations and tunneling rate of dirac electron in graphene

The usual choice of an orthogonal set of four plane-wave solutions of the free-particle Dirac equation does not lend itself readily to direct and complete physical interpretation except in low energy approximation. A different choice of solutions can be made which yields a direct physical interpretation at all energies. Besides the separation of positive and negative energy states there is a further separation of states for which the spin is respectively parallel or anti-parallel to the direction of the momentum vector. This can be obtained from the Maxwell's equation without charges and current in the $E \parallel H$ configuration. Dirac's four-component equation for the relativistic electron is [5-7]. (see our chapter "Chiral Transverse Electromagnetic standing waves with $E \parallel H$ in the Dirac Equation and the spectra of the Hydrogen Atom", section 2,Intech 2011, Behaviour of electromagnetic waves in different media and structures, edited by Ali Akdagli). Here we considerer a bidimensional graphene system so the Dirac's four-component equation for the relativistic electron is:

$$i\hbar\frac{\partial}{\partial t}\psi = \hat{H}^D\psi \tag{45}$$

where:

$$\hat{H}^D = v_F(\bar{\alpha} \cdot \hat{\vec{p}}) + m v_F{}^2\beta \tag{46}$$

$$\alpha_k = \begin{pmatrix} 0 & \sigma_k \\ \sigma_k & 0 \end{pmatrix}, \quad k = 1, 2, 3, \tag{47}$$

$$\beta = \begin{pmatrix} I & 0 \\ 0 & -I \end{pmatrix} = \sigma_z \tag{48}$$

and I is the two-by-two identity matrix and the Fermi velocity v_F is deduced from the chiral electrodynamics with$v_F = c(1 \pm k_0 T)$, where T is the chiral parameter in a metamaterial condition. This result is capital to our approach because we fin a contact point between the graphene system and optical metamaterial making$v_F = c(1 \pm k_0 T)$, no making $c \rightarrow v_F$ as other authors do it. In Figure 5 we have$1.0073 > k_0 T > 0.9927$.

The Hamiltonian commutes with the momentum vector$\hat{p}$. In order to resolve this degeneracy we seek a dynamical variable which commutes with both H and$\hat{p}$. Such a variable is$\hat{\sigma} \cdot \hat{p}$, where$\hat{\sigma}$is the matrix Pauli. The eigenfunctions of the commuting variables p and$\hat{\sigma} \cdot \hat{p}$ are simultaneous:

$$(\hat{\sigma} \cdot)^2 = p^2 \tag{49}$$

Thus for a simultaneous eigenstate of $\hat{p}$ and $\hat{\sigma} \cdot \hat{p}$, the value of $\hat{\sigma} \cdot \hat{p}$ will be $+p$ or $-p$, corresponding to states for which the spin is parallel or antiparallel, respectively, to the momentum vector like a graphene system.

A simultaneous eigenfunction of H and p will have the form of a plane wave

$$\psi_j = u_j \exp[i(\mathrm{p} \cdot \mathrm{r} - Et)/h], \; j=1, 2, 3, 4 \tag{50}$$

where the ψ_j are the four components of the state function and u_j four numbers to be determined. Then E can have either of the two values.

$$E = \pm \varepsilon = \pm \left(m^2 v_F{}^4 + v_F{}^2 p^2\right)^{\frac{1}{2}} \tag{51}$$

We now demand that ψ_j be also an eigenfunction of $\hat{\sigma} \cdot$ belonging to one of the eigenvalues p_E, say, where $p_E = \pm p$.

The eigenvalue equation is

$$\hat{\sigma} \cdot \psi = p_E \psi \tag{52}$$

Since W can be given either of the two values $\pm\varepsilon$ and p_E, the two values $\pm p$, we have found for given p four linearly independent plane wave solutions. It is easily verified that they are mutually orthogonal.

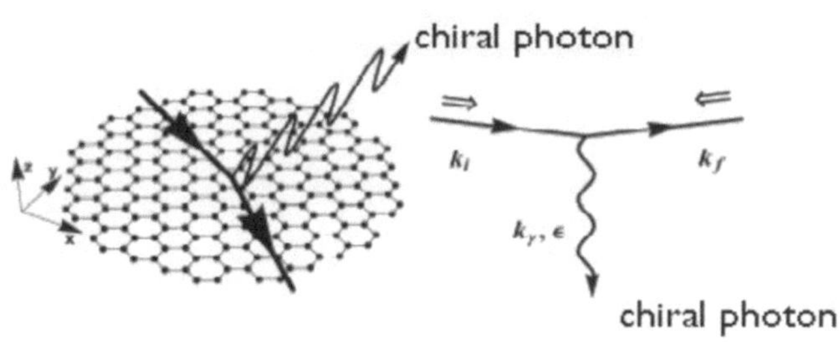

Figure 5. Diagram where the graphene is located in the plane (x, y), the chiral photon is in either direction. The pontos black figure on the left indicates the position of carbon atoms.

The physical interpretation of the solutions is now clear. Each solution represents a homogeneous beam of particles of definite momentum p, of definite energy, either $\pm\varepsilon$, and with the spin polarized either parallel or antiparallel to the direction of propagation. From here we can obtain the well known equation for graphene $H_0 = -i\hbar v_F \sigma \cdot \nabla$.

In this section, we study the tunneling rate of Dirac electrons in graphene through a barrier with an intense electromagnetic field. A one transport phenomenon in graphene is the chiral tunneling [1, 2, 17, 18]. In mono layer graphene a perfect transmission through a potential barrier in the normal direction is expected. This tunneling effect is due to the chirality of the Dirac electrons, which prevents backscattering in general. This kind of reflectionless transmission is independent of the strength of the potential, which limits the development of graphene-based field-effect transistors (FET). The perfect transmission can be suppressed effectively when the chiral symmetry of the Dirac electrons is broken by a laser field, when the n-p junctions in graphene are irradiated by an electromagnetic field in the resonant condition [19,20].We consider a rectangular potential barrier with heightH_0, width D in the X direction, and infinite length in the Y direction. The Fermi level lies in the valence band in the barrier region and in the conduction band outside the barrier. The gray filled areas indicate the occupied states. The optical field under chiral condition is propagated perpendicular to the layer surface and it is circularly polarized along the Z,Y direction with a detuning$\Delta_0=2E_b-\hbar\omega$. We choose $\Delta_0>0$to ensure that there is no inter-band absorption inside the barrier. Meanwhile, $2E_k \gg \hbar\omega$is used to guarantee that the influence of the optical field outside the barrier can be neglected. Thus, neglecting the scattering between different valleys, the scattering process of Dirac electrons in K point is described by the time-dependent Dirac equation, Eq. (45) with$m=0$, so $\psi=\psi_g=(\psi_E, \psi_H)^t$where t means transposed wave function.

$$i\hbar\frac{\partial}{\partial t}\psi_g(r, t)=[H_e+H_0 I+H_{\text{int}}]\psi_g(r, t) \tag{53}$$

where $\psi_g(r, t)=[\phi_A(r, t), \phi_B(r, t)]^t$ is the wave function, $H_e=v_F\hat{\sigma}\cdot p=-i\hbar v_F\hat{\sigma}\cdot\nabla$ is the unperturbed Dirac Hamiltonian obtained from the chiral electrodynamics [17], $\hat{\sigma}=(\sigma_x\sigma_y)$are the Pauli matrices, $v_F\approx 10^6 ms^{-1}$is the Fermi velocity, $H_0(r)$is the height of the potential barrier, I is the unit matrix, and $H_{\text{int}}$is the interaction Hamiltonian with a chiral electromagnetic potential.

$$H_{\text{int}}=-\hbar e v_F[\hat{\sigma}\cdot \mathrm{A}]=\hbar\begin{pmatrix} 0 & H_{12} \\ H_{21} & 0 \end{pmatrix} \tag{54}$$

where e is the electron charge and the chiral potential vector is $[\mathrm{A}]=[A_x e^{i\omega t}, iA_y e^{i\omega t}]$with $i=\sqrt{-1}$ and $|A_x|=|A_y|$ so we have circular polarized wave of the electromagnetic field. When the Dirac electrons perpendicularly collide with the barrier perpendicularly, we can rewrite Eq. (53) as a set of partial differential equations

$$i\hbar\frac{\partial}{\partial t}\phi_A(x, t)=-iv_F\frac{\partial}{\partial x}\phi_B(x, t)+V_0\phi_A(x, t)+H_{12}(t)\phi_B(x, t) \tag{55}$$

$$i\hbar\frac{\partial}{\partial t}\phi_B(x, t)=-iv_F\frac{\partial}{\partial x}\phi_A(x, t)+V_0\phi_B(x, t)+H_{21}(t)\phi_A(x, t) \tag{56}$$

Since the tunneling time is of order of sub-picosecond and the potential $H_{12}(t)$and $H_{21}(t)$ vary as fast as the frequency of incident light beams, this scattering process is strongly time-dependent.

In order to study such a strongly time-dependent scattering process, we employ the finite-difference time-domain (FDTD) method to solve Eq. (47) and Eq. (48) numerically in the time-domain [21]. In the traditional FDTD method, the Maxwell's equations are discretized by using central-difference approximations of the space and time partial derivatives. As a time-domain technique, the FDTD method can demonstrate the propagation of electromagnetic fields through a model in real time. Similar to the discretization of Maxwell's equations in FDTD, we denote a grid point of the space and time as $(i, k)=(i\Delta x, k\Delta t)$ [22]. For the nodal variables we can apply the usual Lee discretization method in 2-D systems with$k_0^2 T^2 \sim 1$ (See Figure 6).

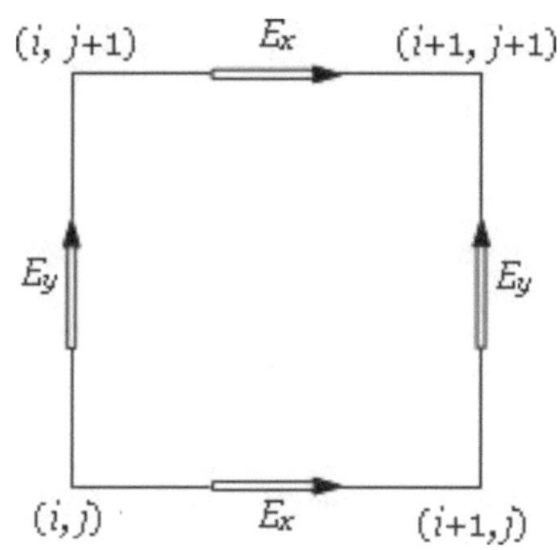

Figure 6. Lee discretization. Here $H_x=(\nabla \times \vec{A})_x E_x=-(\partial(1+T\nabla\times)\vec{A}/\partial t)_x$and with$\vec{J}=0$ (See [21, 22, 24]).

For any function of space and time$G(i\Delta x, k\Delta t)=G^k(i)$, the first order in time or space partial differential can be expressed as:

$$(\frac{\partial}{\partial x}G(x, t))_{x=i\Delta x}\approx\frac{G^k(i+1/2)-G^k(i-1/2)}{\Delta x}, \ldots\ldots, (\frac{\partial}{\partial t}G(x, t))_{t=k\Delta t}\approx\frac{G^{(k+1/2}(i)-G^{(k-1/2}(i)}{\Delta t}$$

These eqs. can be replaced by a finite set of finite differential equations like:

$$\begin{aligned}\phi_A{}^{k+1/2}(i)\left[\frac{1}{\Delta t}-\frac{H_0(i)}{2i}\right]&=\left[\frac{1}{\Delta t}+\frac{H_0(i)}{2i}\right]\phi_A{}^{k-1/2}(i)\\&-\left[\frac{v_F}{\Delta x}-\frac{H_{12}^k(i+1/2)}{2i}\right]\phi_B{}^k(i+1/2)\\&+\left[\frac{v_F}{\Delta x}+\frac{H_{12}^k(i-1/2)}{2i}\right]\phi_B{}^k(i-1/2)\end{aligned}\tag{57}$$

For computational stability, the space increment and the time increment need to satisfy the relation $\Delta x > v_F \Delta t$ [21]. Furthermore, the space increment Δx must far smaller than the wavelength of electrons$\Delta x < \lambda_e / 10$, and the time increment must be far smaller than the period of the electromagnetic fieldT_l.

At the boundary, one-dimensional Mur absorbing boundary conditions are used [21-22]. To compare our results with [25] which use linear polarization for the vector potentialAwhere the chiral parameter T is zero, we considerer at the input boundary, a normalized Gaussian electronic wave packet, where$t_g$and $\tau_g$denote the peak position and the pulse width, respectively. Thus, by solving Eq. (57) directly in the time domain we can demonstrate the propagation of a wave packet through a barrier in real time.

Numerical simulations are shown in Fig. 7. The following parameters are used in our calculation: the peak position $t_0 = 1.5$ ps, the pulse width $\tau_g = 1.0$ ps, the space increment $\Delta x = 0.1$ nm, the time increments$\Delta t = 5 \times 10^{-5} ps$, and the height of the potential barrier $H_0 = 500$ meV.

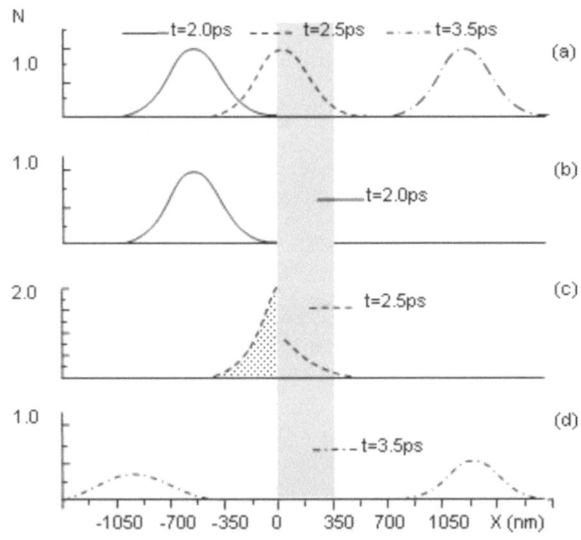

Figure 7. a) Numerical simulations of a wave packet given by$N = |_A|^2 + |_B|^2$, tunneling through a barrier without pump beams. Figures (b)-(d) show the time sequence of a wave packet tunneling through a barrier with pump intensity$I_\omega = 3MW / cm^2$, Δ_0= 5meV, and D = 350 nm. The light grey shows the barrier area.

When there is no pump beams, a perfect chiral tunneling can be found [see Fig. 7 (a)]. This result is consistent with that of Ling et al. [25]. But when the sample is irradiated by an intense non resonant laser beam, a reflected wave packet appears [see Fig. 7 (d)]. The perfect transmission is suppressed. By analyzing the transmitted wave packet and the reflected wave packet, we can obtain the tunneling rate.

To explain the suppression of chiral tunneling, we first investigate the chiral potential wave in the barrier within a rotating wave approximation [23, 24]. Figure 7 (a) shows the renormalized band as a function of momentum k with intensity$I_\omega = 3MW / cm^2$.

Here, the important point is that we make$\pm k_0 T \sim 1 + \alpha \approx 1.0073$, where $\alpha = e^2 / (4\pi\varepsilon_0 \hbar c)$is the fine structure constant, so we verify that our chiral theory is correct. Without external beams, (fig. 7 (a)), we verify the known results on chiral tunneling [11]. Figures 7 (b)-(d) show the time sequence of a wave packet tunneling through a barrier with pump intensity $I_\omega = 3MW / cm^2$, Δ_0= 5meV, and D = 350 nm.

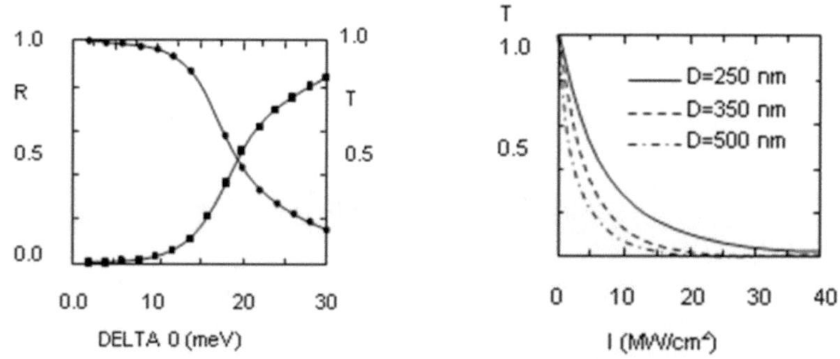

Figure 8. Left: The reflectance R(circles) and the transmittance T (squares) of the barrier as a function of the detuning for $I_\omega = 3MW / cm^2$ and D = 350 nm. Right: The transmittance versus$I_\omega (MW / cm^2)$, having D as parameter. Here, $k_0^2 T^2 \approx 1$, that is the wave electric field is almost parallel to the wave magnetic field within the graphene device.

Under intense light beams, the dressed states are strongly mixed with valence states and conduction states. Therefore, the chiral symmetry of Dirac electrons in graphene can be broken and perfect chiral tunneling is strongly suppressed. Numerical results are shown in Fig. 8 (left) with pump intensity $I_\omega = 3MW / cm^2$ and D = 350 nm. From Fig. 8(left) we can find that the transmission is strongly suppressed, even with laser detuning (e.g., Δ_0= 10 meV, the transmittance is about 0.03).

Figure 8 (left), show that the reflectance decreases, and the transmittance increases as Δ_0 increases. The strong laser field can enhance band mixing and reduce the transmittance. If D increases we can see that the wide barrier can prolong the interaction time between electrons and photons, reduce the tunneling rate, and lower the threshold of the pump laser power (Figure 8 right). Our results can be compared with [25] which use linear polarization ofH_{int}, however we think that inside of a plate of graphene, the eigenvectors of A have to be characterized as chiral waves, because the electrons have a chiral nature in a graphene device. Is for this reason that our numerical results correspond to half times the results shown in [25].

5. *Zitterbewegung* of optical pulses near the Dirac point inside a negative-zero-positive index metamaterial

This optical analog of the *Zitterbewegung* effect is similar to two-dimensional (2DPC) photonic crystal.

The *Zitterbewegung* effect can appear in solids, because the electron at the touching point between two interacting bands in a solid also obeys the massless Dirac equation [26]. In recent years, a growing attention has been devoted to the simulations or demonstrations of the *Zitterbewegung* effects in controllable physical systems, such as, ultra cold atoms, superconductors, semiconductor nanostructures with spin-orbit coupling, a single trapped ion, and graphene.

For a homogenous medium, when a light field is polarized in the z direction, the Beltrami's equation$\nabla \times \vec{E}^{\pm}(\vec{r}, t) = \mp k_{\pm}\vec{E}^{\pm}$can be transformed as a Helmholtz equation written as

$$(\frac{\partial}{\partial x} + i\frac{\partial}{\partial y})(\frac{\partial}{\partial x} - i\frac{\partial}{\partial y})E_z(x, y, \omega) = (ik(\omega))^2 E_z(x, y, \omega) \tag{58}$$

with a wave number $k(\omega)$ This equation can be written in the form of the Dirac equation

$$\begin{pmatrix} 0 & -i(\frac{\partial}{\partial x} - i\frac{\partial}{\partial y}) \\ -i(\frac{\partial}{\partial x} + i\frac{\partial}{\partial y}) & 0 \end{pmatrix}\begin{pmatrix} E_{z1}(x, y, \omega) \\ E_{z2}(x, y, \omega) \end{pmatrix} = k(\omega)\begin{pmatrix} E_{z1}(x, y, \omega) \\ E_{z2}(x, y, \omega) \end{pmatrix} \tag{59}$$

In the case of grapheme near of$\varepsilon_{\|} / \varepsilon_p$, $\mu_{\|} / \mu_p \approx 0$, there are two bands touch each other forming a double-cone structure with a linear dispersion, which can also be called Dirac Point. Near the Dirac point the light transport obeys the massless Dirac equation if we make $v_F = c(1 \pm k_0 T) = v_D$. Also naturally, it would be of great interest to find out what is the condition to have the Dirac dispersion for the light field in a homogenous medium, the wave vector of a medium, $k(\omega)$, can be expanded as $k(\omega) = k(\omega_D) + (\omega - \omega_D) / v_D + \ldots\ldots\ldots$so if$k(\omega_D) = 0$, then we have a linear dispersion$k(\omega) = (\omega - \omega_D) / v_D$, and the Dirac equation is

$$\begin{pmatrix} 0 & -iv_D(\frac{\partial}{\partial x} - i\frac{\partial}{\partial y}) \\ -iv_D(\frac{\partial}{\partial x} + i\frac{\partial}{\partial y}) & 0 \end{pmatrix}\begin{pmatrix} E_{z1}(x, y, \omega) \\ E_{z2}(x, y, \omega) \end{pmatrix} = (\omega - \omega_D)\begin{pmatrix} E_{z1}(x, y, \omega) \\ E_{z2}(x, y, \omega) \end{pmatrix} \tag{60}$$

where$E_{z1}(x, y, \omega)$, $E_{z2}(x, y, \omega)$are two eigenfunctions of the electrical fields (polarized in the z-direction) corresponding to the same frequencyω, and v_D is the group velocity at the Dirac point (ω_D). For media satisfying Eq. (58) with$\omega_D > 0$, the wave number, $k(\omega)$, varies from negative to zero and then to positive with the frequency, and so does the refractive index of the medium with zero refractive index atω_D.

Note that the positive and negative branches of the band structure coexist. Equation (58) is the massless Dirac equation of the light fields in homogenous materials, which is the same as that of electrons in graphene [27]. Therefore, for the NZPI metamaterial [satisfying Eq. (58)], we will have the Dirac point with a double-cone structure for the light field at frequencyω_D. It is expected that the propagation of the light field is analogous to that of the electron in graphene. In fact, in the 2DPC [28], the effective refractive index varies from negative to positive near this Dirac point. Atω_D, Eq. (45) takes the form of diffusion equation,

$$-iv_D\left(\frac{\partial}{\partial x}-i\frac{\partial}{\partial y}\right)E_{z1}(x,\ y,\ \omega)=0,\ -iv_D\left(\frac{\partial}{\partial x}+i\frac{\partial}{\partial y}\right)E_{z2}(x,\ y,\ \omega)=0 \tag{60}$$

which are the same as that of the massless Dirac equation at zero energy. [29]. Therefore, it can be predicted that the behavior of light fields at ω_D has the diffusive properties inside the medium of Eq. (2), like electrons at Dirac point of graphene [27-29]. At or nearω_D, because of $k^2=k_x^2+k_y^2\approx 0$, thek_y component becomes a pure imaginary number for any realk_x, thus the fields along the$y$ direction between a given interval L have the total energy transmittance

$$T_{all}=\int_{-\infty}^{\infty}(\exp(-|k_x|)^2dk_x=1\Big/L \tag{61}$$

which tells us that the propagation of light field at (or near) $\omega_D$exhibits the $L/1$ scaling, a main characteristic of the diffusion phenomenon. This effect can be obtained if $\varepsilon_{||}/\varepsilon_p$, $\mu_{||}/\mu_p\rightarrow 0$with low damping rates.

Here, the Dirac point with the double-cone structure for the light field can also be realized in a homogenous negative-zero-positive index (NZPI) medium [26], in which the two-dimensional Helmholtz equation could be written as the two-dimensional massless Dirac equation. The condition for the realization of the Dirac point in the homogenous optical medium is the index varying from negative to zero and then to positive with frequency, and the light field also obeys eq. (45) near the Dirac point.. Thus, in this work, we show that the *Zitterbewegung* effect with optical pulses appears near the Dirac point in NZPIM slabs. With the realization of the NZPIMs in experiments, we believe that it already has a great possibility to observe *Zitterbewegung* effect with optical pulses in the GHz region.

6. Theoretical description of minimal conductivity in graphene under chiral approach

The first results on the theoretical combination of graphene electron transport properties and a characteristic property of Dirac chiral fermions were obtained by Katsnelson et al, [2]. The intrinsic nature of Dirac fermions gives rise to minimal conductivity even for an ideal crystal, that is, without any scattering processes. The simplest way for the theoretical consideration is the Landauer approach [29]. Assuming that the sample is a ring of length L_y in y direction; author used the Landauer formula to calculate the conductance in the x direction (Fig. 9).

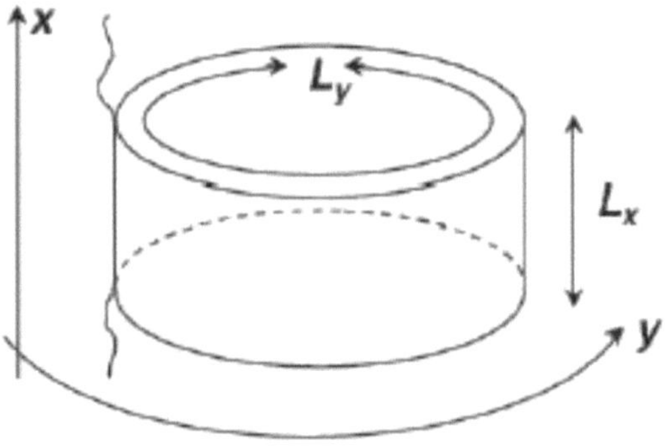

Figure 9. Geometry of the sample. The direction of the current is parallel to the *x* axis [27].

The convenient boundary conditions are not physical, but to get finite transparency one should choose$L_x L_y$.

In the coordinate representation the Dirac equation (59) at zero energy takes the form $(\Psi_i = v_D E_{zi})$

$$-i\left(\frac{\partial}{\partial x} - i\frac{\partial}{\partial y}\right) v_D E_{z1}(x,\ y,\ \omega) = -i\left(\frac{\partial}{\partial x} - i\frac{\partial}{\partial y}\right)\Psi_1 = 0 \tag{62}$$

$$-i\left(\frac{\partial}{\partial x} + i\frac{\partial}{\partial y}\right) v_D E_{z2}(x,\ y,\ \omega) = -i\left(\frac{\partial}{\partial x} + i\frac{\partial}{\partial y}\right)\Psi_2 = 0 \tag{63}$$

The solutions of these equations are$\Psi_1(x - iy), \Psi_2(x + iy)$, which are complex conjugated functions which are general solutions.

Due to periodicity in y direction both wave functions should be proportional to$\exp(ik_y y)$ where$k_y = 2\pi n / L_y$, $n = 0, \pm 1, \pm 2, \ldots$ This means that the dependence on the x is also fixed:

the wave functions are proportional to$\exp(2\pi n x / L_y)$. The introduced boundary conditions at the sample edges are $x = 0$ $x = 0$ and$x = L_x$. The assumption is that the sample is doped graphene with the Fermi energy$E_F = v k_F = -V_0$. The wave functions in the sample are supposed to have the same y-dependence, that is,$\Psi_{1,2}(x,\ y) = \Psi_{1,2}(x)\exp(ik_y y)$.

Requiring continuity of the each wave function at the edges of sample, one can find the transmission coefficient:

$$T_n = \frac{\cos^2\phi}{\cosh^2(k_y L_x).\sin^2\phi} \tag{64}$$

Where $\sin\phi = k_y / k_F$,

$$k_x = \sqrt{k^2 - k_y^2} \tag{74}$$

Further, one should assume that $k_F L_x \gg 1$ and put $\phi = 0$ in equation. Thus, the trace of the transparency which is just the conductance per valley per spin is

$$g = \frac{e^2}{h} tr(\sum T_n) = \frac{e^2}{h}(\sum_{n=-\infty}^{\infty} \frac{1}{\cosh^2(k_y L_{x\approx})}) = \frac{e^2}{h}\frac{L_y}{\pi L_x} \tag{65}$$

The conductance then equals $\sigma L_y / L_x$ and the conductivity is $e^2 / \pi h$. Experimentally, it is close to e^2/h [31], that is in π times larger than present estimation.

Zitterbewegung – circular motion of elementary particles caused by an interference between positive and negative energy states – leads to the fluctuation of the position of an electron. This relativistic "jittering" of an electron in graphene could be interpreted in terms of classical physics as an interaction of electron with some potential caused by the presence of disorder in crystal. Therefore, the Zitterbewegung plays a role of "intrinsic" disorder in the system which appears in the presence of minimal conductivity of the ideal crystal (without scattering) even at zero temperatures.

Under this chiral approach, we show that the optical transparency of suspended graphene can be defined by the fine structure constant, $\alpha = e^2 / \hbar c$, the parameter that describes coupling between light and relativistic electrons and is traditionally associated with quantum electrodynamics rather than condensed matter physics. Despite being only one atom thick, graphene is found to absorb a significant ($\alpha = 0.78\%$ to $\pi\alpha = 2.3\%$) fraction of incident white light, which is a consequence of graphene's unique electronic structure. This value translates directly into universal dynamic conductivity $G = e^2 / 4\pi\hbar$ within a few % accuracy.

There is a small group of phenomena in condensed matter physics, which are defined only by the fundamental constants and do not depend on material parameters. Examples are the resistivity quantum h / e^2 that appears in a variety of transport experiments, including the quantum Hall effect and universal conductance fluctuations, and the magnetic flux quantum h / e playing an important role in the physics of superconductivity (here h is the Planck constant and e the elementary charge). By and large, it requires sophisticated facilities and special measurement conditions to observe any of these phenomena. Here, we show that such a simple observable as the visible transparency of graphene [1] is defined by the fine structure constant, α. Our results are in agreement with the theory of ideal two-dimensional (2D) Dirac fermions [2,3] and its recent extension into visible optics [4], which takes into account the triangular warping and nonlinearity of graphene's electronic spectrum, but our result is π^{-1}times the result of [2,3].

Optical properties of thin films are commonly described in terms of dynamic or optical conductivity G. For a 2D Dirac spectrum with a conical dispersion relation $\varepsilon = \hbar v_F |k|$ ($k$the wavevector). G was theoretically predicted [32, 33] to exhibit a universal value$G = e^2 / 4\hbar$, our result is $G_0 = e^2 / 4\pi\hbar$if the photon energy E is much larger than both temperature and Fermi energyε_F. Both conditions are stringently satisfied in our visible-optics experiments. The universal value of G also implies that all optical properties of graphene (its transmittance T, absorption $A_{abs}$and reflection R) can be expressed through fundamental constants only (T, $A_{abs}$and R are unequivocally related to G in the 2D case). In particular, it was noted by Kuzmenko *et al* [34] that $T_{trasp} = (1 + 2\pi G_0 / c)^{-2} \approx (1 + 0.5\pi\alpha) \approx 1 - \pi\alpha$for the normal light incidence. Our result is given by$T_{trasp} \approx 1 - \alpha$. We emphasize that – unlike G – both T and R are observable quantities that can be measured directly by using graphene membranes.

7. Absorption of light by quasi 2D Dirac fermions

Here, following [35], we show how the universal value of graphene's opacity can be understood qualitatively, without calculating its dynamic conductivity. Let a light wave with electric field E and frequency ω fall perpendicular to a graphene sheet of a unit area. The incident energy flux is given by $W_i = c / 4\pi(|E|^2)$

Taking into account the momentum conservation$k$ for the initial $|i\rangle$ and final $|f\rangle$ states, only the excitation processes pictured in Fig. 10 contribute to the light absorption. The absorbed energy $W_a = n\hbar\omega$ is given by the number n of such absorption events per unit time and can be calculated by using Fermi's golden rule as $n = (2\pi / \hbar)(|L|^2)D$ where L is the matrix element for the interaction between light and Dirac fermions, and D is the density of states at $\varepsilon = E / 2 = \hbar\omega / 2$ (see Fig. A1). For quasi 2D Dirac fermions, $D(\hbar\omega / 2) = \hbar\omega / \pi\hbar^2 v_F^2$and is a linear function of ε.

The interaction between light and Dirac fermions is generally described by the Hamiltonian
$H = v_F \hat{\sigma} \cdot \hat{p} = v_F \hat{\sigma} \cdot (\hat{p} - e / cA) = H_0 + H_A$

where the first term is the standard Hamiltonian for 2D Dirac quasiparticles in graphene [1] and
$H_A = -v_F \hat{\sigma} \cdot (e / c)A = v_F \hat{\sigma} \cdot (e / i\omega)E$describes their interaction with electromagnetic field.

Here $A = (ic / \omega)E$ is the vector potential and σ the standard Pauli matrices. Averaging over all initial and final states and taking into account the valley degeneracy, our calculations yield $L^2 = |\langle f | v_F \hat{\sigma} \cdot (e / i\omega)E | i \rangle|^2 = (1/8)e^2 v_F^2 E^2 / \omega^2$

This results in $W_a=(e^2/4\hbar)\mid E\mid^2$and, consequently, absorption $A_{abs}=W_a/W_i=\pi e^2/c\hbar)=\pi\alpha$ both of which are independent of the material parameter v_F that cancels out in the calculations ofW_a. Also note that the dynamic conductivity $G=W_a/\mid E\mid^2$is equal to$e^2/4\hbar$. Our result is$A_{abs}\approx W_a/W_i=\pi e^2/\pi c\hbar)=\alpha$ in correspondence with $v_F=c(1-k_0T)$ Because graphene practically does not reflect light (R<<1 as discussed above), its opacity (1 – T) is dominated by the derived expression forA_{abs}.

In the case of a zero-gap semiconductor with a parabolic spectrum (e.g., bilayer graphene at low ε), the same analysis based on Fermi's golden rule yields$A_{abs}=2\pi\alpha$. This shows that the fact that the optical properties of graphene are defined by the fundamental constants is related to its 2D nature and zero energy gap and does not directly involve the chiral properties of Dirac fermions.

On a more general note, graphene's Hamiltonian H has the same structure as for relativistic electrons (except for coefficient vF instead of the speed of light c). The interaction of light with relativistic particles is described by a coupling constant, a.k.a. the fine structure constant. The Fermi velocity is only a prefactor for both HamiltoniansH_0 andH_A and, accordingly, one can expect that the coefficient may not change the strength of the interaction, as indeed our calculations show.

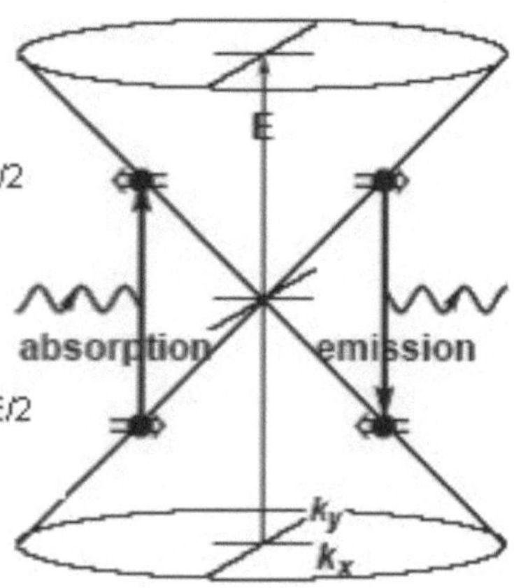

Figure 10. Excitation processes responsible for absorption of light in graphene.

Thus, we have found that the visible opacity of suspended graphene is given by $\pi\ \alpha$ within a few percent accuracy and increases proportionally to the number of layers N for few-layer graphene. Its dynamic conductivity at visible frequencies is remarkably close to the universal value of$(e^2/4\hbar)N$. The agreement between the experiment and theory is particularly striking because it was believed the universality could hold only for low energies (< 1eV) beyond which the electronic spectrum of graphene becomes strongly warped and nonlinear and the approximation of Dirac fermions breaks down.

Electrons from the valance band (bottom) are excited into empty states in the conduction band with conserving their momentum and gaining the energy $E= \hbar\omega$.

The approximation of 2D Dirac fermions is valid for graphene only close to the Dirac point and, for higher energies ε, one has to take into account such effects in graphene's band structure as triangular warping and nonlinearity.

Most theories suggest$4e^2/\pi h$, which is about π times smaller than the typical values observed experimentally. The reason of disagreement is still an open question – "mystery of a missing pie". A value of conductivity $\sim e^2/h$ at the Dirac point can emerge in case of preservation of chiral symmetry by disorder or when the dominant disorder does not scatter electrons between two valleys. Furthermore, macroscopic inhomogeneity (on the scale larger than the mean free path) is also important for measurements ofσ_{min}.

8. Conclusions

In this chapter we presented a short review on the chiral propagation of electron waves in monolayer graphene and optical simulation with optical field in the negative-zero-positive index metamaterial NZPIM and its close connection. Section II presented an enhanced vector diagram of Maxwell's equations for chiral media with quasi parallel electromagnetic fields,$E \parallel H$. Chiral waves in graphene acting as metamaterial media were discussed in section III. In section IV, two equations and tunneling rate of Dirac electron in graphene were derived. In Section V, *Zitterbewegung* of optical pulses near the Dirac point inside a negative-zero-positive index metamaterial, showed that the chiral field near the Dirac point becomes a diffusive wave. The last sections described the theoretical description of minimal conductivity in graphene under chiral approach and absorption of light by quasi 2D Dirac fermions.

We reviewed the formulation of graphene's massless Dirac Hamiltonian, under the chiral electromagnetism approach, like a metamaterial medium, hopefully demystifying the material's unusual chiral, relativistic, effective theory. The novel result here was that in our theory we did not make $c \to v_F$[1, 2, 27-35], but we obtained $v_F$as $v_F = c(1-k_0T)$ if$k_0T > 0$ or $v_F = c(1+k_0T)$ if$k_0T < 0$. These results were derived of the Chiral Electrodynamics with T as the chiral parameter and$k_0 = \omega / c$ [5, 6, 7]. This situation corresponds to chiral electrons where the Efield is almost parallel to the$\boldsymbol{H}$ field, and the movement of electron is helical in the valence band of the membrane graphene. Semi classically, for an electron in hydrogen atom, this condition was shown by Huang (2006) for E wave perpendicular to $\boldsymbol{H}$ wave with radiation ($E = H$) and by Torres-Silva for E parallel to$\boldsymbol{H}$, without radiation ($E = iH$), see refs ([5, 36, 37]). Here, we have extended this result to the electron bound to a carbon atom.

There is a small group of phenomena in condensed matter physics, which are defined only by the fundamental constants and do not depend on material parameters. Examples are the resistivity quantum h / e^2that appears in a variety of transport experiments, including the quantum Hall effect and universal conductance fluctuations, and the magnetic flux quantum h / e playing an important role in the physics of superconductivity. In this chapter we founded that under this chiral approach, the optical transparency of suspended graphene can be defined by the fine structure constant,$\alpha = e^2 / \hbar c$, that is$T_{trasp} \approx 1 - \alpha$, and $v_F / c = \alpha = (1 - k_0 T)$.

Also we give a support to the similitude of the band structure of a macroscopic photonic crystal with the electronic band structure of graphene, which is experimentally much more difficult to access, allows the experimental study of various relativistic phenomena. With our analytical and numerical analysis we hope to verify that, similar behaviors exist to electrons in graphene treated as mass-less particle, ie, electron wave discovered only 8 years ago graphene is already one of the most studied carbon allotropes. But this material still poses a lot of theoretical and experimental questions, which have to be answered.

Acknowledgements

I wish to thank to colleague Jorge Benavides Silva of the EIEE for many useful discussions on particles and photons

Author details

H. Torres-Silva*

Address all correspondence to: htorres@uta.cl

Escuela de Ingeniería Eléctrica y Electrónica Universidad de Tarapacá- Arica, Chile

References

[1] Novoselov, S., et al. (2005). *Nature*, 438, 197-200.

[2] Katsnelson, M. I., et al. (2006). *Nature Physics*, 2, 620-624.

[3] Wilton, D. (1995). *IEEE Antennas and Propagation Magazine.*, 37, 7-11.

[4] Uckun, S. (1998). *Electrotechnical Conference. Melecon 98*, 283-286.

[5] Torres-Silva, H. (2011). Chiral Transverse Electromagnetic standing waves with E||H in the Dirac Equation and the spectra of the Hydrogen Atom. ", section 2. Intech, vol Behavior of electromagnetic waves in different media and structures, edited by Ali Akdagli

[6] Torres-Silva, H. (2000). Internal publication on Chiral Electrodynamics. (in spanish), PCI:00.1036/97 UTA-CHILE

[7] Torres-Silva, H. (2008). *Ingeniare*, 16(1), pp. 43-47, 48-52.

[8] Shelby, R. A., et al. (2001). *Science* [5514], 292, 77-79.

[9] Veselago, V. G. (1968). *Soviet Physics Uspekhi*, 10(4), 509-514.

[10] Smith, D. R., et al. (2000). *Phys. Rev. Lett*, 84(18), 4184-4187.

[11] Monzon, C., & Forester, D.W. (2005). *Phys. Rev. Lett*, 95, 123904.

[12] Torres-Silva, H. (2008). *Ingeniare*, 16, 119-122.

[13] Lindell, I., Sihvola, A. H., Tretyakov, S. A., & Viitanen, y. A. J. (1994). Electromagnetic Waves in Chiral and Bi-Isotropic Media. Artech House, Boston.

[14] Plum, E., Zhou, J., Dong, J., Fedotov, V. A., Koschny, T., Soukolis, C. M., & Zheledev, y. N. I. (2009). *Physical Review B.*, 79(3), 0354071-0354076.

[15] Tretyakov, S., Nefedov, I., Sihvola, A., Maslovski, S., & Simovski, C. (2003). *J. Electromagn. Waves Appl*, 17, 695.

[16] Zhou, J., Dong, J., Wang, B., Koschny, T., Kafesaki, M., & Soukoulis, C. (2009). *Phys. Rev. B*, 79, 121104.

[17] Hannes, W. R., & Titov, M. (2010). *Europhysics Letters*, 89, 47007.

[18] Abanin, D. A., & Levitov, L. S. (2007). *Science*, 317, 641.

[19] Elias, D. C., et al. (2009). *Science*, 323, 610.

[20] Fistul, M. V., & Efetov, K. B. (2007). Phys. Rev. Lett ., 98, 256803.

[21] Yee, K. S. (1966). IEEE Transactions on Antennas and Propagation, ., 14, 302.

[22] Mur, G. (1981). *IEEE Transactions on Electromagnetic Compatibility*, EMC-23, 377.

[23] Fistul, M. V., & Efetov, K. B. (2007). *Phys. Rev. Lett*, 98, 256803.

[24] Torres-Silva, H. (2011). *Ingeniare*, 19, 67-75.

[25] Ling, Jiang-Tao, Su, Fu-Hai, Huang, Hai, & Xin, Hai. (2007). Arxiv:1007.3430 , 3

[26] Wang, Li-Gang, et al. (2009). *Opt. Lett*, 34, 1510-1512.

[27] Nomura, K., & Mac Donald, A. H. (2007). *Phys. Rev. Lett*, 98, 076602.

[28] Notomi, M. (2000). *Phys. Rev. B*, 62, 10696.

[29] Katsnelson, M. I. (2006). *Eur. Phys. J. B*, 51, 157.

[30] Zhang, F., et al. (2008). *J. Appl. Phys*, 103, 084312.

[31] Novoselov, K. S., et al. (2005). *Nature*, 438, 197-200.

[32] Gusynin, V. P., et al. (2006). *Phys. Rev. Lett.*, 96, 256802.

[33] Falkovsky, L. A., et al. (2007). *Phys. Rev. B*, 76, 153410.

[34] Kuzmenko, A. B., et al. (2008). *Phys. Rev. Lett.*, 100, 117401.

[35] Nair, R., et al. (2008). *Science*, 320, 1308.

[36] Huang, X. Q. (2006). arxiv:physics/0601169.

[37] Torres-Silva, H. (2008). *Ingeniare*, 16, 24-30.

Section 2

Numerical and Analytical Solutions

Chapter 3

Numerical Modeling of Electromagnetic Wave Propagation Through Bi-Isotropic Materials

I. Barba, A. Grande, A.C.L. Cabeceira, A. Gómez,
J.A. Pereda and J. Represa

Additional information is available at the end of the chapter

http://dx.doi.org/10.5772/50680

1. Introduction

As part of the present wave of interest in electromagnetic propagation in complex materials (left-handed, chiral, nonreciprocal, nonlinear, etc.), a special mention may be made to bi-isotropic materials. After a period of relevance during the nineties [1-4], nowadays they enjoy a renewed interest, because of new fabrication techniques that allow the use of mature technologies (printed or integrated circuit) [5,6], and also the discovery of new properties associated to the bi-isotropy, such us the presence of negative refraction index [6-11]. In order to study electromagnetic wave propagation in such materials, different traditional numerical methods, both in time and frequency domain, have been extended so that they may deal with additional constitutive parameters, as well as their inherently dispersive. This last point is especially important when dealing with time-domain methods, in which the frequency dispersion must be translated to time-domain.

In this work, we want to present a review of different techniques developed during the last years:

- Time domain techniques: Finite Differences in the Time Domain (FDTD), Transmission Line Matrix (TLM) and Multiresolution in Time Domain (MRTD). All of them have been modified to treat with bi-isotropic media, and with several methods to include frequency dispersion.
- In the frequency domain, the extension of the coupled mode method to cover the electromagnetic wave propagation in closed structures containing these new media.

INTECH
open science | open minds

© 2012 Barba et al.; licensee InTech. This is an open access article distributed under the terms of the Creative Commons Attribution License (http://creativecommons.org/licenses/by/3.0), which permits unrestricted use, distribution, and reproduction in any medium, provided the original work is properly cited.

2. Bi-Isotropic materials. Constitutive Relations

In a general bi-isotropic material, the electric and magnetic fields are coupled. There are different ways of formulating its constitutive relations in frequency domain [1]. Among them, we use:

$$\begin{aligned} \vec{D} &= \varepsilon \vec{E} + \xi \vec{H} \\ \vec{B} &= \zeta \vec{E} + \mu \vec{H} \end{aligned} \tag{1}$$

In which we assume the medium is linear. $\vec{D}$, $\vec{E}$, $\vec{B}$, and $\vec{H}$ are the electromagnetic field vectors. ▪ and ▪ are the permittivity and permeability, ▪ and ▪ would be the cross-coupling parameters: the four parameters may depend on frequency (dispersive behavior). In a bi-anisotropic medium, they have tensor form, and may contain up to 36 scalar material parameters. ▪ and ▪ are usually expressed in the following way:

$$\begin{aligned} \zeta &= \frac{\chi + j\kappa}{c} \\ \xi &= \frac{\chi - j\kappa}{c} \end{aligned} \tag{2}$$

c being the light speed in vacuum. The parameter is known as "Pasteur" parameter, while ▪ is the "Tellegen" parameter; thanks to the introduction of c in the equation, both parameters are dimensionless; at the same time, in case of non-dispersive behavior, they are also real [1]. Substituting these values into Eq. 2, it leads to:

$$\begin{aligned} \vec{D} &= \varepsilon \vec{E} + \frac{\chi - j\kappa}{c} \vec{H} \\ \vec{B} &= \mu \vec{H} + \frac{\chi + j\kappa}{c} \vec{E} \end{aligned} \tag{3}$$

Using these relations, instead of Eq. 1, the constitutive parameters have a more clear physical meaning: the Pasteur parameter describes the "chirality" of the medium, i.e. if the medium is "non chiral" (that is, it is equivalent to its mirror image), then ▪ ▪ = 0. In other case, the medium is said to be "chiral". The Tellegen parameter represents the nonreciprocity of the medium. A medium may be nonreciprocal due to other reasons (for example, a magnetized ferrite), but only a cross coupling like the one described in Eq. 3 may be nonreciprocal and isotropic (that is "bi-isotropic") at the same time [2].

While reciprocal chiral materials have been extensively researched during the last years, and different geometries and fabrication techniques have been designed [5,6], nonreciprocal materials have stayed mainly in the plane of theoretical discussion. In [12,13], Lakhtakia and Weiglhofer affirmed, using a covariance requirement deduced by Post (the so-called "Post constraint" [14]), that a bi-isotropic medium must be reciprocal, so the Tellegen parameter should always be 0. This opened a discussion on the theoretical possibility of the existence of such materials that lasted more than a decade, mainly between Lakhtakia and Weiglhofer [12,13,15-18] and Lindell, Sihvola, Tretyakov et al. [3,4,19-23]. Finally, Tretyakov et al. [24] built a "Tellegen particle" and verified its non-reciprocal behavior. At the same time, it has been found that some natural media (Fe_2TeO_6 and Cr_2O_3) though anisotropic, violate the Post constraint [25], and it may be demonstrated that only in uniform and unbounded media it may be stated that the Tellegen parameter vanishes [26-28], so "the Post constraint as a general dogma should be buried with all due honors" [29]. Other authors, nevertheless, consider the question still open [30].

An alternative formulation of Eq. 3 was proposed by Kong [31]:

$$\begin{aligned} \vec{D} &= \varepsilon \vec{E} + \frac{1}{c}\left(\chi \vec{H} - \kappa \frac{d\vec{H}}{dt}\right) \\ \vec{B} &= \mu \vec{H} + \frac{1}{c}\left(\chi \vec{E} - \kappa \frac{d\vec{E}}{dt}\right) \end{aligned} \tag{4}$$

Note that this equation is formulated in time domain, instead of frequency domain, and that the chirality parameter □ is not equivalent to the one present in Eq. 3. In addition, this equation does not take into account the intrinsically dispersive behavior of this class of materials [1,6,9-11]. To deal with that dispersion, already in 1937, E.U. Condon [32] proposed a second-order resonant model for optical chiral materials. Though in 1999 Kristensson showed that such model is non-physical [33], it is still useful as an approximation: experimental measurements show that the chirality parameter follows indeed a second order-resonant behavior, that may be modeled as a combination of two orthogonal oscillators [34]. Electric permittivity and magnetic permeability also follow a second-order resonant behavior (Lorentz model [35]) with a resonant frequency virtually identical to ω_0 [11, 36]

Electromagnetic waves in general dispersive bi-isotropic media show some interesting behavior [1]. First, the existence of the Pasteur parameter leads to:

- Optical/electromagnetic activity (OA): it is the turning of the plane of a linearly polarized electromagnetic wave about the direction of motion as the wave travels through the material. This is caused by the fact that a chiral material transmits right- and left- handed circularly polarized waves with different phase velocities.

- Optical/electromagnetic rotatory dispersion (ORD) is the variation of the optical/electromagnetic activity with frequency. Usually, that dispersion leads to an OA that changes its sign at a resonance frequency (Cotton effect).

- Circular dichroism (CD) is a modification of the nature of field polarization. It refers to the different absorption of left- and right-handed circularly polarized electromagnetic waves. As a consequence, linear polarization of a wave changes into an elliptical polarization while travelling through a chiral material.

Second, the existence of the Tellegen parameter leads to:

- Non-orthogonallity of the electric and magnetic field vectors

- Non-reciprocal scattering parameters
- Isotropic polarization and magnetization induced in it by either a uniform electric or magnetic field [19]
- Broken Time-reversal and space-inversion symmetries, that would lead with different phenomena, still under discussion [30]

Finally, in certain cases, the cross-coupling relation may lead to a negative refraction index [7, 8, 30], which has been confirmed experimentally in chiral media, as mentioned before [9-11]

3. Time Domain Modelling

3.1. Finite Differences in Time Domain (FDTD)

The FDTD method is a time-domain numerical technique for solving Maxwell's equations. Time-domain techniques produce wideband results with a single code execution and provide deeper physical insight into the phenomena under study. In addition, nonlinear or time-varying media are more easily handled in time-domain.

In the classical formulation of the FDTD approach the computational space is divided into cells where the electric and magnetic field components are distributed according to the Yee lattice [37]. Then, the space and time derivatives in Maxwell's curl equations are approximated by means of second-order central differences, obtaining a system of explicit finite-difference equations that allows us compute the transient electric and the magnetic fields as function of time. The resulting algorithm is a marching-in-time procedure that simulates the actual electromagnetic waves in a finite spatial region.

The original FDTD scheme has been successfully extended to the modeling of different kinds of materials [38]. However, the magnetoelectric coupling that characterizes the bi-isotropic media makes the extension of the FDTD method to incorporate bi-isotropic media a challenging problem. First attempts can be found in [39-41]. In these works extensions of Berenger's PML for bi-isotropic and bi-anisotropic media are developed assuming that the constitutive parameters are constant with frequency.

The schemes presented in [42-43] to incorporate bi-isotropic media are based on the field decomposition into the right- and the left-handed circularly polarized eigenwaves. These eigenwaves are uncoupled and propagate in an equivalent isotropic media with effective material parameters. Thus, the resulting sets of update equations do not differ in their general forms from the conventional FDTD update equations. Unfortunately, these schemes are only valid in some particular cases, like normal incidence on a bi-isotropic medium or free wave propagation, where the eigenwaves are not coupled. Problems like oblique incidence on a chiral slab cannot be tackled using this technique.

The formulations developed in [44-45] consider the constitutive relations proposed by Kong as governing equations. These approaches may be suitable for problems involving narrow-band signals. In [44] the algorithm is formulated for the particular case of Tellegen media. In [45] the scheme is generalized for bi-isotropic media and a closed-form stability criterion, with the same physical meaning than the conventional Courant's condition, is derived for Tellegen media.

The first formulation that considered the dispersive nature of the constitutive parameters and allowed full transient modeling of general dispersive bi-isotropic media was presented in [46-47]. In this formulation the frequency dependence of the permittivity and the permeability is represented by Lorentzian models while a single-resonance Condon model [1] [32] is adopted to represent the dispersion in the chirality parameter. In [46] the dispersion is handled using the piecewise linear recursive convolution (PLRC) technique, whereas in [47] the auxiliary differential equation (ADE) method is employed.

One important feature of the schemes given in [46-47] is the use of a new FDTD lattice developed to deal with the magnetoelectric coupling [44]. The strategy followed is to overlap two Yee cells making the *x*-, *y*-, and *z*-components to coincide at the same point, respectively, and, therefore, defining *x*-, *y*-, and *z*-nodes staggered in space and time. The advantage of this new mesh is that it simplifies the formulation since interpolation of the values of the fields is not needed. However it presents drawbacks: the standard FDTD boundary conditions are not applicable and it requires implementing a specific interface with Yee's original lattice.

Other FDTD models for dispersive chiral media have been developed in [48-49]. These schemes use Yee's conventional mesh and employ the *Z*-transform technique to convert frequency domain constitutive relations to Z-domain. However, these formulations lose the second-order accuracy in time, since in order to decouple the updating equations backward differences are used to approximate some time derivatives.

In this work we focus on an extension of the FDTD method for modeling frequency-dispersive chiral media presented in [50]. The approach consists of discretizing Maxwell's curl equations according to Yee's scheme and employing the Mobius transformation [51] to discretize the constitutive relations. This scheme overcomes the main limitations found in other approaches as it is a full-dispersive Model and is based on a modification of Yee's original cell. Moreover, the scheme preserves the second-order accuracy and the explicit nature of the original FDTD formulation.

3.2. FDTD Formulation

A Differential model

In the Laplace domain, Maxwell's curl equations can be written as

$$\begin{aligned} s\vec{B}(\vec{r}, s) &= \nabla \times \vec{E}(\vec{r}, s) \\ s\vec{D}(\vec{r}, s) &= \nabla \times \vec{H}(\vec{r}, s) \end{aligned} \tag{5}$$

and the constitutive relations for chiral media are

$$\begin{aligned}\vec{D}(\vec{r}, s) &= \varepsilon(s)\vec{E}(\vec{r}, s) - \frac{1}{c} j\kappa(s)\vec{H}(\vec{r}, s)\\ \vec{B}(\vec{r}, s) &= \mu(s)\vec{H}(\vec{r}, s) + \frac{1}{c} j\kappa(s)\vec{E}(\vec{r}, s)\end{aligned} \tag{6}$$

where the permittivity and permeability are characterized by a Lorentz model [35] and the chirality parameter by a Condon model [32]

$$\varepsilon(s) = \varepsilon_\infty + \frac{(\varepsilon_s - \varepsilon_\infty)\omega_e^2}{\omega_e^2 + 2\delta_e s + s^2} \tag{7}$$

$$\mu(s) = \mu_\infty + \frac{(\mu_s - \mu_\infty)\omega_h^2}{\omega_h^2 + 2\delta_h s + s^2} \tag{8}$$

$$\kappa'(s) = j\kappa(s) = \frac{\tau_\kappa \omega_\kappa^2 s}{\omega_\kappa^2 + 2\delta_\kappa \omega_\kappa s + s^2} \tag{9}$$

being ε_∞ (μ_∞) the permittivity (permeability) at infinite frequency, $\varepsilon_s(\mu_s)$ the permittivity (permeability) at zero frequency, ω_e, ω_h and ω_κ the resonance frequencies, δ_e, δ_h and δ_κ ▪ the loss factors and τ_κ a time constant. Introducing (6) into (5) we obtain

$$\begin{aligned}\mu_\infty s\vec{H}(\vec{r}, s) &= -\nabla \times \vec{E}(\vec{r}, s) - \vec{J}_{hh}(\vec{r}, s) - \vec{J}_{he}(\vec{r}, s)\\ \varepsilon_\infty s\vec{E}(\vec{r}, s) &= \nabla \times \vec{H}(\vec{r}, s) - \vec{J}_{ee}(\vec{r}, s) - \vec{J}_{eh}(\vec{r}, s)\end{aligned} \tag{10}$$

where the terms $\vec{J}_{hh}$, $\vec{J}_{ee}$, $\vec{J}_{eh}$ and $\vec{J}_{he}$ are auxiliary current densities defined as

$$\vec{J}_{hh}(\vec{r}, s) = \sigma_h(s)\vec{H}(\vec{r}, s) \tag{11}$$

$$\vec{J}_{eh}(\vec{r}, s) = \varsigma_h(s)\vec{H}(\vec{r}, s) \tag{12}$$

$$\vec{J}_{ee}(\vec{r}, s) = \sigma_e(s)\vec{E}(\vec{r}, s) \tag{13}$$

$$\vec{J}_{he}(\vec{r}, s) = \varsigma_e(s)\vec{E}(\vec{r}, s) \tag{14}$$

with $\sigma_h = s(\mu - \mu_\infty)$, $\varsigma_h = -s\hat{\kappa}/c$, $\sigma_e = s(\varepsilon - \varepsilon_\infty)$ and $\varsigma_e = -\varsigma_h$. $\vec{J}_{hh}$, $\vec{J}_{eh}$ and $\vec{J}_{ee}$, $\vec{J}_{he}$ are evaluated at the same point and at the same time step than $\vec{H}$ and $\vec{E}$, respectively.

B Difference model

http://dx.doi.org/10.5772/50680

A discrete time-domain version of (10) is obtained by applying the property $sF(s) \leftrightarrow dF(t)/dt$. Then, using central differences for the derivatives, average in time for $\vec{J}_{hh}$, $\vec{J}_{ee}$ and average in space for $\vec{J}_{he}$, $\vec{J}_{eh}$, the following expressions are obtained [50]

$$\begin{aligned}\vec{H}^{n+\frac{1}{2}}\left(k+\tfrac{1}{2}\right) = \vec{H}^{n-\frac{1}{2}}\left(k+\tfrac{1}{2}\right) - \frac{\Delta t}{\mu_\infty}\Big\{ [\nabla\times\vec{E}]^{n}\left(k+\tfrac{1}{2}\right) \\ + \frac{1}{2}\left[\vec{J}_{hh}^{n+\frac{1}{2}}\left(k+\tfrac{1}{2}\right)+\vec{J}_{hh}^{n-\frac{1}{2}}\left(k+\tfrac{1}{2}\right)\right] + \frac{1}{2}\left[\vec{J}_{he}^{n}(k)+\vec{J}_{he}^{n}(k+1)\right]\Big\}\end{aligned} \tag{15}$$

$$\begin{aligned}\vec{E}^{n+1}(k) = \vec{E}^{n}(k) + \frac{\Delta t}{\varepsilon_\infty}\Big\{ [\nabla\times\vec{H}]^{n+\frac{1}{2}}(k) \\ - \frac{1}{2}\left[\vec{J}_{ee}^{n+1}(k)+\vec{J}_{ee}^{n}(k)\right] - \frac{1}{2}\left[\vec{J}_{eh}^{n+\frac{1}{2}}\left(k-\frac{1}{2}\right)+\vec{J}_{eh}^{n+\frac{1}{2}}\left(k+\frac{1}{2}\right)\right]\Big\}\end{aligned} \tag{16}$$

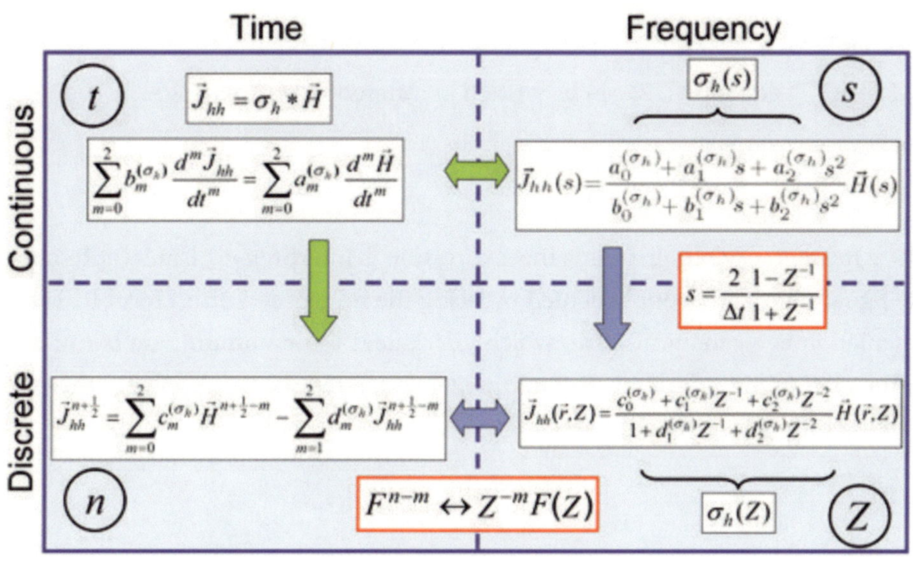

Figure 1. Transformation of the constitutive equations into the discrete-time domain.

We should now derive discrete-time expressions for the constitutive relations (11)-(14). The procedure is outlined in Figure 1. One attempt may consist on first transforming them into the continuous-time domain by again using the property $sF(s) \leftrightarrow dF(t)/dt$, and then approximating the resulting second-order differential equations by using finite differences. However, there are different ways to perform this discretization and each of them leads to different algorithms with different numerical properties. In this study, the constitutive relations (11)-(14) are discretized following an alternative way, first transforming them into the Z-domain and then obtaining the discrete-time domain version. Thus, for (11) we obtain

$\vec{J}_{hh}(\vec{r}, Z)=\sigma_h(Z)\vec{H}(\vec{r}, Z)$

where $\sigma_h(Z)$ is the Z-domain magnetic conductivity. To obtain $\sigma_h(Z)$ we adopt Mobius transformation technique [51]. This approach involves the following change of variable

$$s=\frac{2}{\Delta t}\frac{1-Z^{-1}}{1+Z^{-1}}$$

Applying this change to $\sigma_h(Z)$ yields

$$\vec{J}_{hh}(\vec{r}, Z)=\frac{c_0^{(\sigma_h)}+c_1^{(\sigma_h)}Z^{-1}+c_2^{(\sigma_h)}Z^{-2}}{1+d_1^{(\sigma_h)}Z^{-1}+d_2^{(\sigma_h)}Z^{-2}}\vec{H}(\vec{r}, Z) \tag{17}$$

where $c_m^{(\sigma_h)}$ and $d_m^{(\sigma_h)}$ are real-valued coefficients. Now considering the Z-transform property $Z^{-m}F(Z)\leftrightarrow F^{n-m}$ equation (17) can be written in difference form, as follows

$$\vec{J}_{hh}^{n+\frac{1}{2}}=\sum_{m=0}^{2}c_m^{(\sigma_h)}\vec{H}^{n+\frac{1}{2}-m}-\sum_{m=1}^{2}d_m^{(\sigma_h)}\vec{J}_{hh}^{n+\frac{1}{2}-m}$$

To reduce the memory requirements this expression is interpreted as an infinite-impulse response digital filter and is implemented by using the transposed direct form II. This type of implementation is a canonical form, which guarantees the minimum number of delay elements, thus the minimum number of additional storage variables. Applying this approach, the resulting difference equations are

$$\vec{J}_{hh}^{n+\frac{1}{2}}=\vec{W}_{hh,1}^{n-\frac{1}{2}}+c_0^{(\sigma_h)}\vec{H}^{n+\frac{1}{2}} \tag{18}$$

$$\vec{W}_{hh,1}^{n+\frac{1}{2}}=\vec{W}_{hh,2}^{n-\frac{1}{2}}-d_1^{(\sigma_h)}\vec{J}_{hh}^{n+\frac{1}{2}}+c_1^{(\sigma_h)}\vec{H}^{n+\frac{1}{2}} \tag{19}$$

$$\vec{W}_{hh,2}^{n+\frac{1}{2}}=c_2^{(\sigma_h)}\vec{H}^{n+\frac{1}{2}}-d_2^{(\sigma_h)}\vec{J}_{hh}^{n+\frac{1}{2}} \tag{20}$$

These equations are evaluated at$z=(k+(1/2))\Delta z$. Note, that they are coupled to (15). Decoupling them an eliminating $\vec{W}_{hh,2}$ we obtain

$$\vec{H}^{n+\frac{1}{2}}\left(k+\frac{1}{2}\right)=\frac{1}{2\mu_{\infty}+\Delta t c_0^{(\sigma_h)}}\left(2\mu_{\infty}\vec{H}^{n-\frac{1}{2}}\left(k+\frac{1}{2}\right)\right.$$
$$\left.-\Delta t\left\{2\,[\nabla\times\vec{E}\,]^{n}\left(k+\frac{1}{2}\right)+\vec{W}_{hh}^{n-\frac{1}{2}}\left(k+\frac{1}{2}\right)+\vec{J}_{hh}^{n-\frac{1}{2}}\left(k+\frac{1}{2}\right)+\vec{J}_{he}^{n}(k)+\vec{J}_{he}^{n}(k+1)\right\}\right) \tag{21}$$

and

$$\begin{aligned}\vec{J}_{hh}^{n+\frac{1}{2}}&=\vec{W}_{hh}^{n-\frac{1}{2}}+c_0^{(\sigma_h)}\vec{H}^{n+\frac{1}{2}}\\ \vec{W}_{hh}^{n+\frac{1}{2}}&=c_2^{(\sigma_h)}\vec{H}^{n-\frac{1}{2}}-d_2^{(\sigma_h)}\vec{J}_{hh}^{n-\frac{1}{2}}-d_1^{(\sigma_h)}\vec{J}_{hh}^{n+\frac{1}{2}}+c_1^{(\sigma_h)}\vec{H}^{n+\frac{1}{2}}\end{aligned} \tag{22}$$

where $\vec{W}_{hh}\equiv\vec{W}_{hh,1}$. Using the same procedure to discretize (12), we obtain the following equation, which is evaluated at $z=(k+(1/2))\Delta z$

$$\begin{aligned}\vec{J}_{eh}^{n+\frac{1}{2}}&=\vec{W}_{eh}^{n-\frac{1}{2}}+c_0^{(\xi_h)}\vec{H}^{n+\frac{1}{2}}\\ \vec{W}_{eh}^{n+\frac{1}{2}}&=c_2^{(\xi_h)}\vec{H}^{n-\frac{1}{2}}-d_2^{(\xi_h)}\vec{J}_{eh}^{n-\frac{1}{2}}-d_1^{(\xi_h)}\vec{J}_{eh}^{n+\frac{1}{2}}+c_1^{(\xi_h)}\vec{H}^{n+\frac{1}{2}}\end{aligned} \tag{23}$$

After that (13) is discretized in the same way than (11) y (12). The resulting equations are coupled to (16). Decoupling them we get to the next expression

$$\vec{E}^{n+1}(k)=\frac{1}{2\varepsilon_{\infty}+\Delta t c_0^{(\sigma_e)}}\left(2\varepsilon_{\infty}\vec{E}^{n}(k)\right.$$
$$\left.+\Delta t\left\{2\,[\nabla\times\vec{H}\,]^{n+\frac{1}{2}}(k)-\vec{W}_{ee}^{n}(k)-\vec{J}_{ee}^{n}(k)-\vec{J}_{eh}^{n+\frac{1}{2}}\left(k-\frac{1}{2}\right)-\vec{J}_{eh}^{n+\frac{1}{2}}\left(k+\frac{1}{2}\right)\right\}\right) \tag{24}$$

and

$$\begin{aligned}\vec{J}_{ee}^{n+1}&=\vec{W}_{ee}^{n}+c_0^{(\sigma_e)}\vec{E}^{n+1}\\ \vec{W}_{ee}^{n+1}&=c_2^{(\sigma_e)}\vec{E}^{n}-d_2^{(\sigma_e)}\vec{J}_{ee}^{n}-d_1^{(\sigma_e)}\vec{J}_{ee}^{n+1}+c_1^{(\sigma_e)}\vec{E}^{n+1}\end{aligned} \tag{25}$$

Finally, (14) is discretized using the Mobius transformation, as was previously done for (11)–(13), and implemented as

$$\begin{aligned}\vec{J}_{he}^{n+1}&=\vec{W}_{he}^{n}+c_0^{(\xi_e)}\vec{E}^{n+1}\\ \vec{W}_{he}^{n+1}&=c_2^{(\xi_e)}\vec{E}^{n}-d_2^{(\xi_e)}\vec{J}_{he}^{n}-d_1^{(\xi_e)}\vec{J}_{he}^{n+1}+c_1^{(\xi_e)}\vec{E}^{n+1}\end{aligned} \tag{26}$$

Equations (25) and (26) are evaluated at $z = k\Delta z$.

Thus, the resulting computational procedure comprises the following calculations on each time step:

1. $\vec{H}^{n+(1/2)}$ is calculated by means of (21)
2. $\vec{J}_{hh}^{n+(1/2)}$ and $\vec{W}_{hh}^{n+(1/2)}$ are updated using (22), where $\vec{H}^{n+(1/2)}$ is known from step 1.
3. $\vec{J}_{eh}^{n+(1/2)}$ and $\vec{W}_{eh}^{n+(1/2)}$ are obtained by using (23), where $\vec{H}^{n+(1/2)}$ has been obtain in 1.
4. $\vec{E}^{n+1}$ is evaluated by using (24)
5. $\vec{J}_{ee}^{n+1}$ and $\vec{W}_{ee}^{n+1}$ are calculated by means of (25), where $\vec{E}^{n+1}$ is known from step 4.
6. $\vec{J}_{he}^{n+1}$ and $\vec{W}_{he}^{n+1}$ are updated by using (26), where $\vec{E}^{n+1}$ has been obtained in 4.

This algorithm preserves the second-order accuracy and the explicit nature of the conventional FDTD formulation with only 4 additional back-stored variables per field component and cell.

3.3. Results

Wave propagation in a chiral slab

FDTD schemes emulate the progression of the fields as they actually evolve in space and time. This feature of time-domain simulators will allow us to visualize the noteworthy properties of wave propagation in chiral media.

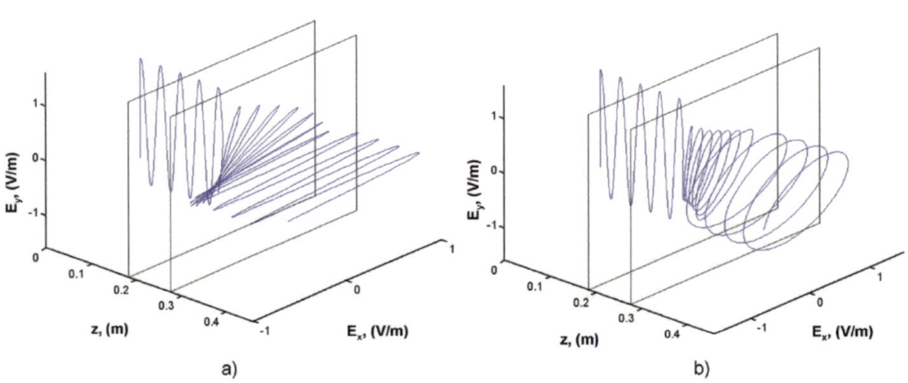

Figure 2. Evolution of the polarization of a wave traveling in a chiral slab: (a) lossless case: optical activity, (b) lossy case: optical activity and circular dichroism.

Thus, in a first simulation we have considered a time-harmonic electric field that propagates in the z positive direction and impinges on a slab of chiral material. Figure 2.a shows a snapshot of the electric field in the entire simulation domain. It can be seen how, due to the optical activity, the plane of polarization rotates as the wave travels through the chiral slab.

When losses are present in a chiral medium circular dichroism appears. The circular dichroism modifies the nature of the field polarization. To visualize this phenomenon we performed the same simulation than in the previous case but now considering a lossy chiral slab. Figure 2.b shows how the polarization rotates as the wave propagates through the chiral slab and how the linearly polarized wave degenerates progressively into an elliptically polarized wave.

Reflection and transmission from a chiral slab

Consider the propagation of an electromagnetic wave that travels in air and impinges on a slab of chiral medium. In a chiral medium, the reflected and transmitted waves can be obtained through the following reflection and transmission dyadics

$$\overline{\overline{R}} = \begin{pmatrix} R_{co} & 0 \\ 0 & R_{co} \end{pmatrix} \quad \overline{\overline{T}} = \begin{pmatrix} T_{co} & T_{cr} \\ T_{cr} & T_{co} \end{pmatrix}$$

being R_{co} the copolarized reflection coefficient and T_{co}, T_{cr} the co- and crosspolarized transmission coefficients, respectively. The crosspolarized reflection coefficient equals zero.

The dispersive behavior of the constitutive parameters were assumed to follow the models given in (7)-(9) with the following values $\varepsilon_s = 4.2\varepsilon_0$, $\varepsilon_\infty = 3.3\varepsilon_0$, $\mu_s = 1.1\mu_0$, $\mu_\infty = \mu_0$, $\omega_e = \omega_h = \omega_\kappa = 2\pi \times 7.5GHz$, $\delta_e = 0.05\omega_e$, $\delta_h = 0.05\omega_h$, $\delta_\kappa = 0.07$and $\tau_\kappa = 1ps$.

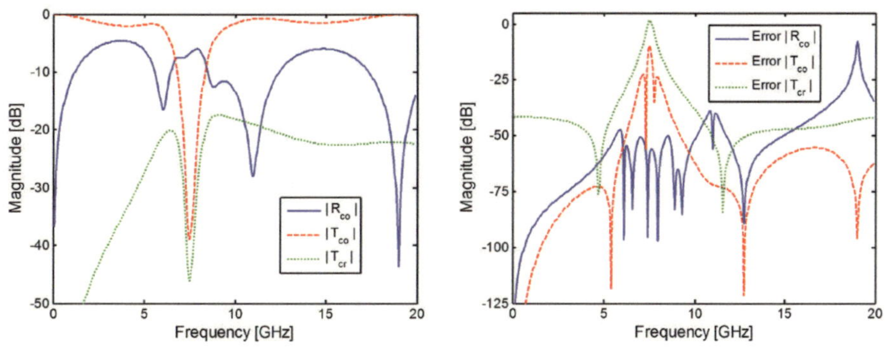

Figure 3. Reflection and transmission coefficients: theoretical values and relative errors.

A band limited Gaussian pulse was injected at a source point in the air region. The time step was $\Delta t = 1$ps and the space discretization$\Delta z = 0.3$mm. The thickness of the slab was $d = 30\Delta z$. Figure 3 depicts the theoretical reflection and transmission coefficients [1] and the relative error of the coefficients computed by the present FDTD formulation. These results demonstrate the accuracy and the full transient capability of this FDTD approach.

4. Multiresolution in Time Domain (MRTD)

Multiresolution in Time Domain (MRTD) technique is a new approach to reduce the simulation time of FDTD schemes (actually, it may be considered a generalization of FDTD [52]), keeping the same degree of accuracy. In 1996, Krumpholz showed that Yee's FDTD scheme could be derived by applying the method of moments for the discretization of Maxwell's equations, using pulse basis functions for the expansion of the unknown fields [53]. In a MRTD scheme the expansion is completed by means of a twofold expansion in scaling and wavelet functions with respect to time/space. Scaling functions takes into account smoothly varying fields. In regions or time periods characterized by strong field variations or field singularities, higher resolution is enhanced by incorporating wavelet functions in the field expansions. The major advantage of the use of MRTD in the time domain is the capability to develop time and space adaptive grids. The scheme is, then, a generalization of Yee's FDTD and can extend this technique's capabilities by improving computational efficiency and substantially reducing computer resources.

The MRTD approach has been successfully applied to finite-difference time-domain (FDTD) simulations of passive structures [53-56]. In those cases, multiresolution expansion has been usually applied to the space dependence of the electromagnetic fields. In [57], the expansion was applied to the time dependence of the fields. In this case, such expansion has the purpose of improving the resolution in the calculation of convolutions in dispersive media. As we have seen previously, the chirality parameter, present in eq. (3) is intrinsically dispersive (eqs. (7)-(9)), so we expand here a similar technique suitable for chiral media.

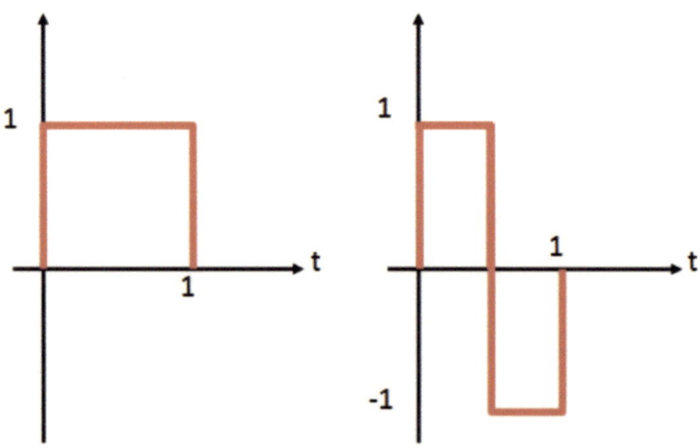

Figure 4. Haar wavelet functions: left, scale function $\Phi_n(t)$. Right, first order wavelet function $\psi_n(t)$. Both cases are particularized for $n = 1$.

To discretize the constitutive equations (2), we must develop all the components of the field vectors ($\vec{E}$, $\vec{D}$, $\vec{B}$, $\vec{H}$) in scale and wavelet functions:

$$F=\sum_{i}\phi_i(z)\sum_{n}\left({}_nF_i^{\phi}\phi_n(t)+{}_nF_i^{\psi}\psi_n(t)\right) \tag{27}$$

F may be any component of any of the field vectors above mentioned. We have included two levels of resolution in the time dependence of the function, while the space discretization is represented only in one dimension, assuming dependence on z. The extension to three dimensions is straightforward: if the scale (Φ) and wavelet (ψ) functions are Haar functions [58], as shown in Figure 4, the space discretization is equivalent to the FDTD discretization explained in the previous section. That means that the discretization of Maxwell curl equations will lead to two parallel Yee's networks, one for scale coefficients (${}_nF_i^{\phi}$), and a second one for wavelet coefficients (${}_nF_i^{\psi}$).

The results for both networks are combined in the constitutive relations (Eq. (3)). If we discretize them in time domain, using a recursive convolution method (in a similar way to FDTD [46]), we obtain the following expressions (x component of $\vec{D}$, $\vec{B}$):

$$\begin{aligned}
D^x(z,t)&=\sum_i\left\{\begin{array}{c}\sum_n \varepsilon\left({}_nE_i^{x,\phi}\phi_n(t)+{}_nE_i^{x,\psi}\psi_n(t)\right)\\ -\sum_m\left\{\frac{1}{c}\int_0^t\left({}_{m'}H_i^{x,\phi}\phi_m(t-\tau)+{}_mH_i^{x,\psi}\psi_m(t-\tau)\right)\kappa'(\tau)d\tau\right\}\end{array}\right\}\phi_i(z)\\
B^x(z,t)&=\sum_i\left\{\begin{array}{c}\sum_n \mu\left({}_nH_i^{x,\phi}\phi_n(t)+{}_nH_i^{x,\psi}\psi_n(t)\right)\\ +\sum_m\left\{\frac{1}{c}\int_0^t\left({}_mE_i^{x,\phi}\phi_m(t-\tau)+{}_mE_i^{x,\psi}\psi_m(t-\tau)\right)\kappa'(\tau)d\tau\right\}\end{array}\right\}\phi_i(z)
\end{aligned} \tag{28}$$

where κ′(τ) is the inverse Laplace transform of the chirality parameter in frequency domain κ(ω). In this equation${}_n E_i^{x,\phi}$, for example, would represent the coefficient for the scale function of the *x*-component of the electric field$\vec{E}$, when $t = n\ t$ and $z = i$ □ z. Note that the equation includes only the dispersion in the chirality response: dielectric and magnetic response could be included as in [57]. If we are working with Haar functions, this equation may be simplified [59], obtaining:

$$\begin{aligned}
D^x(z,t)&=\sum_{i,n}\left\{\begin{array}{c}\varepsilon\left({}_nE_i^{x,\phi}\phi_n(t)+{}_nE_i^{x,\psi}\psi_n(t)\right)\\ -\frac{1}{c}\phi_n(t)\sum_m\left\{{}_{n-m}H_i^{x,\phi}\kappa'^{\phi}_m+{}_{n-m}H_i^{x,\psi}\kappa'^{\psi}_m\right\}\end{array}\right\}\phi_i(z)\\
B^x(z,t)&=\sum_{i,n}\left\{\begin{array}{c}\mu\left({}_nH_i^{x,\phi}\phi_n(t)+{}_nH_i^{x,\psi}\psi_n(t)\right)\\ +\frac{1}{c}\phi_n(t)\sum_m\left\{{}_{n-m}E_i^{x,\phi}\kappa'^{\phi}_m+{}_{n-m}E_i^{x,\psi}\kappa'^{\psi}_m\right\}\end{array}\right\}\phi_i(z)
\end{aligned} \tag{29}$$

where $\kappa'^{\phi}_{n-m}=\int_{(n-m)\Delta t}^{(n-m+1)\Delta t}\kappa'(\tau)d\tau,\kappa'^{\psi}_{n-m}=\int_{(n-m+1/2)\Delta t}^{(n-m+1)\Delta t}\kappa'(\tau)d\tau-\int_{(n-m)\Delta t}^{(n-m+1/2)\Delta t}\kappa'(\tau)d\tau$. If we compare these equations with (27), in order to separate scale and wavelet coefficients of D^x and B^x, we find:

$$
\begin{aligned}
{}_nD_i^{x,\phi} &= \varepsilon_n E_i^{x,\phi} - \frac{1}{c_0}\sum_m \left\{ {}_{n-m}E_i^{x,\phi}\kappa'^{\phi}_m + {}_{n-m}E_i^{x,\psi}\kappa'^{\psi}_m \right\} \\
{}_nB_i^{x,\phi} &= \mu_n H_i^{x,\phi} + \frac{1}{c_0}\sum_m \left\{ {}_{n-m}E_i^{x,\phi}\kappa'^{\phi}_m + {}_{n-m}E_i^{x,\psi}\kappa'^{\psi}_m \right\}
\end{aligned}
\tag{30}
$$

$$
\begin{aligned}
{}_nD_i^{x,\psi} &= \varepsilon\cdot{}_nE_i^{x,\psi} \\
{}_nB_i^{x,\psi} &= \mu\cdot{}_nH_i^{x,\psi}
\end{aligned}
\tag{31}
$$

As we may see, the convolutions sums appear in eq. (30), i.e., they affect only the coefficients of the scale functions of $\vec{D}$ and $\vec{B}$ (${}_nD_i^{x,\phi}$ and ${}_nB_i^{x,\phi}$): if the wavelet functions give account of the faster variations of the fields, it is logic that they do not depend on the long term values stored in the integral is reasonable.

So, we may follow a procedure similar to the one described in [57] for isotropic dispersive materials, i.e. the algorithm would be composed by two different steps:

1. Propagation of the field, following the discretized Maxwell curl equations. We obtain updated values of vectors $\vec{D}$ and $\vec{B}$, using the previous values of $\vec{E}$ and $\vec{H}$.
2. Updating of the values of $\vec{E}$ and $\vec{H}$, using the constitutive relations of the medium; we should update first the wavelet (ψ) coefficients, following Eq. (31), because their value is necessary to solve Eq. (30) and obtain the new values for the scale (ϕ) coefficients.

4.1. Results

To validate numerically the accuracy of our approximation, we have simulated the propagation of a plane wave linearly polarized through a chiral medium, with $\varepsilon=2\varepsilon_0$, $\mu=\mu_0$, and κ following the Condon model: ω_0=1 GHz, τ=3 ps and a damping factor δ=0 (medium without losses). To simplify the results, we have not taken into account the dielectric and magnetic dispersions (they do not affect the electromagnetic activity). The discretization is Δt = 1 ps and Δx = 0.375 mm. We have introduced a harmonic source, with a frequency of 7 GHz, and after 8000 time steps, we have calculated the rotation of the polarization plane in different points: the result is compared with the one obtained using a FDTD scheme [46]. The results are shown in Table 1: [59]

	Theoretical value	FDTD		MRTD	
		Value	Error	Value	Error
1000Δz	11.55°	11.54°	0.09%	11.54°	0.09%
2000Δz	23.10°	23.09°	0.04%	23.08°	0.09%
3000Δz	34.64°	34.56°	0.2%	34.65°	0.03%
4000Δz	46.21°	46.11°	0.2%	46.04°	0.15%

Table 1. Rotation of the polarization plane during the propagation of a plane wave through a chiral medium [59]

5. Transmission Line Matrix (TLM)

The Transmission Line Matrix (TLM) method is a numerical technique that solves the wave equation of a propagation phenomenon in time domain. As in the previous cases, the basic TLM network [60] must be modified to include the cross terms emerging in the Maxwell equations, thus in the wave equation, when dealing with bi-isotropic media. We consider the propagation of waves in the xy–plane: although it is a bi-dimensional problem, the medium presents cross coupling on all the field components on every plane perpendicular to propagation direction, so the number of unknown field components cannot be reduced less than twelve.

After a simple manipulation, Maxwell equations can be arranged in four sets of equations as shown in the Table 2, related by the constitutive relations described in equation (4).

where Φ is defined as$\Phi^2 = \varepsilon\mu - \chi^2\varepsilon_0\mu_0$, and must have a non–zero value. This mathematical condition, verified by all bi-isotropic media [1] means that any wave solution has associated a finite value of the phase velocity.

{1}	$\frac{\partial E_z}{\partial x} = \frac{\partial B_y}{\partial t}$,	$\frac{\partial E_z}{\partial y} = -\frac{\partial B_x}{\partial t}$,	$\frac{\partial B_y}{\partial x} - \frac{\partial B_x}{\partial y} = {}^2\frac{\partial E_z}{\partial t} - \frac{\kappa}{c}\frac{\partial^2 Q_{mz}}{\partial t^2}$
{2}	$\frac{\partial H_z}{\partial x} = -\frac{\partial D_y}{\partial t}$,	$\frac{\partial H_z}{\partial y} = \frac{\partial D_x}{\partial t}$,	$\frac{\partial D_y}{\partial x} - \frac{\partial D_x}{\partial y} = -\Phi^2\frac{\partial H_z}{\partial t} - \frac{\kappa}{c}\frac{\partial^2 Q_{ez}}{\partial t^2}$
{3}	$\frac{\partial D_z}{\partial x} = \Phi^2\frac{\partial H_y}{\partial t} + \frac{\kappa}{c}\frac{\partial^2 Q_{ey}}{\partial t^2}$,	$\frac{\partial D_z}{\partial y} = -\Phi^2\frac{\partial H_x}{\partial t} - \frac{\kappa}{c}\frac{\partial^2 Q_{ex}}{\partial t^2}$,	$\frac{\partial H_y}{\partial x} - \frac{\partial H_x}{\partial y} = \frac{\partial D_z}{\partial t}$
{4}	$\frac{\partial B_z}{\partial x} = -\Phi^2\frac{\partial E_y}{\partial t} - \frac{\kappa}{c}\frac{\partial^2 Q_{my}}{\partial t^2}$,	$\frac{\partial B_z}{\partial y} = \Phi^2\frac{\partial E_x}{\partial t} + \frac{\kappa}{c}\frac{\partial^2 Q_{mx}}{\partial t^2}$,	$\frac{\partial E_y}{\partial x} - \frac{\partial E_x}{\partial y} = -\frac{\partial B_z}{\partial t}$

Table 2. The Four Sets of Electromagnetic Field Vectors Equations [62]

To obtain a simplified representation, two next vectors, with no physical meaning, have been introduced:

$$\vec{Q}_m = \mu\vec{H} + \frac{X}{c}\vec{E} + \vec{B}, \qquad \vec{Q}_e = \varepsilon\vec{E} + \frac{X}{c}\vec{H} + \vec{D} \tag{32}$$

For every set of resulting equations, we may derive an independent wave equation describing the propagation, and having the form:

$$\frac{\partial^2 A}{\partial x^2} + \frac{\partial^2 A}{\partial y^2} - \phi^2\frac{\partial^2 A}{\partial t^2} + \frac{\kappa}{c}\frac{\partial^3}{\partial t^3}\left[R + \frac{\partial S}{\partial t}\right] = 0 \tag{33}$$

A	R	S
E_z	Q_{mz}	0
H_z	$-Q_{ez}$	0
D_z	$\varepsilon Q_{mz} - \frac{X}{c}Q_{ez}$	Q_{ez}
B_z	$\frac{X}{c}Q_{mz} - \mu Q_{ez}$	Q_{mz}

Table 3. Symbols*A*, *R*and *S* in wave equations [62]

The symbol *A* represents whatever *z*-component of the electromagnetic field vectors, and each symbol *R* or *S* equals the *z*-component of a linear combination of $\vec{Q}_m$ and $\vec{Q}_e$ (both introduced in (32)) depending on the component *A*, as detailed in Table 3.

5.1. TLM Modeling

The classical TLM method works with interconnected transmission–lines along the coordinate axes establishing a network. The points at which the transmission–lines intersect are referred to as "nodes". Then, space and time are discretized, and voltage and current pulses scatter from point to point in space, $\Delta\ell$, in a fixed time stepΔt, [60]. The starting point to obtain any TLM model is to discretize the electromagnetic field equations (either the Maxwell equations in Table 2 or the wave equations (33)), and to compare it with the equations modeled by the TLM algorithm, that is, the equations relating the voltages and currents in the transmission–line network. For a parallel connection of the transmission–line in the *xy*–plane, the voltages and currents are related by the differential equations:

$$\frac{\partial V_z}{\partial x} = -L\frac{\partial I_x}{\partial t} \qquad \frac{\partial V_z}{\partial y} = -L\frac{\partial I_y}{\partial t} \qquad \frac{\partial I_x}{\partial x} + \frac{\partial I_y}{\partial y} = -2C\frac{\partial V_z}{\partial t} \tag{34}$$

where *L* and *C* represent the inductance and the capacitance per unit length of the transmission–lines, respectively. In these equations, the voltage V_z, and the currents I_x and I_y symbolize the standard electric quantities in the transmission–line network [60].

A look at Table 2 allows us to implement the model of every set of equations with a classical TLM mesh of shunt-connected nodes. Furthermore, the twelve field vectors components are required in the new model due to the existing magneto–electric coupling, even in the 2D case. In this way, a single set is not enough, but four meshes are needed to simulate each of the twelve field components. Every TLM network will be labeled {1}, {2}, {3}, or {4}, following their equivalent set of Maxwell equations in Table 2. This leads to a specific analogy between each TLM network electric quantity and its respective electromagnetic field vectors component, as shown in Table 4 where$\phi_r = \phi / c$.

	{1}	{2}	{3}	{4}
$V_z \equiv ...$	E_z	$-\mu_0 c H_z$	D_z / ε_0	$-cB_z$
$-I_x \equiv ...$	B_y / μ_0	cD_y	$\Phi_r^2 H_y$	$\Phi_r^2 \varepsilon_0 c E_y$
$I_y \equiv ...$	B_x / μ_0	cD_x	$\Phi_r^2 H_x$	$\Phi_r^2 \varepsilon_0 c E_x$

Table 4. Analogies between TLM meshes quantities and field vectors [62].

{1}	$\frac{\partial V_z}{\partial x} = -\mu_0 \frac{\partial I_x}{\partial t},$	$\frac{\partial V_z}{\partial y} = -\mu_0 \frac{\partial I_y}{\partial t},$	$\frac{\partial I_x}{\partial x} + \frac{\partial I_y}{\partial y} = -\frac{\varphi^2}{\mu_0}\frac{\partial V_z}{\partial t} + \frac{\kappa}{\eta_0}\frac{\partial^2 Q_{mz}}{\partial t^2}$
{2}	$\frac{\partial V_z}{\partial x} = -\mu_0 \frac{\partial I_x}{\partial t},$	$\frac{\partial V_z}{\partial y} = -\mu_0 \frac{\partial I_y}{\partial t},$	$\frac{\partial I_x}{\partial x} + \frac{\partial I_y}{\partial y} = -\frac{\varphi^2}{\mu_0}\frac{\partial V_z}{\partial t} + \kappa\frac{\partial^2 Q_{ez}}{\partial t^2}$
{3}	$\frac{\partial V_z}{\partial x} = -\mu_0 \frac{\partial I_x}{\partial t} + \kappa\eta_0 \frac{\partial^2 Q_{ey}}{\partial t^2},$	$\frac{\partial V_z}{\partial y} = -\mu_0 \frac{\partial I_y}{\partial t} - \kappa\eta_0 \frac{\partial^2 Q_{ex}}{\partial t^2},$	$\frac{\partial I_x}{\partial x} + \frac{\partial I_y}{\partial y} = -\varphi_r^2 \varepsilon_0 \frac{\partial V_z}{\partial t}$
{4}	$\frac{\partial V_z}{\partial x} = -\mu_0 \frac{\partial I_x}{\partial t} - \kappa \frac{\partial^2 Q_{my}}{\partial t^2},$	$\frac{\partial V_z}{\partial y} = -\mu_0 \frac{\partial I_y}{\partial t} + \kappa \frac{\partial^2 Q_{mx}}{\partial t^2},$	$\frac{\partial I_x}{\partial x} + \frac{\partial I_y}{\partial y} = -\varphi_r^2 \varepsilon_0 \frac{\partial V_z}{\partial t}$

Table 5. The differential equations for the four TLM meshes [62].

The corresponding TLM equivalences of the vectors $\vec{Q}_m$ and$\vec{Q}_e$, introduced in (32), are:

$$\left.\begin{aligned} Q_{mx} &= \frac{\mu_0}{\phi_r^2}\left[\mu_r I_y\{3\} + \chi I_y\{4\} + \phi_r^2 I_y\{1\}\right] \\ Q_{my} &= -\frac{\mu_0}{\phi_r^2}\left[\mu_r I_x\{3\} + \chi I_x\{4\} + \phi_r^2 I_x\{1\}\right] \\ Q_{mz} &= -\frac{1}{c}\left[\mu_r V_z\{2\} + \chi V_z\{1\} + V_z\{4\}\right] \end{aligned}\right\} \quad \left.\begin{aligned} Q_{ex} &= \frac{1}{\phi_r^2 c}\left[\varepsilon_r I_y\{4\} + \chi I_y\{3\} + \phi_r^2 I_y\{2\}\right] \\ Q_{ey} &= -\frac{1}{\phi_r^2 c}\left[\varepsilon_r I_x\{4\} + \chi I_x\{3\} + \phi_r^2 I_x\{2\}\right] \\ Q_{ez} &= -\varepsilon_0\left[\varepsilon_r V_z\{1\} - \chi V_z\{2\} + V_z\{3\}\right] \end{aligned}\right\} \tag{35}$$

Again, these vectors have no special physical meaning. Please note that whatever the relationship between field components and TLM variables is, these networks have the same wave propagation characteristics of the classic TLM network, well studied in the literature, [60]. Therefore, the numerical dispersion and consequently the velocity error of the bi-isotropic TLM network are well known.

To take into account the propagation velocity (the same in the four meshes following (3)), a permittivity stub is shunt connected at all the nodes. The normalized admittance of these stubs is $y_o=4(\phi_r^2-1)$. Then, propagation velocities into each TLM network depend onϕ, different from the velocity in the modeled bi-isotropic medium. The same transmission–lines characteristics are imposed for all the meshes, and consequently the TLM meshes impedances Z_{TLM} equalη_o / ϕ_r, where η_o denotes the vacuum intrinsic impedance.

Additionally, a new circuit element is needed at every node in order to incorporate the terms in Table 2 with no equivalence in Table 4. The technique developed in [60] to model the waves propagation in complex media, based on the voltage source connection, is used for this purpose. It is important to remark that, since these voltages sources V_s^k will introduce the adequate signal correction at any time iteration, we obtain a TLM algorithm that models the correct propagation velocity.

The simulation begins with the excitation of one or several field vectors components, depending on the problem (symmetry, boundaries, output, ...). Regarding the equivalences in Table 4, voltage impulses are initially introduced into the system according to the desired field excitation. The scattering and incidence processes of these voltage pulses follow the classical TLM method procedures [60], but, for a bi-isotropic 2D–TLM model, the algorithm is not yet completed, since at every time iteration *k*, the voltages at each node must be transferred from a pair of networks to the other pair:

$$\begin{aligned}
V_n\{3\} &= \varepsilon_r V_n\{1\} - \chi V_n\{2\} + \frac{\kappa}{\Delta t} \Delta_t^1 V_n\{2\} \\
V_n\{4\} &= \mu_r V_n\{2\} - \chi V_n\{1\} - \frac{\kappa}{\Delta t} \Delta_t^1 V_n\{1\} \\
V_n\{1\} &= \frac{1}{\phi_r^2}\left[\mu_r V_n\{3\} + \chi V_n\{4\} - \frac{\kappa}{\Delta t} \Delta_t^1(\mu_r V_n\{2\} - \chi V_n\{1\})\right] \\
V_n\{2\} &= \frac{1}{\phi_r^2}\left[\varepsilon_r V_n\{4\} + \chi V_n\{3\} + \frac{\kappa}{\Delta t} \Delta_t^1(\varepsilon_r V_n\{1\} - \chi V_n\{2\})\right]
\end{aligned} \tag{36}$$

Here the index n denotes the branch number (n = 15) of the node, and the symbol Δ_t^d represents the d–order time variation of the pulses. The values of the new elements to be connected at the TLM node equivalent circuit are obtained from the discretized form of wave equations (33). The TLM method deals with the total voltage at the node, so it is more useful to write the wave equations in terms of such voltages. For example, the wave equation for the *z*- component of the electric field become for the mesh {1}:

$$\frac{\partial^2 {}_k V_T\{1\}}{\partial x^2} + \frac{\partial^2 {}_k V_T\{1\}}{\partial y^2} - \phi_r^2 \varepsilon_0 \mu_0 \frac{\partial^2 {}_k V_T\{1\}}{\partial t^2} + \frac{\kappa}{c}\frac{\partial^3}{\partial t^3}\left[-\frac{\mu}{\mu_0} {}_k V_T\{2\} + \chi_k V_T\{1\} - {}_k V_T\{4\} \right] = 0 \tag{37}$$

The voltage of the four sources V_s^k at every time iteration k can be obtained directly from these TLM wave equations. The resulting expressions for each TLM mesh are:

$$\begin{aligned} V_s^{k+1}\{1\} &= V_s^{k-1}\{1\} - \frac{1}{\phi_r^2}\frac{\kappa}{\Delta t}[\gamma]^k \\ V_s^{k+1}\{2\} &= V_s^{k-1}\{2\} + \frac{1}{\phi_r^2}\frac{\kappa}{\Delta t}[\rho]^k \\ V_s^{k+1}\{3\} &= V_s^{k-1}\{3\} - \frac{1}{\phi_r^2}\frac{\kappa}{\Delta t}\left[\varepsilon_r \gamma + \chi \rho - \frac{\kappa}{\Delta t}\Delta_t^1 \rho\right]^k \\ V_s^{k+1}\{4\} &= V_s^{k-1}\{4\} + \frac{1}{\phi_r^2}\frac{\kappa}{\Delta t}\left[\mu_r \rho + \chi \gamma + \frac{\kappa}{\Delta t}\Delta_t^1 \gamma\right]^k \end{aligned} \tag{38}$$

where the symbols and indicate:

$$\gamma = \Delta_t^3\left[\mu_{rk} V_T\{2\} - \chi_k V_T\{1\} + {}_k V_T\{4\}\right] \qquad \rho = \Delta_t^3\left[\varepsilon_{rk} V_T\{1\} - \chi_k V_T\{2\} + {}_k V_T\{3\}\right] \tag{39}$$

The sources are introduced into the algorithm by adding them to the voltage impulses, before the scattering process, as follows:

$${}_k V_T\{m\} = \frac{2}{4 + y_0}\left(\sum_{n=1}^{4} {}_k V_n^i\{m\} + y_{0k} V_5^i\{m\}\right) + V_s^k\{m\} \tag{40}$$

Finally, the information to get out any field vectors component, at any point of the network and at any time $k\,\Delta t$, may be found in Table 4. The voltage V_z at the node represents the total voltage V_T (10), and the currents I_x and I_y are proportional to the voltage pulses difference. For example, the electric field strength components can be expressed in terms of the incident voltage pulses upon the node:

$$\begin{aligned} {}_k E_x &= \frac{{}_k V_3^i\{4\} - {}_k V_1^i\{4\}}{\Delta \ell \sqrt{2}\phi_r^2} \qquad {}_k E_y = \frac{{}_k V_2^i\{4\} - {}_k V_4^i\{4\}}{\Delta \ell \sqrt{2}\phi_r^2} \\ {}_k E_z &= \frac{\sum_{n=1}^{4} {}_k V_n^i\{1\} + y_{0k} V_5^i\{1\}}{\Delta \ell 2 \phi_r^2} + \frac{V_s^k\{1\}}{\Delta \ell} \end{aligned} \tag{41}$$

Finally, all these steps (pulse scattering, node connection, signal transfer between meshes (36), voltage sources updating (38), voltage impulses summation (40), output of the fields values (41)…) are repeated as many time steps as required.

5.2. Results

To validate the TLM model, we have modeled again the plane wave propagation in bi-isotropic media and compared the results with known theoretical ones [1]. In order to simulate a plane wave propagating along the *x*–direction, the mesh boundaries are modeled with absorbing conditions [62] for limits in *x*–direction to avoid not desired reflections, and magnetic walls for the *y*–direction. It is to be noted that in the four meshes (Table 4), the total voltage at the node is either equivalent to an electric component ({1} and {3}) or to a reversed–sign magnetic component ({2} and {4}) of the electromagnetic field, thus the same boundary conditions in the *y*–direction have been modeled at all the meshes.

The propagation of a plane wave propagation in a bi-isotropic medium with relative permittivity and permeability ε_r = 5.0 and μ_r=1.5, Tellegen parameter χ = 1.2, and Pasteur parameter (Kong model) κ= –0.01 ps is now computed. In the simulation, the computational space domain is n_x = 5,000 nodes, and n_y = 10 nodes along the *x*– and *y*–axes, respectively, with a space step $\Delta\ell$ = 1.00 mm. A monochromatic excitation of frequency 4.00 GHz is applied at the plane $x = 500\Delta\ell$. The excitation is introduced with linear polarization in the *z*–direction of the electric field strength$E_z = E_0\sin(\omega t)$, then:

$$ {}_kV_n\{1\} = {}_{k-1}V_n\{1\} + \frac{1}{2}E_0\sin(\omega t)\ \mathit{for}\, n = 1, \ldots, 5 \tag{42}$$

	Theoretical values	TLM results	Relative Error (%)
θ_{EH}	115.99°	116.09°	–0.09
θ_{DB}	64.01°	64.12°	–0.17
$10^2\,\eta_{EH}/_o$	54.77	54.63	+0.26
$10^2\,\eta_{DB}/_o$	54.77	54.92	–0.27
$10\,\varepsilon_r$	44.94	44.87	+0.16
$10\mu_r$	13.48	13.46	+0.15

Table 6. Angles, impedances and effective parameters [62]

In Table 6, the angle tilt between electric and magnetic field vectors θ_{EH} and θ_{DB}, the intrinsic impedances η_{EH} and η_{DB}, and the effective relative permittivity and permeability are shown after *10,000* time iterations (Δt= 2.36 ps) of the algorithm. In Table 7, the rotation of the angle of the polarization plane at different distances from the source is presented. The respective theoretical values [62] and relative errors are included also in these two tables of results: they are in good agreement with the theoretical values, with relative errors less than *0.3* %.

Distance to the source (Δℓ)	Theoretical values	TLM results	Relative Error (%)
500	90.60º	90.74º	–0.15
1000	91.21º	91.36º	–0.17
1500	91.81º	91.97º	–0.18
2000	92.41º	92.60º	–0.20
2500	93.02º	93.26º	–0.26

Table 7. Rotation of the polarization plane [62]

6. Frequency Domain Modelling: The Coupled Mode Method (CMM)

The Coupled Mode Method is one of the available numerical methods in the frequency domain used for analyzing the electromagnetic wave propagation inside waveguides that contain any isotropic, anisotropic or complex medium [63-69]. Originally, this method was formulated by Schelkunoff [63] for uniform closed structures involving heterogeneous and non-isotropic media. Afterwards, it was extended to cover other structures such us open dielectric guides [64], waveguides containing bi-isotropic media [65-69], and free propagation in unbounded bi-isotropic media [70].

Moreover, the application of this numerical method is not only restricted to the analysis of uniform structures. In fact, one of the main advantages of the coupled mode method is that it can be combined with the Mode Matching Method (MMM) in order to analyze discontinuities between two waveguides that contain isotropic, anisotropic or complex media. Thus, joining the coupled mode method and the mode matching method a hybrid numerical technique capable of analyzing 3D structures implemented in a waveguide that presents discontinuities in the energy propagation direction is obtained [71-72]. Due to the mode matching method properties, there is no limitation in the number of discontinuities that we can analyze. Therefore 3D periodic structures can also be examined with this numerical tool.

6.1. Theory

In few words, the coupled mode method is a method of moments [63] which consists on expanding the electromagnetic field components inside a uniform waveguide partially filled with any of the former media in terms of a set of base functions previously defined. Thus, the general expression of any electromagnetic field component is expressed as

$$G(x,\ y,\ z)=\sum_{n=1}^{\infty} c_n g_n(x,\ y,\ z)=\sum_{n=1}^{\infty} c_n \tilde{g}_n(x,\ y) e^{-\Gamma z} \tag{43}$$

where c_n are the expansions coefficients, $\tilde{g}_n$the selected base functions and the propagation constant of the *proper modes*, i.e., the modes of the partially filled waveguide. In many cases these base functions are the electric and magnetic fields of the different modes supported by the *basis structure,* i.e., the empty waveguide. Therefore, they are called *basis modes*. When the guiding structure is a rectangular waveguide these basis modes correspond to the *TE* and *TM* modes of the empty rectangular waveguide [63].

The obtention of the expansions coefficients and the propagation constants of the proper modes constitutes the mathematical procedure of the coupled mode method. Mainly, it consists on selecting two electromagnetic field vectors as primary expansions[1] and then substituting them into the x- and y- components of the Maxwell Equations.

Next, the other two field vectors are expressed in terms of the primary ones through the constitutive relations (3). Thus, the obtained equations only depend of the primary fields expansions. In order to solve the problem, the Galerkin method is applied [73] and a set of differential equations, named the Generalized Telegraphist Equations, is obtained [63].

Finally, after some analytical work involving the z-components of the Maxwell Equations [63,69] and the constitutive relations, we reach to an eigenvalue problem whose solution provides the expansion coefficients and the propagation constants of the proper modes

$$\{[A]-\Gamma[U]\}[C]=0. \tag{44}$$

In Eq. (44) $[A]$ is a coupling matrix between the basis modes taken into account in the procedure, $[U]$ is an unitary matrix, is a vector related to the propagation constants of the electromagnetic field and [C] is a matrix containing the coefficients of the electromagnetic field expansions. In following sections this procedure is analyzed in more detail.

6.1.1. A. The different formulations of the coupled mode method:

As it is know, the electromagnetic field is described by four field vectors $\vec{H}$and, $\vec{B}$which are related among themselves through the constitutive relations. Thus, only the expansions of two primary fields have to be proposed: one electric ($\vec{E}$or$\vec{H}$) and another magnetic ($\vec{E}$or$\vec{B}$). The other two field vectors are obtained by substituting the primary fields into the constitutive relations. Hence, for the same problem, the coupled mode method can be formulated in different ways giving rise to different formulations that produce similar solutions.

So far, in the literature two alternative formulations of the coupled mode method can be found: the EH- and EB-formulations [67-69]. In the first one the primary fields are $\vec{E}$ and$\vec{H}$, whereas in the second one the primary fields are $\vec{E}$ and$\vec{B}$. As we will see later, the main difference between the EH- and EB-formulations of the coupled mode method resides in the satisfaction or not of all boundary conditions of the electromagnetic field components over the perfect electric conductor that define the waveguide. In particular, the issue is related to

1 As it will be shown later these primary fields can be $\vec{E}$ and $\vec{H}$, $\vec{D}$ or $\vec{B}$ and $\vec{E}$.

the normal component of the magnetic field over the walls in contact with the chiral material, as they must be no null-valued [69].

EH-formulation of the coupled mode method.

In the EH-formulation, the primary expansions, particularized for a uniform rectangular waveguide [63], are

$$E_x(x, y) = \sum_{q=1}^{nte} V_{[q]} \frac{\partial T_{[q]}(x, y)}{\partial y} + \sum_{p=1}^{ntm} V_{(p)} \frac{\partial T_{(p)}(x, y)}{\partial x} \tag{45}$$

$$E_y(x, y) = -\sum_{q=1}^{nte} V_{[q]} \frac{\partial T_{[q]}(x, y)}{\partial x} + \sum_{p=1}^{ntm} V_{(p)} \frac{\partial T_{(p)}(x, y)}{\partial y} \tag{46}$$

$$E_z(x, y) = \sum_{p=1}^{ntm} k_{c(p)} V_{(p)}^{z} T_{(p)}(x, y) \tag{47}$$

$$H_x(x, y) = \sum_{q=1}^{nte} I_{[q]} \frac{\partial T_{[q]}(x, y)}{\partial x} - \sum_{p=1}^{ntm} I_{(p)} \frac{\partial T_{(p)}(x, y)}{\partial y} \tag{48}$$

$$H_y(x, y) = \sum_{q=1}^{nte} I_{[q]} \frac{\partial T_{[q]}(x, y)}{\partial y} + \sum_{p=1}^{ntm} I_{(p)} \frac{\partial T_{(p)}(x, y)}{\partial x} \tag{49}$$

$$H_z(x, y) = \sum_{q=1}^{nte} k_{c_{[q]}} I_{[q]}^{z} T_{[q]}(x, y) + h_0 \tag{50}$$

Here, *V*'s and *I*'s represent the electric and magnetic field's expansion coefficients, and the *T*-functions are the *T*-potentials that give the electromagnetic field components of the basis modes *TE* and *TM* in the empty waveguide [63,68]. For rectangular waveguides they are

- *TE* modes

$$T_{[n,m]} = \frac{\eta_{[n,m]}}{k_{c_{[n,m]}} \sqrt{ab}} \cos\left(\frac{n\pi}{a} x\right) \cos\left(\frac{m\pi}{b} y\right), \quad n, m = 0, 1, 2, \ldots \text{ but not } n = m = 0 \tag{51}$$

- *TM* modes

$$T_{(n,m)} = \frac{\eta_{(n,m)}}{k_{c_{(n,m)}}\sqrt{ab}} \sin\left(\frac{n\pi}{a}x\right)\sin\left(\frac{m\pi}{b}y\right) \tag{52}$$
$$n, m = 1, 2, 3, \ldots.$$

with

$$\eta_{\{n,m\}} = \begin{cases} 2 & if \quad n \neq 0,\ m \neq 0 \\ \sqrt{2} & if \quad n = 0 \ or \ m = 0 \end{cases} \tag{53}$$

Moreover, in Eqs. (45) -(52), $k_{c(p)}$ and $k_{c[q]}$ are the cutoff wavenumber of the basis modes; the term h_0 is a constant that must be taken into account in some media in order to obtain the correct solution of the electromagnetic field. Also, *nte* and *ntm* correspond, respectively, to the number of basis modes *TE* and *TM* taken into account in the expansions. In all expressions, parentheses refer to *TM*-modes and brackets to *TE*-modes.

Following the procedure of the coupled mode method previously mentioned at the beginning of this section and detailed in [63,68], the Generalized Telegraphist Equations for the EH-formulation turn out as:

$$\frac{dV_{(n)}}{dz} = j\omega \int_s \left(B_x \frac{\partial T_{(n)}}{\partial y} - B_y \frac{\partial T_{(n)}}{\partial x} \right) ds + k_{c_{(n)}} V^z_{(n)} \tag{54}$$

$$\frac{dV_{[m]}}{dz} = -j\omega \int_s \left(B_x \frac{\partial T_{[m]}}{\partial x} + B_y \frac{\partial T_{[m]}}{\partial y} \right) ds \tag{55}$$

$$\frac{dI_{(n)}}{dz} = -j\omega \int_s \left(D_x \frac{\partial T_{(n)}}{\partial x} + D_y \frac{\partial T_{(n)}}{\partial y} \right) ds \tag{56}$$

$$\frac{dI_{[m]}}{dz} = -j\omega \int_s \left(D_x \frac{\partial T_{[m]}}{\partial y} - D_y \frac{\partial T_{[m]}}{\partial x} \right) ds + k_{c_{[m]}} I^z_{[m]} \tag{57}$$

Inspecting the set of equations (54)-(57) it can be seen that it is constituted by 4 equations and 6 unknowns. Therefore, in order to find their solution, two additional expressions relating the longitudinal coefficients with the transversal ones are needed. These relations are provided by the longitudinal components of the Maxwell equations and the constitutive relations; they can be obtained following two alternative ways that give two different formulations of the EH-formulation.

EHD-formulation

http://dx.doi.org/10.5772/50680

The first presented way to obtain the additional relations between the longitudinal and transversal coefficients it the one proposed by Ogusu [64]. Although it is not the original procedure, it provides the more intuitive way to relate these coefficients. Therefore, it is firstly introduced.

In this formulation the additional relations are obtained by rewriting the constitutive relations (3) as

$$E_z = \frac{1}{\varepsilon_0} \frac{\mu_r}{\left[\mu_r \varepsilon_r - \left(\chi^2 + \kappa^2\right)\right]} D_z - c \frac{(\chi - j\kappa)}{\left[\mu_r \varepsilon_r - \left(\chi^2 + \kappa^2\right)\right]} B_z \tag{58}$$

$$H_z = c \frac{(\chi + j\kappa)}{\left[\left(\chi^2 + \kappa^2\right) - \mu_r \varepsilon_r\right]} D_z - \frac{1}{\mu_0} \frac{\varepsilon_r}{\left[\left(\chi^2 + \kappa^2\right) - \mu_r \varepsilon_r\right]} B_z \tag{59}$$

Then, after introducing the longitudinal components of the Maxwell equations and the expansions (45)-(50) into (58)-(59), and applying the Galerkin mode [73], we get the desired expressions

$$\begin{aligned} k_{c_{(n)}} V^z_{(n)} = {} & \frac{c}{j\omega} \sum_{j=1}^{nte} V_{[j]} k^2_{c_{[j]}} k^2_{c_{(n)}} \int_S \frac{(\chi - j\kappa)}{\left[\mu_r \varepsilon_r - \left(\chi^2 + \kappa^2\right)\right]} T_{[j]} T_{(n)} ds \\ & - \frac{1}{j\omega\varepsilon_0} \sum_{i=1}^{ntm} I_{(i)} k^2_{c_{(i)}} k^2_{c_{(n)}} \int_S \frac{\mu_r}{\left[\mu_r \varepsilon_r - \left(\chi^2 + \kappa^2\right)\right]} T_{(i)} T_{(n)} ds \end{aligned} \tag{60}$$

$$\begin{aligned} k_{c_{[m]}} I^z_{[m]} = {} & \frac{1}{j\omega\mu_0} \sum_{j=1}^{nte} V_{[j]} k^2_{c_{[j]}} k^2_{c_{[m]}} \int_S \frac{\varepsilon_r}{\left[\left(\chi^2 + \kappa^2\right) - \mu_r \varepsilon_r\right]} T_{[j]} T_{[m]} ds \\ & - \frac{c}{j\omega} \sum_{i=1}^{ntm} I_{(i)} k^2_{c_{(i)}} k^2_{c_{[m]}} \int_S \frac{(\chi + j\kappa)}{\left[\left(\chi^2 + \kappa^2\right) - \mu_r \varepsilon_r\right]} T_{(i)} T_{[m]} ds \end{aligned} \tag{61}$$

Note that in (60)-(61) the longitudinal coefficients are directly related with the transversal ones, i.e., no matrix inversion is required. Therefore we call it *direct EH*-formulation of the coupled mode method or, simply, *EHD*-formulation.

Once the longitudinal and transversal coefficients are known, only the coefficient h_0 have to be determined. Following a similar procedure, the value of h_0 can be obtained from

$$h_0 \int_S ds = \frac{1}{j\omega\mu_0} \sum_{j=1}^{nte} V_{[j]} k^2_{c_{[j]}} \int_S \frac{\varepsilon_r}{\left[\left(\chi^2 + \kappa^2\right) - \varepsilon_r\right]} T_{[j]} ds - \frac{c}{j\omega} \sum_{i=1}^{ntm} I_{(i)} k^2_{c_{(i)}} \int_S \frac{(\chi + j\kappa)}{\left[\left(\chi^2 + \kappa^2\right) - \varepsilon_r\right]} T_{(i)} ds \tag{62}$$

EHI-formulation

The alternative process to obtain the relations between the longitudinal and transversal coefficients is the original one formulated by Schelkunoff [63]. Now, the longitudinal components of the Maxwell equations and the expansions (45)-(50) are directly introduced in the constitutive relations (3). After applying the Galerkin method [73], the desired equations are obtained

$$
\begin{aligned}
I_{(n)} &= -j\omega\varepsilon_0 \sum_{i=1}^{ntm} k_{c_{(i)}} V_{(i)}^{z} \int_S \varepsilon_r T_{(i)} T_{(n)} ds \\
&- j\frac{\omega}{c} \sum_{j=1}^{nte} k_{c_{[j]}} I_{[j]}^{z} \int_S (\chi - j\kappa) T_{[j]} T_{(n)} ds \\
&- j\frac{\omega}{c} h_0 \int_S (\chi - j\kappa) T_{(n)} ds
\end{aligned}
\tag{63}
$$

$$
\begin{aligned}
V_{[m]} &= -j\omega\mu_0 \sum_{j=1}^{nte} k_{c_{[j]}} I_{[j]}^{z} \int_S \mu_r T_{[j]} T_{[m]} ds - j\omega\mu_0 h_0 \int_S \mu_r T_{[m]} ds \\
&- j\frac{\omega}{c_0} \sum_{i=1}^{ntm} k_{c_{(i)}} V_{(i)}^{z} \int_S (\chi + j\kappa) T_{(i)} T_{[m]} ds
\end{aligned}
\tag{64}
$$

Inspecting (62) and (63) two main differences with the EHD-formulation can be seen. The first one is that in order to isolate the longitudinal coefficients, matrix inversions are required. Therefore, as the longitudinal coefficients cannot be directly expressed in terms of the transversal ones, we call this formulation "Indirect EH-formulation" of the coupled mode method or, simply, EHI-formulation. The second difference deals with the obtention of the constant h_0. As it influences over the expansions coefficients, it must be taken into account in the eigenvalue problem. Thus a third relation must be included. Following a similar procedure, the value of h_0 can be obtained from

$$
h_0 \int_S \mu_r ds = -\sum_{j=1}^{nte} k_{c_{[m]}} I_{[m]}^{z} \int_S \mu_r T_{[m]} ds - \sqrt{\frac{\varepsilon_0}{\mu_0}} \sum_{i=1}^{ntm} k_{c(i)} V_{(i)}^{z} \int_S (\chi + j\kappa) T_{(i)} ds
\tag{65}
$$

EB-formulation

We have seen that, in the EH formulation of the coupled mode method, both electric and magnetic fields inside the partially filled waveguide are expressed in terms of the electric and magnetic fields of the empty waveguide. Therefore they satisfy the same boundary conditions over the perfect electric conductor. As long as the media contained in the waveguide allow the proper modes and the basis modes to satisfy the same boundary conditions, the EH-formulation provides correct results. This happens when the media are dielectric, isotropic or anisotropic. The problem appears when the medium produces a discontinuity in any component of the electromagnetic field that cannot be reproduced with the basis modes of the waveguide. This occurs, for example, when a bi-isotropic medium fills the waveguide. If we focus our attention on the normal component of the magnetic field over the perfect electric conductor, we see that it must be not null valued, whereas the normal component of the magnetic induction does. However, when the problem is analyzed with the coupled mode method, it can be seen that, if the magnetic field satisfies the same boundary conditions than the basis modes, the normal component of the magnetic field will be zero over the PEC, which is incorrect. In order to avoid this inconvenience we can choose as primary fields those that satisfy the same boundary conditions over the perfect electric conductors than the basis modes fields. These are $\vec{E}$ and $\vec{B}$. Therefore, the proposed primary expansions for these fields are

$$E_x = \sum_{m=1}^{nte} V_{[m]} \frac{\partial T_{[m]}}{\partial y} + \sum_{n=1}^{ntm} V_{(n)} \frac{\partial T_{(n)}}{\partial x} \tag{66}$$

$$E_y = -\sum_{m=1}^{nte} V_{[m]} \frac{\partial T_{[m]}}{\partial x} + \sum_{n=1}^{ntm} V_{(n)} \frac{\partial T_{(n)}}{\partial y} \tag{67}$$

$$E_z = \sum_{n=1}^{ntm} k_{c_{(n)}} V^z_{(n)} T_{(n)} \tag{68}$$

$$B_x = \mu_0 \sum_{m=1}^{nte} b_{[m]} \frac{\partial T_{[m]}}{\partial x} - \mu_0 \sum_{n=1}^{ntm} b_{(n)} \frac{\partial T_{(n)}}{\partial y} \tag{69}$$

$$B_y = \mu_0 \sum_{m=1}^{nte} b_{[m]} \frac{\partial T_{[m]}}{\partial y} + \mu_0 \sum_{n=1}^{ntm} b_{(n)} \frac{\partial T_{(n)}}{\partial x} \tag{70}$$

$$B_z = \mu_0 \sum_{m=1}^{nte} k_{c_{[m]}} b^z_{[m]} T_{[m]} + \mu_0 b_0 \tag{71}$$

Following a similar procedure than the presented for the conventional formulation of the coupled mode method, an eigenvalue problem of Eq. (54)-(57) type is obtained. As it happened in the previous formulations, in order to solve this set of equations, two new equivalent relations between longitudinal and transversal coefficients and another third one to obtain b_0 are needed. When the medium filling the waveguide is an isotropic chiral medium, the coefficient b_0 turns to be null.

Solving the Telegraphist equation eigenvalue problem, the coefficients expansions and the propagation constants for H and $\vec{H}$ can be obtained in a way similar to the EH-formulation.

Magnetic Field

When only the knowledge of the electric field or the constant propagations of any proper modes is required, the expansions of $\vec{H}$ and $\vec{E}$ obtained using the EB- formulation are enough to obtain the desired solution. However, if we need to characterize a discontinuity, an expansion for $\vec{H}$ is also necessary. As it was previously mentioned, the obtention of this new expansion for $\vec{E}$ involves Eqs. (66)-(71) and the constitutive relations (3). Once the $\vec{B}$ expansions are known, the mode matching method can be easily extended [71-72, 75] and any discontinuity can be analyzed.

6.2. Results

In order to check the behavior of the different formulations of the coupled mode method previously presented, the analysis of electromagnetic wave propagation inside a uniform rectangular waveguide filled with an isotropic chiral medium is proposed (Figure 5). It is important to note that there is no analytical solution available of this problem. Hence, a priori one could not establish which formulation provides the most accurate results. However, according to the conclusions extracted from the analysis of parallel-plate waveguide containing an isotropic chiral media [67], and the fact that the EB-formulation was developed to ensure the satisfaction of all boundary conditions of the electromagnetic field over the perfect electric conductor, we can infer that the EB- is the optimal formulation for analyzing this problem. Thus, along this section this statement is verified with numerical results. For that purpose we focus our attention in two properties of the electromagnetic field: the propagation constant values of the proper modes and their magnetic field components profile. In the first case it can be seen how the convergence behavior of each formulation is. In the second one it is shown how the magnetic field obtained by the EB-formulation satisfies the boundary conditions of the normal component of the magnetic field whereas that extracted from the EH-formulation does not.

Figure 6 shows the variation of the phase constant of the fundamental proper mode normalized to the vacuum wavenumber versus the number of basis TE and TM modes (convergence diagram) for the structure of Figure 5 provided by the mentioned formulations of the coupled mode method. Here, the basis modes TE and TM have been taken by increasing their cutoff frequency and the simulation data is displayed in the figure caption. An inspection of this figure evidences that the final values of the propagation constant provided by the EB- and EHI-formulations are very similar but slightly different than the EHD ones. Moreover, the convergences of the EB- and EHI- are faster than the EHD- one as they need a smaller number of basis modes TE and TM to reach to the final value. Therefore, with these evidences, which can be extended to the rest of proper modes, we can infer that the EHI-formulation gives more accurate results than the EHD-.

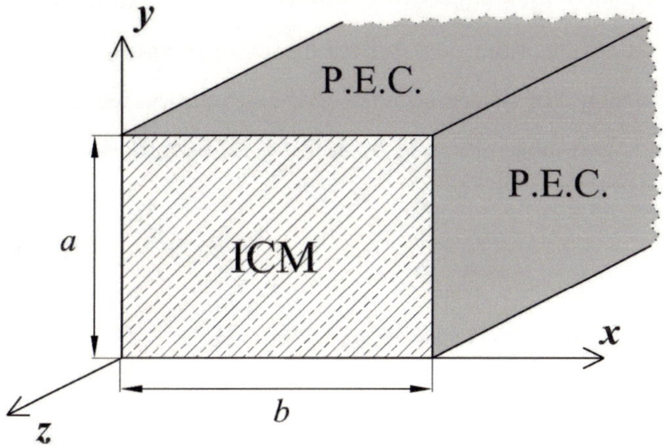

Figure 5. Cross section of a uniform rectangular waveguide fully loaded with an isotropic chiral media (ICM). P.E.C. stands for perfect electric conductor

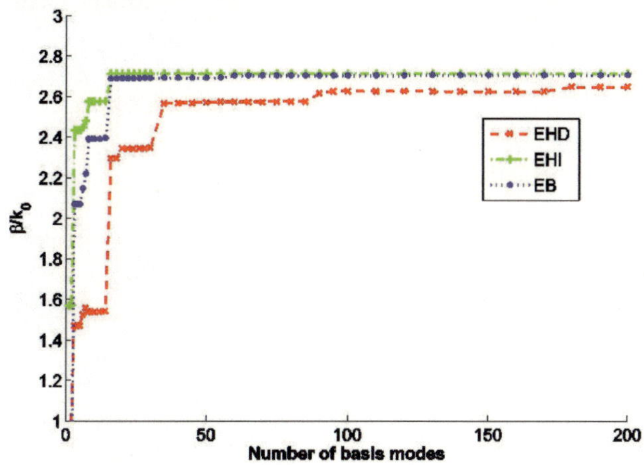

Figure 6. Convergence diagram of the fundamental mode for a uniform waveguide completely filled with an isotropic chiral media. Dimensions in mm: $a \times b = 22.86 \times 10.16$;$\varepsilon_r = 3$, $\mu_r = 1$, $\chi = 0$ and $\kappa = 1.5$.

As it was previously mentioned, since the EHI-formulation implies a matrix inversion that introduces a numerical error, one could think that the results provided by the EHD-formulation should be more accurate than those from the EHI. However, this statement is false and the EHI- results are closer to the EB- than the EHD- ones. The explanation of this disagreement is related to the fact that in the EHD-formulation the continuous E_z component is written as the product of discontinuous functions. This situation, which is right analytically, is not correct when is obtained numerically, as it is impossible to reproduce a discontinuous function in terms of a finite number of sines and cosines. In the other hand, in the EHI-formulation, although matrix inversions are required, a discontinuous function is expressed in terms of the product of discontinuous and continuous functions, which is correct both analytically and numerically. For this reason, the EHI-formulation results agree with the EB-formulation ones better than those obtained from the EHD-formulation.

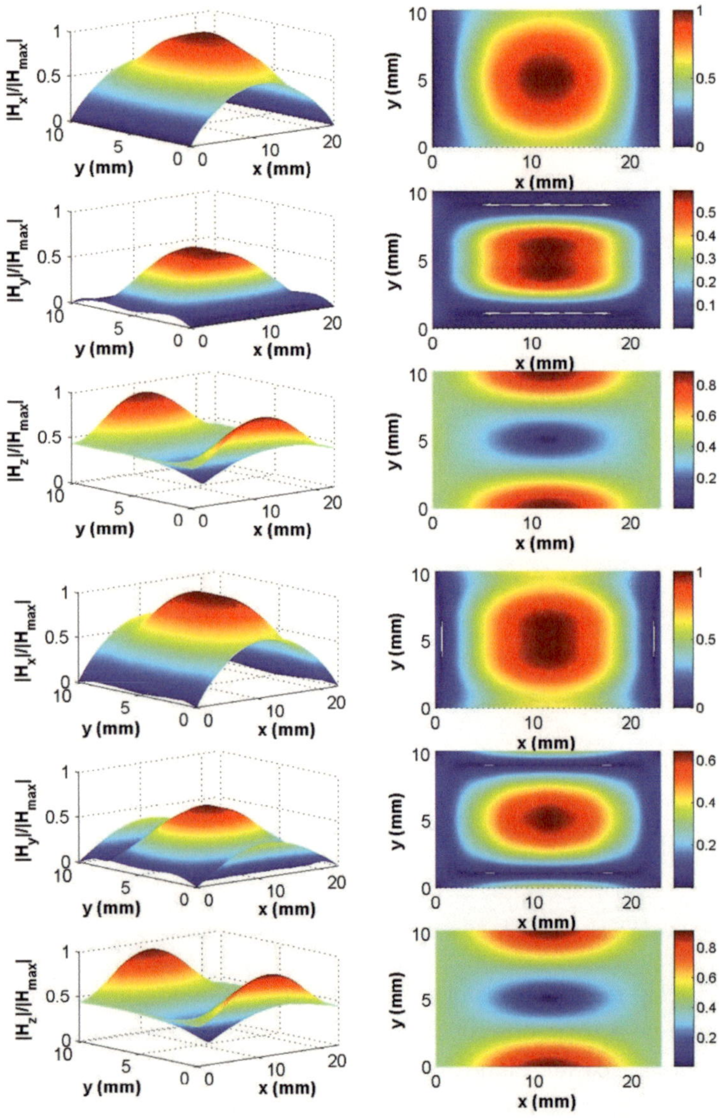

Figure 7. Module of the magnetic field components of the fundamental mode, normalized to their maximum value, as a function of the x- and y- coordinates provided by (a) EHI-formulation and (b) EB-formulation. Dimensions in mm: a × b = 22.86 × 10.16. $\varepsilon_r = 3$, $\mu_r = 1$, $\chi = 0$ and $\kappa = 0.75$. Number of basis modes 120 for the EB-formulation and 150 for the EH-.

Finally, in Figure 7 the module of the three components of the magnetic field normalized to their maximum values, as a function of the *x*- and *y*- coordinates are presented. As it is clear that the EHI-formulation is more appropriate that the EHD when analyzing this problem, only results for EHI- and EB-formulations are shown. On inspecting Figure 7 it is easy to prove that the magnetic field obtained with the EB-formulation does satisfy the boundary conditions of the normal component of the magnetic field whereas that extracted from the EH-formulation does not. It can be seen in Figure 7 how the normal component of the magnetic field over the perfect electric conductor walls is null in the EHI-formulation meanwhile it is not null in the EB- one.

7. Conclusions

A review of different numerical techniques for the modeling of electromagnetic wave propagation through bi-isotropic materials has been presented. In the time domain, different FDTD models have been mentioned, with emphasis in one based on Mobius transformation, as well as one MRTD approach, suitable to improve recursive convolution techniques and one TLM model. In the frequency domain, two different formulations of the coupled mode method, the EH- and the EB-formulations, have been presented. It has been seen that the EB-formulation is the optimal for analyzing electromagnetic wave propagation inside a rectangular waveguide that contains an isotropic chiral media because the solution that it gives verify all the boundary conditions of the electromagnetic field over the perfect electric walls that constitute the waveguide.

Acknowledgements

Partial support for this work was provided by the Spanish MICINN through the TEC2010-21496-C03-01, subprogram "Ramón y Cajal" (RYC-2010-06922) and CONSOLIDER CSD2008-00066 projects.

Author details

I. Barba[1*], A. Grande[1], A.C.L. Cabeceira[1], A. Gómez[2], J.A. Pereda[2] and J. Represa[1]

*Address all correspondence to: ibarba@ee.uva.es

1 University of Valladolid,, Spain

2 University of Cantabria,, Spain

References

[1] Lindell, I. V., Sihvola, A. H., Tretyakov, S. A., & Viitanen, A. J. (1994). Electromagnetic waves in Chiral and Bi-Anisotropic Media. Boston, Artech House.

[2] Sihvola, A. H. (1994). Electromagnetic Modeling of Bi-Isotropic Media. *Progress in Electromagnetic Research, PIER*, 9, 45-86.

[3] Sihvola, A. H., & Lindell, I. Material effects on bi-anisotropic electromagnetics. 1995, *IEICE Trans. Electron. (Tokyo) E78-C:*, 1383-1390.

[4] Tretyakov, S. A., Sihvola, A. H., Sochava, A. A., & Simovski, C. R. (1998). Magnetoelectric interactions in bi-anisotropic media. *J. Electromag. Waves Appl. JEMWA*, 12, 481-497.

[5] Soukoulis, C. M., & Wegener, M. (2011). Past achievements and future challenges in the development of three-dimensional photonic metamaterials. *Nature Photonics*, 5, 523-530.

[6] Barba, I., Cabeceira, A. C. L., García-Collado, A. J., Molina-Cuberos, G. J., Margineda, J., & Represa, J. (2011). Quasi-planar Chiral Materials for Microwave Frequencies. In: Kishk A, editor. Electromagnetic Wave Propagation in Complex Matter. Rijeka: Intech , 97-116.

[7] Pendry, J. B. (2004). A Chiral Route to Negative Refraction. *Science*, 306, 1353-1355.

[8] Tretyakov, S. A., Sihvola, A. H., & Jylhä, L. (2005). Backward-wave regime and negative refraction in chiral composites. *Photonics and Nanostructures: Fundamentals and Applications,*, 3, 107-115.

[9] Plum, E., Zhou, J., Dong, J., Fedotov, V. A., Koschny, T., Soukoulis, C. M., & Zheludev, N. I. (2009). Metamaterial with negative index due to chirality,. *Physical Review B*, 79, 035407-1.

[10] Zhou, J., Dong, J., Wang, W., Koschny, T., Kafesaki, M., & Soukoulis, C. M. (2009). , Negative refractive index due to chirality. Physical Review B , 79, 121104-1.

[11] Molina-Cuberos, G. J., García-Collado, A. J., Barba, I., & Margineda, J. (2011). Chiral Metamaterials With Negative Refractive Index Composed by an Eight-Cranks Molecule. *IEEE Antennas and Wireless Propagation Letters*, 10, 1488-1490.

[12] Lakhtakia, A. (1994). Ten years past the Post. In McCall MW, Dewar G, editors. Complex Mediums V: Light and Complexity., *Proc. SPIE (Int. Soc. for Optical Engineering),*, 5508, 85-94.

[13] Weiglhofer, W. S. (1994). On a medium constraint arising directly from Maxwell's equations. *J. Phys.*, A27:, L871-L874.

[14] Post, E. J. (1962). Formal Structure of Electromagnetics. Amsterdam, North-Holland.

[15] Lakhtakia, A., & Weiglhofer, W. S. (1995). On a constraint on the electromagnetic constitutive relations of nonhomogeneous linear media. *IMA J. of Appl. Math.*, 54, 301-306.

[16] Lakhtakia, A., & Weiglhofer, W. S. (1996). Lorentz covariance, Occam's razor, and a constraint on linear constitutive relations. Phys. Lett. A Erratum A222: 459., 213, 107-111.

[17] Weiglhofer, W. S., & Lakhtakia, A. (1998). The Post constraint revisited Arch. *Elektron. Uebertrag.*, 52, 276-279.

[18] Lakhtakia, A. (2004). On the genesis of the Post constraint in modern electromagnetism. *Optik-International Journal of Light and Electron Optics*, 115, 151-158.

[19] Raab, R. E., & Sihvola, A. H. (1997). On the existence of linear non-reciprocal bi-isotropic (NRBI) media. JPhys. A: Math. Gen. , 30, 1335-1344.

[20] Lindell, I. V., & Sihvola, A. H. (2004). Perfect electromagnetic conductor. *Journal of Electromagnetic waves and Applications JEMWA*, 19, 861-869.

[21] Lindell, I. V., & Wallén, K. H. (2004). Differential-form electromagnetics and bi-anisotropic Q-media. *J. Electromagn. Waves and Appl. JEMWA*, 18, 957-968.

[22] Lindell, I. V., & Wallén, K. H. (2004). Generalized Q-media and field decomposition in differential-form approach. *J. Electromagn. Waves and Appl. JEMWA*, 18, 1045-1056.

[23] Lindell, I. V., & Sihvola, A. H. (2005). Transformation method for problems involving perfect electromagnetic conductor (PEMC). *IEEE Trans.Antennas Propag.*, 53, 3005-3011.

[24] Tretyakov, S. A., Maslovski, S. I., Nefedov, I. S., Viitanen, A. J., Belov, P. A., & Sanmartin, A. (2003). Artificial Tellegen particle,. *Electromagnetics*, 23, 665-680.

[25] de Lange, O. L., & Raab, R. E. (2001). Post's constraint for electromagnetic constitutive relations,. *J. Optics A3*, L23-L26.

[26] Serdyukov, A. N., Semchenko, I. V., Tretyakov, S. A., & Sihvola, A. H. (2001). Electromagnetics of Bi-anisotropic Materials: Theory and Applications. Amsterdam, Gordon and Breach.

[27] Canto, J. R., Paiva, C. R., & Barbosa, A. M. (2011). Modal Analysis of Bi-isotropic H-Guides. *Progress In Electromagnetics Research PIER*, 111, 1-24.

[28] Matos, S. A., Paiva, C. R., & Barbosa, A. M. (2009). A Spacetime Algebra Approach to Moving Bi-isotropic Media. *Proc IEEE AP-S/URSI International Symp. APSURSI 09.*

[29] Hehl, F. W., & Obukhov, Y. N. (2005). Linear media in classical electrodynamics and the Post constraint. *Phys. Lett. A*, 334, 249-259.

[30] Kamenetskii, E. O., Sigalov, M., & Shavit, R. (2009). Tellegen particles and magnetoelectric metamaterials. *Journal of Applied Physics*, 105, 013537-1.

[31] Kong, J. A. (2008). Electromagnetic Wave Theory. Cambridge, USA EMW Publishing., 1016.

[32] Condon, E. U. (1937). Theories of optical rotatory power. *Rev. Modern. Phys.*, 9, 432-457.

[33] Kristensson, G. (1999). Condon's model on optical rotatory power and causality- a scientific trifle. In: Eds. A. Karlsson and G. Kristensson. Festskrift till Staffan Ström: Lund, Sweden, KF Sigma. Available: http://www.lunduniversity.lu.se/o.o.i.s?id=12683&postid=530495. Accesed 2012 May 30., 97-119.

[34] Rogacheva, A. V., Fedotov, V. A., Schwanecke, A. S., & Zheludev, N. I. (2006). Giant Gyrotropy due to Electromagnetic-Field Coupling in a Bilayered Chiral Structure. *Physical Review Letters*, 97, 177401-1.

[35] Balanis, C. A. (1989). Advanced Engineering Electromagnetics. New York, USA, Wiley.

[36] Bahr, A. J., & Clausing, K. R. (1994). An approximate model for artificial chiral material. *IEEE Trans. Antennas Propag.*, 42, 1592-1599.

[37] Yee, K. S. (1966). Numerical solution of initial boundary value problems involving Maxwell's equation is isotropic media. *IEEE Trans. Antennas Propag.*, 14, 302-307.

[38] Taflove, A. (1995). Computational Electrodynamics: The Finite-Difference Time-Domain Method. Norwood, MA, Artech House.

[39] García, S., Perez, I., Martín, R., & Olmedo, B. (1998). Extension of Berenger's PML for bi-anisotropic media. *IEEE Microwave Guided Wave Lett.*, 8, 297-299.

[40] Teixeira, F., & Chew, W. (1998). General closed-form PML constitutive tensors to match arbitrary bi-anisotropic and dispersive linear media. *IEEE Microwave Guided Wave Lett.*, 8, 223-225.

[41] Garcia, S., Perez, I., Martin, R., & Olmedo, B. (1999). BiPML: A PML to match waves in bi-anisotropic media. *Microwave Opt. Technol. Lett.*, 20, 44-48.

[42] Akyurtlu, A., & Werner, D. H. (2004). BI-FDTD: A novel finite-difference time-domain formulation for modeling wave propagation in bi-isotropic media. *IEEE Trans. Antennas Propag.*, 52, 416-425.

[43] Akyurtlu, A., & Werner, D. H. (2004). A novel dispersive FDTD formulation for modeling transient propagation in chiral Metamaterials. *IEEE Trans. Antennas Propag.*, 52, 2267-2276.

[44] Barba, I., Grande, A., Cabeceira, A. C. L., & Represa, J. (2003). Time domain modeling of electromagnetic wave propagation in Tellegen media. *Microwave Opt. Technol. Lett.*, 38, 167-168.

[45] Grande, A., Barba, I., Cabeceira, A. C. L., Represa, J., Kärkkäinen, K., & Sihvola, A. H. (2004). Analyzing the stability of the new FDTD technique for non-dispersive bi-iso-

tropic media using the von Neumann method. *VIII ACES*, Symp. Syracuse., New York.

[46] Grande, A., Barba, I., Cabeceira, A. C. L., Represa, J., So, P. M. P., & Hoefer, W. J. R. (2004). FDTD modeling of transient microwave signals in dispersive and lossy bi-isotropic media. *IEEE Trans. Microwave. Theory Tech.*, 52, 773-784.

[47] Grande, A., Barba, I., Cabeceira, A. C. L., Represa, J., Kärkkäinen, K., & Sihvola, A. H. (2005). Two dimensional extension of a novel FDTD technique for modeling dispersive lossy bi-isotropic media using the auxiliary differential equation method. *IEEE Microwave Wireless Components Lett.*, 15, 375-377.

[48] Demir, V., Elsherbeni, A. Z., & Arvas, E. (2005). FDTD formulation for dispersive chiral media using the Z transform method. *IEEE Trans. Antennas Propag.*, 53, 3374-3384.

[49] Nayyerir, V., Soleimani, M., Rashed-Mohassel, J., & Dehmollaian, M. (2011). FDTD modeling of dispersive bianisotropic media using Z-transform method. *IEEE Trans. Antennas Propag.*, 59, 2268-2278.

[50] Pereda, J. A., Grande, A., Gonzalez, O., & Vegas, A. (2006). FDTD modeling of chiral media by using the Mobius transformation technique. *IEEE Antennas Wireless Propag. Lett.*, 5, 327-330.

[51] Pereda, J. A., Vegas, A., & Prieto, A. (2002). FDTD modeling of wave propagation in dispersive media by using the Mobius transformation technique. *IEEE Trans. Microwave Theory Tech.*, 506, 1689-1695.

[52] Bushyager, N., & Tentzeris, E. M. (2006). *MRTD (Multi Resolution Time Domain) Method in Electromagnetics (Synthesis Lectures in Computational Electromagnetics).*, San Francisco, USA:, Morgan & Claypool,, 116.

[53] Krumpholz, M., & Katehi, L. P. B. (1996). MRTD: New time-domain schemes based on multiresolution analysis. *IEEE Trans. Microwave Theory Tech.*, 44, 555-571.

[54] Tentzeris, E. M., Cangellaris, A., Katehi, L. P. B., & Harvey, J. (2002). Multiresolution time-domain (MRTD) adaptive schemes using arbitrary resolutions of wavelets,. *IEEE Trans. Microwave Theory Tech.*, 50, 501-516.

[55] Fujii, M., & Hoefer, W. J. R. (1998). A three-dimensional Haar-wavelet-based multiresolution analysis similar to the FDTD method-Derivation and application. *IEEE Trans. Microwave Theory Tech.*, 46, 2463-2475.

[56] Grivet-Talocia, S. (2000). On the accuracy of Haar-based multiresolution time- domain schemeS. *IEEE Microwave Guided Wave Lett.*, 10, 397-399.

[57] Barba, I., Represa, J., Fujii, M., & Hoefer, W. J. R. (2003). Multiresolution model of electromagnetic wave propagation in dispersive materials. *International Journal of Numerical Modelling*, 16, 239-247.

[58] Daubechies, I. (1992). *Ten Lectures on Wavelets.Philadelphia*, USA, SIAM, 357.

[59] Barba, I., Grande, A., Cabeceira, A. C. L., & Represa, J. (2006). A Multiresolution Model of Transient Microwave Signals in Dispersive Chiral Media. *IEEE Transactions on Antennas & Propagation*, 54, 2808-1812.

[60] Hoefer, W. J. R. (1985). The Transmission-Line Matrix method- Theory and Applications. *IEEE Trans. Microwave Theory Tech.*, 33, 882-893.

[61] Barba, I., Cabeceira, A. C. L., Represa, J., & Panizo, M. (1996). Modelling dispersive dielectrics for HSCN TLM method. *Int. J. Numer. Modeling*, 14, 15-30.

[62] Cabeceira, A. C. L., Barba, I., Grande, A., & Represa, J. (2006). A Time-Domain Modelling for EM wave propagation in Bi-Isotropic media based on the TLM methoD. *IEEE Trans. Microwave Theory Tech.*, 54, 2780-2789.

[63] Schelkunoff, S. A. (1952). Generalized telegraphist's equations for waveguides. *Bell Syst. Tech. J.*, 3, 784-801.

[64] Ogusu, K. (1977). Numerical analysis of rectangular dielectric waveguide and its modifications. *IEEE Trans Microwave Theory Tech MTT-25*, 441-446.

[65] Xu, Y., & Bosisio, R. G. (1995). An efficient method for study of general bi-anisotropic waveguides. *IEEE Trans Microwave Theory Tech MTT-43*, 873-879.

[66] Xu, Y., & Bosisio, R. G. (1997). A study on the solutions of chirowaveguides and bianisotropic waveguides with the use of coupled-mode analysis. *Microwave Opt Technol Lett*, 14, 308-311.

[67] Gómez, Á., Vegas, A., & Solano, M. A. (2004). A brief discussion on the different formulations of the coupled mode method in chiral media: Application to the parallel-plate chirowaveguide. *Microwave and Optical Technology Letters*, 42, 181-185.

[68] Gómez, Á., Vegas, A., & Solano, M. A. (2006). On two different formulations of the Coupled Mode Method: Application to 3D rectangular chirowaveguides. *AEU-International Journal of Electronics and Communications*, 60, 690-704.

[69] Gómez, Á., Lakhtakia, A., Vegas, A., & Solano, M. A. (2010). Hybrid technique for analysing metallic waveguides containing isotropic chiral materials. *IET Microwaves, Antennas and Propagation*, 4, 305-315.

[70] Gómez, Á., Barba, I., Cabeceira, A. C. L., Represa, J., Vegas, A., & Solano, M. A. (2008). Application of Schelkunoff's method for simulating isotropic chiral free propagation: Clarifying some common errors. *Journal of Electromagnetic Waves and Applications*, 22, 861-871.

[71] Solano, M. A., Vegas, A., & Prieto, A. (1994). Numerical analysis of discontinuities in a rectangular waveguide loaded with isotropic or anisotropic obstacles by means of the coupled-mode method and the mode matching method. *Int J Numer Model Electron Networks Devices Fields*, 7, 433-452.

[72] Solano, M. A., Vegas, A., & Prieto, A. (1998). Theoretical and experimental study of two ports structures composed of different size rectangular waveguides partially filled with isotropic dielectrics. *Int J Electron*, 84, 521-528.

[73] Harrington, R. F. (1993). Field computation by moment methods. *IEEE Press Series on Electromagnetic Waves*, New Jersey, Chapter 1.

[74] Solano, M. A., Vegas, A., & Gómez, A. (2006). Comments on "a comprehensive study of discontinuities in chirowaveguides". *IEEE Trans. on Microwave Theory and Techniques*, 54, 1297-1298.

[75] Gómez, Á., Lakhtakia, A., Margineda, J., Molina-Cuberos, G. J., Núñez, M. J., Saiz, Ipiña. J. A., Vegas, A., & Solano, M. A. (2008). Full-wave hybrid technique for 3-D isotropic-chiral-material discontinuities in rectangular waveguides: Theory and experiment. *IEEE Trans. on Microwave Theory and Techniques*, 56, 2815-2825.

Chapter 4

Integral Equation Analysis with Characteristic Basis Functions

Jaime Laviada, Fernando Las-Heras and Marcos R. Pino

Additional information is available at the end of the chapter

http://dx.doi.org/10.5772/50502

1. Introduction

Full-wave electromagnetic analysis algorithms dependence on the number of unknowns N. The complexities ranges from $O(N \log N)$ for certain iterative schemes such as the fast multipole method [8,42] or the adaptive integral method [5] to $O(N^3)$ for the conventional method of moments (MoM) [20].

Thus, the reduction of the number of unknowns results very attractive since it reduces the demanded computational resources in terms of memory and CPU time. In the method of moments context, the choice of the basis functions has a strong impact on the degrees of freedom that are required to solve the problem and, therefore, strategies for unknowns reduction are based on choosing the basis functions as good as possible.

Conventional piecewise basis functions such as Rao-Wilto-Glisson (RWG) [40] or rooftops [15], involve an expansion that is able to model almost any kind of current. Thus, it could be convenient to limit the set of basis functions so that is is only made of that ones which can model currents with a physical meaning and, therefore, to reduce the number of degrees of freedom of the problem under analysis. It is well-known that entire-domain basis functions enable a very efficiently modeling of the currents. Nonetheless, the computation of a set of entire-domain basis functions it is only possible for certain canonical geometries [44,31,4] or via numerical techniques which involve solving the entire problem [19,18] with the corresponding computational cost.

The basis functions to be used for solving our problem should be able to model any kind of current that could be induced in the geometry under analysis. In a general analysis, it is not possible to obtain a finite and complete set of basis functions unless the conditions of the

INTECH
open science | open minds

© 2012 Laviada et al.; licensee InTech. This is an open access article distributed under the terms of the Creative Commons Attribution License (http://creativecommons.org/licenses/by/3.0), which permits unrestricted use, distribution, and reproduction in any medium, provided the original work is properly cited.

problem are strongly restricted. Nonetheless, it will be shown in this book chapter that it is possible to compute a set of basis functions that can be progressively increased to reach an accuracy similar to the accuracy that would be achieved by means of low-level basis functions (RWG, rooftops, etc.).

The entire-domain basis functions are usually expressed as linear combinations of low-level basis functions. This definition enables an easy and efficient computation of the system of equations matrix since it is possible to exploit the linear properties of the integral equations to carry out the computation.

Among the pioneering works on this field, we find the theory of*characteristic modes* that has been carry out by Harrington [19,18]. This modes are computed by solving the following generalized eigenvalue problem:

$$\overline{\overline{X}}\bar{I}_n=\lambda_n\overline{\overline{R}}\bar{I}_n \tag{1}$$

where the matrices $\overline{\overline{R}}$ and $\overline{\overline{X}}$ are the real and imaginary part of the impedance matrix, respectively. The impedance matrix should have been computed from the EFIE with the Galerkin's method. Thus, the eigenvalues, λ_n, and eigenvectors, I_n, will be real due to the symmetries. The choice of the eigenvectors as basis functions yields a diagonal matrix with the corresponding computational advantages. If the geometry under analysis is modeled with N conventional low-level basis functions (e.g., RWG or rooftop), the number of eigenvalues will be also N. This eigenvalues are a complete base that can model any current with the same accuracy as the one that could be achieved with the N low-level basis functions. Nonetheless, the eigenvector with highest eigenvalues will have a more significant contribution to the radiation so that the eigenvectors under a certain threshold can be discarded as basis functions. Thus, it is possible to find a set of basis functions with very good accuracy whose size is smaller than N.

During the eighties and nineties, the research on this kind of basis functions was diffused. Nevertheless, the interest on these techniques has reemerged due to the possibilities in the application to the design of antennas [6]. In this research line, the physical meaning of the characteristic modes is exploited for the design of antennas with certain characteristics. The convergence of the accuracy as a function of the number of basis functions is improved by including a source mode that is computed by exciting the antenna and attaching the induced current with the eigenvectors.

Computation of characteristic modes involves the solution of an eigenvalue problem that limits the size of the elements where the basis functions can be defined. For this reason, this technique has been limited to the analysis of arrays with elements that are not connected each other. In addition, once the basis functions has been computed for an element, they can be copied for the rest of the elements and, therefore, to save CPU time. An example of this technique has been shown in [32] where the authors analyze a 3 × 3 antenna array by defining basis functions for each element. In particular, they consider three kinds of elements depending on the position of the element (center, corner or border) and they reuse the same basis function for each kind of element.

A similar problematic happens in the adaptive basis function method that has been proposed in [45]. In this method, a diagonally dominant moment matrix is achieved by building macrobasis functions for a certain body. This macrobasis functions are chosen so that they are orthogonal to the testing functions on the body. This method was also extended in [14,12,13] for the solution of arrays of equally-spaced disconnected elements.

A special case of basis functions defined as aggregation of low-level basis functions is the *Multilevel Method of Moments* [22]. The authors developed an iterative system based on the generation of hierarchical basis functions for the analysis of two-dimensional scattering problems. In this scheme, the solution of the problem was computed by iteratively solving it with basis functions of different sizes so that it enables to refine the solution. Although the aim of these basis functions was not an efficient modeling of the current, it was one of the pioneering on nesting basis functions which will be one of the key point of the techniques presented in this book chapter.

The previous scheme would be modified in [38] for the analysis of printed circuits. In this case, the technique was modified to work with high level basis functions without the need of going back to the low-level basis functions in the intermediate stages. In this method, the circuits were split into blocks and artificial ports were introduced at the end of the blocks. The basis functions were obtained by feeding the natural and the artificial ports and considering the continuity equation for the currents. The extension of this technique for geometries different from printed circuits (e.g. a three-dimensional scatterer) was not accomplished to the authors' best knowledge.

A technique similar to the previous one is the *Subdomain Multilevel Approach* (SMA) [43] for the analysis of certain kind of microstrip circuits. In this method, the geometry is decomposed into several subdomains (e.g. the feeding network and the antennas) and these subdomains are fed by a generator. The induced currents are used as the basis functions which are referred to as macrobasis functions. The rest of the geometry, which is not decomposed into subdomains, is analyzed with conventional basis functions (weak compression) or through new macrobasis functions computed by feeding each port (strong compression). The accuracy of the method is improved by decomposing each macrobasis functions into two basis functions corresponding to the real and imaginary parts. In addition, the joint between macrobasis functions is improved by including rootops between them. The main drawback of this method is also the difficulty to be extended to arbitrary geometries.

The *characteristic basis function method* (CBFM) [39] laid the foundations to generate macrobasis functions for three-dimension arbitrary-shaped scattering and radiation problems. In this first work, the generation of the characteristic basis functions (CBFs) yields two kinds of basis functions primary and secondary ones. This method will be later presented and compared with some new techniques for generating CBFs.

Another method which shares many points in common with the CBFM is the *Synthetic-Functions eXpansion* (SFX) [35]. The different versions of the CBFM [33,11] as well as the SFX can be considered as a generalization of the Multilevel method of moments and SMA to three-dimensional problems where plane-waves or external dipoles are used instead of artificial ports. This point will be deeply discussed later in section 2.2.

This chapter is structured as follows. Firstly, the theory of the analysis with characteristic basis functions is presented. This first section includes the general basis of any method based on macrobasis functions defined as linear combinations of low-level basis functions, the theoretical consideration of generating CBFs with the external sources method, a comparison of methods for the computation of the reduced matrix and the recursive application of the characteristic basis functions. In the next section, several applications of the CBFM are described including monostatic radar cross section computation, synthesis of phased arrays and analysis of partial modifications on large structures. Finally, the conclusions of the previous sections are summarized.

2. Analysis with characteristic basis functions

2.1 General basis

The characteristic basis function method pursues to find a set of efficient basis functions so that a smaller number of degrees of freedom is required to solve the problem with the method of moments. The aforementioned basis functions are defined on domains grouping several contiguous low-level basis functions (e.g., RWG or rooftops). In the CBFM context, the domains of definitions of the characteristics basis functions are referred to as *blocks*. Thus, the first step in order to apply the CBFM is to split the geometry under analysis into B blocks (see Figure 1).

Figure 1. Block decomposition for an airplane-like geometry.

On each block, a set of basis functions, which are referred to as characteristic basis functions, are defined. Therefore, the current is expressed as follows:

$$\mathbf{J} = \sum_{i=1}^{B} \sum_{n=1}^{K_i} I_{i,n} \boldsymbol{\mathcal{F}}_{i,n}, \tag{2}$$

where K_i is the number of CBFs in the i-th block; $\mathcal{F}_{i,n}$ is the n-th CBF in the i-th block; and, the coefficients $I_{i,n}$ are the weights of the CBFs in order to express the final solution as a linear combination of the new basis functions. This coefficients are the unknowns to be solved by the method of moments. The CBFs are defined as linear combination of the low-level basis functions:

$$\mathcal{F}_{i,n} = \sum_{m=1}^{N_i} I_{i,m}^{(n)} \mathbf{f}_{i,m} \tag{3}$$

where N_i is the number of low-level basis functions in the i-th block; and $f_{i,m}$ is the m-th low-level basis functions belonging to the i-th block. The definition of the CBFs as a linear combination enables a more simple computation of certain terms such as the reaction terms in the impedance matrix and, therefore, it enables an easy adaptation of previously implemented method of moments codes.

In order to achieve a compact notation, it is useful to define matrices grouping the coefficients of the CBFs for each block. This matrix is defined for the i-th block as follows:

$$\bar{\bar{J}}_i = \begin{bmatrix} I_{i,1}^{(1)} & I_{i,1}^{(2)} & \cdots & I_{i,1}^{(K_i)} \\ I_{i,2}^{(1)} & I_{i,2}^{(2)} & \cdots & I_{i,2}^{(K_i)} \\ \vdots & \vdots & \ddots & \vdots \\ I_{i,N_i}^{(1)} & I_{i,N_i}^{(2)} & \cdots & I_{i,N_i}^{(K_i)} \end{bmatrix}. \tag{4}$$

2.2 External illumination method

The generation of the CBFs is usually accomplished by means of one of the two following techniques: a) primary and secondary CBFs method; b) external illumination method.

The first method was proposed in the pioneering technique about the CBFM [39] and it was mainly designed for the analysis of arrays of M elements not connected between them. The primary basis functions were obtained by computing the current on each isolated element as a consequence of feeding the port connected to it. Once the primary CBFs are computed for each element, the secondary basis functions for a given element are computed by considering the current induced due to the radiation of the primary CBFs on the rest of the elements. Thus, one primary CBF and $M-1$ secondary basis functions are defined for each element.

Although the primary and secondary CBFs method has shown an excellent behavior for the analysis of arrays, the accuracy is not so good for scattering problems where the primary CBFs are generated by solving the isolated blocks for the illumination under analysis. In addition, the CBFs generated by this method depend on the employed incident field and, therefore, they have to be recomputed again if the excitations are changed.

The *external illumination method* was introduced in [33,11] to solve the previous problems. As it will be next shown, the CBFs computed by this technique are a complete set, what has not been proved to be true for the primary and secondary CBFs method.

Let us consider the electric field integral equation for perfect electric conductors (PECs) in terms of the currents induced on each block:

$$0 = \mathbf{E}_i + \sum_{n=1}^{B} \mathcal{L}(\mathbf{J}_n)\Big|_t , \tag{5}$$

where $\mathbf{J}_n$ is the current induced in the n-th block and E_i is the impressed field. The linear operator $\mathsf{L}(\cdot)$ relates the currents and the electric field radiated by them and it is defined in [7].

If the currents on all the blocks but the m-the are known, then the integral equation can be expressed as:

$$-\mathcal{L}(\mathbf{J}_m)\big|_t = \mathbf{E}_i + \sum_{\substack{n=1 \\ n\neq m}}^{B} \mathcal{L}(\mathbf{J}_n)\Bigg|_t , \tag{6}$$

the solution of (6) yields the current on the block. It is important to observe that if the superposition principle is considered, then the final current on the block can be obtained as a linear combination of the currents induced due to the field radiated by the rest of the blocks and the impressed field.

The field outside a certain block can be expressed as a linear combination of a set of modes which are a complete base for the electromagnetic field. The most common expansions of the field are *spherical-wave expansion* and the *plane-wave expansion*. Another possible expansion is possible by enclosing the block with a surface supporting equivalent magnetic and electric currents which model the radiation due to the sources outside the block. This last expansion is the one used in the SFX [35] where the RWG basis functions are used to model the equivalent currents.

If it is assumed that *any* field impinging the block can be expanded as a discrete and finite set of waves (or as the field radiated by a discretized equivalent current enclosing the block[35]), what can be carried out with arbitrary accuracy by using spherical-or plane-waves (or low-level basis functions), then *any* current induced in the block can be expressed as a linear combination of the currents induced by each wave of the aforementioned set (or by the field radiated by each low-level basis functions).

The approach presented in [33,11] is based on approximating the external illumination by plane-waves. In order to understand this approach, it is important to consider the well-known expressions for plane-wave expansion. These expressions enable us to expand the electric field under a certain plane $z = z_0$ as [17, chapter 3]:

$$\mathbf{E}(\mathbf{r}) = \int_{-\infty}^{\infty}\int_{-\infty}^{\infty} \mathbf{T}(k_x, k_y) e^{-j\mathbf{k}\cdot\mathbf{r}} dk_x dk_y \qquad z \geq z_0, \tag{7}$$

where the vector $\mathbf{T}(k_x, k_y)$ is the *Plane Wave Spectrum* (PWS); and the propagation vector is defined as $k = k_x x + k_y y + k_z z$ with:

$$k_z = \begin{cases} \sqrt{k^2 - k_x^2 - k_y^2}, & k_x^2 + k_y^2 \leq k^2 \\ -j\sqrt{k_x^2 + k_y^2 - k^2}, & k_x^2 + k_y^2 > k^2 \end{cases} \tag{8}$$

It is also possible to use analogous expressions for the cases of sources above or under other planes (e.g., $x = x_0$). Since it is possible to enclose the block by a surface with flat faces, then the field outside can be expanded by employing a plane-wave expansion for each face of the enclosing surface.

In order to illustrate it, let us consider the two-dimensional case depicted in Figure 2. In this example, the body is enclosed by a rectangle so that the sources can be decomposed into four groups. The first and second groups are the sources above and under the rectangle, respectively, whereas the third and four groups are the sources at the right and left of the rectangle which do not belong to any of the previous groups. Thus, it is possible to carry out expansions similar to the one in (7) for each plane and, therefore, to express the field inside the rectangle as a superposition of plane-waves.

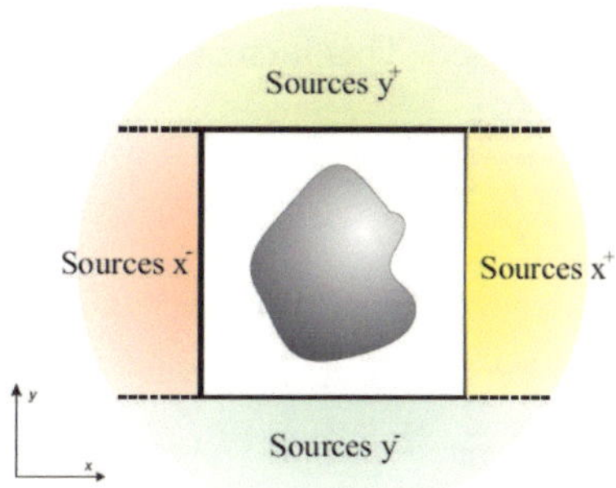

Figure 2. Decomposition of the outer sources into regions for the expansion into plane-waves.

The PWS is *infinite* and *unbounded* as inferred from (7). Hence, several numerical approximations must be carried out. The first one consist on sampling the PWS; guidelines to choose the number of plane-waves depending on the electrical size of the block are given in [28].

In order to truncate the PWS, it is important to observe that it can be split into two regions. The visible region corresponds to the area where the propagation vector is real and, therefore, the wave propagates in that direction. On the other hand, the invisible region corresponds to propagation vectors where at least one of the components is complex and, therefore, this kind of waves is attenuated when it travels away from the sources. Thus, if we assume that the block is far enough from the sources, then it is possible to neglect the contribution from the invisible spectrum. Similar approximations for the PWS were carried out for the analysis of open cavities in [41].

If the blocks are close (or touching each other), then the approximation of neglecting the invisible spectrum can slow down the convergence of the solution in terms of the number of the CBFs. In order to overcome it, when illuminating with plane-waves, the blocks are not considered isolated. Instead of that, they are solved together with a small portion of the continuous blocks. Typical block extensions are between 0.1λ and λ.

Taking into account these two approximations, the coefficients for the CBFs can be computed as:

$$\tilde{\bar{\bar{Z}}}_{ii}\bar{\bar{J}}'_i=\tilde{\bar{\bar{V}}}_i \quad by\ i=1,\ \ldots,\ M \tag{9}$$

where is the matrix containing the field tested by the basis functions in the extended *i*-th block due to each plane waves. Each column of the matrix corresponds to an excitation vector and, therefore, its dimensions are $\tilde{N}_i\times N_{PW}$, where $\tilde{N}_i$ is the number of low-level basis functions in the extended block and N_{PW} is the number of considered plane-waves.

After discarding the currents on the extensions, a complete set of basis functions, $\bar{\bar{J}}'_i$, is achieved. Nonetheless, the linear dependency of these CBFs is not warranted and, hence, if they were used as basis functions, it would yield a MoM matrix that could be ill-conditioned or even singular. In order to detect the number of degrees of freedom in the matrix $\bar{\bar{J}}'_i$, a *singular value decomposition* (e.g., see [16]) is used. This factorization yields the following result:

$$\bar{\bar{J}}'_i=\bar{\bar{U}}_i\bar{\bar{D}}_i\bar{\bar{W}}_i \tag{10}$$

where $\bar{\bar{W}}_i$ and $\bar{\bar{U}}_i$ are orthogonal matrices with dimensions $N_i\times N_i$ and $N_{PW}\times N_{PW}$, where as $\bar{\bar{D}}_i$ is a diagonal matrix with dimensions $N_i\times N_{PW}$ that contains the singular values σ. It will be assumed that the singular values in the diagonal of $\bar{\bar{D}}_i$ are arranged from the highest one to the smaller one without loss of generality.

The columns of $\bar{\bar{U}}_i$ are orthogonal each other and, therefore, they are a complete base of the vector subspace generated by the column vectors in $\bar{\bar{U}}_i$. In addition, the ratio σ_k/σ_1 is a measurement of the dependency of the *k*-column of $\bar{\bar{U}}_i$ with respect to the first $k-1$ first columns. Hence, if the singular values under a given threshold are approximated by zero, then it is only necessary to retain the columns in $\overline{\overline{U}}$ with the singular values over the previous threshold. Finally, the matrix containing the coefficients by columns of the CBFs is:

$$\overline{\overline{J}}_i = \overline{\overline{U}}_i(:,\ 1:K_i) \tag{11}$$

where K_i is the number of singular values normalized by σ_1 that are over a given threshold t_{svd}. The Matlab [36] notation $U_i(:,\ 1:K_i)$ denotes all the columns in U_i from the first one to the K_i-th one.

The inclusion of the SVD provides an efficient way to compute CBFs that are independent each other and, furthermore, it enables an adjust of the quality of the solution by means of the threshold t_{svd}. On the contrary, one of the drawbacks of employing a SVD is that the CBFs lose the physical meaning since they do not correspond anymore to the currents induced by each plane-wave but they are a linear combination of them.

2.3 Reduced matrix computation

In the CBFM, the coefficients of the matrix of the system of equations are the reaction terms between CBFs. Thus, considering the EFIE, the Galerkin's method and a symmetric product for the testing, then the reaction term between the m-th basis function in the i-th block and the n-th basis function in the j-th block is given by:

$$\overline{\overline{Z}}_{ij}^{(1)}(m,n) = \left\langle \mathcal{F}_{i,m}, \mathcal{L}\left(\mathcal{F}_{j,n}\right)\right\rangle, \tag{12}$$

where the index (1) denotes that the matrix contains the interaction between the CBFs instead of the interaction between low-level basis functions. This superindex will be deeply exploited when considering the multilevel version of the CBFM.

If (3) is introduced in (12):

$$\begin{aligned}\overline{\overline{Z}}_{ij}^{(1)}(m,n) &= \sum_{p=1}^{N_i}\sum_{q=1}^{N_j} I_{i,p}^{(m)} I_{j,q}^{(n)} \left\langle \mathbf{f}_{i,p}, \mathcal{L}\left(\mathbf{f}_{j,q}\right)\right\rangle = \sum_{p=1}^{N_i}\sum_{q=1}^{N_j} I_{i,p}^{(m)} I_{j,q}^{(n)} \overline{\overline{Z}}_{ij}^{(0)}(p,q) \\ &= \left[I_{i,1}^{(m)}\ I_{i,2}^{(m)} \cdots I_{i,N_i}^{(m)} \right] \cdot \overline{\overline{Z}}_{ij}^{(0)} \cdot \begin{bmatrix} I_{j,1}^{(m)} \\ I_{j,2}^{(m)} \\ \vdots \\ I_{j,N_j}^{(m)} \end{bmatrix},\end{aligned} \tag{13}$$

where $\overline{\overline{Z}}_{ij}^{(0)}$ is the matrix that contains the reaction terms between the low-level basis functions in the i-th observation block and the j-th source block.

Equation (13) enables us to express the reaction terms between the CBFs of two blocks as a simple product between matrices:

$$\overline{\overline{Z}}_{ij}^{(1)} = \overline{\overline{J}}_i^T \overline{\overline{Z}}_{ij}^{(0)} \overline{\overline{J}}_j \tag{14}$$

where $\overline{\overline{Z}}_{ij}^{(1)}$ is the matrix containing the reaction terms between the CBFs in the i-th observation block and j-th source block.

The matrix of the system of equations for the CBFM, $\overline{\overline{Z}}^{(1)}$, is usually referred to as the *reduced matrix* and it can be computed by means of matrix-matrix products involving the coefficients of the CBFs and the reaction terms of the low-level basis functions:

$$\overline{\overline{Z}}^{(1)} = \begin{bmatrix} \overline{\overline{J}}_1^T \overline{\overline{Z}}_{11}^{(0)} \overline{\overline{J}}_1 & \overline{\overline{J}}_1^T \overline{\overline{Z}}_{12}^{(0)} \overline{\overline{J}}_2 & \cdots & \overline{\overline{J}}_1^T \overline{\overline{Z}}_{1B}^{(0)} \overline{\overline{J}}_B \\ \overline{\overline{J}}_2^T \overline{\overline{Z}}_{21}^{(0)} \overline{\overline{J}}_1 & \overline{\overline{J}}_2^T \overline{\overline{Z}}_{22}^{(0)} \overline{\overline{J}}_2 & \cdots & \overline{\overline{J}}_2^T \overline{\overline{Z}}_{2B}^{(0)} \overline{\overline{J}}_B \\ \vdots & \vdots & \ddots & \vdots \\ \overline{\overline{J}}_B^T \overline{\overline{Z}}_{B1}^{(0)} \overline{\overline{J}}_1 & \overline{\overline{J}}_B^T \overline{\overline{Z}}_{B2}^{(0)} \overline{\overline{J}}_2 & \cdots & \overline{\overline{J}}_B^T \overline{\overline{Z}}_{BB}^{(0)} \overline{\overline{J}}_B \end{bmatrix}. \tag{15}$$

It is important to remark that the computation of the reduced matrix involves the entire MoM matrix. Nonetheless, it is not necessary to store the entire matrix in memory at the same time. Instead of that, it is only required to store the block of the original MoM matrix that is employed for the computation of each submatrix in the reduced matrix.

Similar steps can be followed in order to computed the feeding vector (or feeding matrix if multiple excitations are considered) by means of the following expression:

$$\overline{\overline{V}}^{(1)} = \begin{bmatrix} \overline{\overline{J}}_1^T \overline{\overline{V}}_1^{(0)} \\ \overline{\overline{J}}_2^T \overline{\overline{V}}_2^{(0)} \\ \vdots \\ \overline{\overline{J}}_B^T \overline{\overline{V}}_B^{(0)} \end{bmatrix} \tag{16}$$

where $V_i^{(0)}$ is the feeding matrix containing the low-level feeding vectors of the i-th block.

2.3.1 Fast computation of the reduced matrix

Since the reduced matrix involves all the terms in the MoM matrix, the computational times can be considerably high for electrically large problems. Several approaches are available in the literature to mitigate this drawback. One of the most successful approaches has been the one presented in [10] for the fast computation of reaction terms between the CBFs of an array of repeated elements. This method is based on a *multipolar expansion* of the free space Green's function as in the FMM.

Another method is the one presented in [34] based on the *adaptive cross approximation* (ACA) [2,3]. In this method, the matrices containing the reaction terms between two blocks are modeled by a matrix product of two matrices that can be efficiently computed. In addition, this efficiency improves with the distance between blocks.

A third technique has been proposed in [30]. In this approach, the field radiated by the CBFs of a source block is computed on the observation block by carrying out an interpolation so that it avoids integrating the currents on the source block for each point of the observation block.

Table 1 contains a comparison of the main characteristic of these methods. The points that have been considered are the following ones:

	ACA	Multipolar expansion	Interpolation
Setup time	✓	X	✓
Additional memory requirements	✓	X	✓
Dependency on the Green''s function	✓	X	✓
Distance adaptation	✓	X	✓
Continuous blocks	–	X	✓
Arbitrarily-shaped blocks	✓	✓	X

Table 1. Comparison of several fast methods for computing the reaction terms between CBFs.

- Setup time: the multipolar expansion requires the precomputation of certain parameters. On the hand, the interpolation scheme and the ACA does not require this step.
- Additional memory requirements: the precomputations carried out by the multipolar expansion on the setup stage must be kept in memory so that introduces an additional penalty that does not exist in the other methods.
- Dependency on the Green's function: the ACA and interpolation scheme does not depend on the Green's function that is considered (e.g., free space, multilayer, periodic structure, etc.). Nevertheless, different formulations of the multipolar scheme have to be considered depending on the Green''s function.
- Distance adaptation: the ACA becomes faster depending on the distance between the blocks. Similarly, the interpolation scheme reduces the number of evaluation points so that it also becomes faster if the distance between source and observation blocks is increased. The conventional multipolar expansion does not provide a mechanism to take advantage of the distance between the blocks.
- Contiguous blocks: multipolar expansion can only be applied to blocks in the far-field so that it is not used for contiguous blocks. Although, the interpolation scheme and the adaptive cross approximation can work for contiguous blocks, it results not so efficient for these blocks due to a considerable slowdown of the computations.

- Arbitrary-shaped blocks: the interpolation scheme is designed for planar blocks and, therefore, it only works for two-dimensional blocks.

According to the previous considerations, the ACA seems to be the most complete and versatile method for the combination of it with the CBFM. Thus, it will be used in the rest of the paper for the computation of the reduced matrix unless otherwise indicated.

2.4 Solution of the system of equations

Once the matrix and feeding vector have been computed, the following system of equations must be solved:

$$\overline{\overline{Z}}^{(1)}\overline{I}^{(1)} = \overline{V}^{(1)} \tag{17}$$

where $I^{(1)}$ is the vector containing the unknowns for the expansion of the currents in (2):

and $I_i^{(1)}$ are the coefficients of the expansions of the CBFs in the i-th block:

$$\overline{I}^{(1)} = \begin{bmatrix} \overline{I}_1^{(1)} \\ \overline{I}_2^{(1)} \\ \vdots \\ \overline{I}_B^{(1)} \end{bmatrix} \tag{18}$$

and $I_i^{(1)}$ are the coefficients of the expansions of the CBFs in the i-th block:

$$\overline{I}_i^{(1)} = \begin{bmatrix} I_1^{(1)} \\ I_2^{(1)} \\ \vdots \\ I_i, N_i \end{bmatrix} \tag{19}$$

Once (17) has been solved, it is possible to retrieved the currents vector for the low-level basis functions by expanding the coefficients of the CBFs:

$$\overline{I} = \begin{bmatrix} \overline{\overline{J}}_1 I_1^{(1)} \\ \overline{\overline{J}}_2 I_2^{(1)} \\ \vdots \\ \overline{\overline{J}}_B \overline{I}_B^{(1)} \end{bmatrix} \tag{20}$$

If multiple excitations are considered, then $\widetilde{V}^{(1)}$ becomes a matrix a previously discussed. As a consequence, $\tilde{I}^{(1)}$ become also a matrix containing the solution for each excitation on each column.

2.5 Nested characteristic basis functions

The most important feature of the CBFM is the capacity to solve a problem with a number of unknowns smaller than the number required by the conventional MoM. It has been empirically proved that the compression rate of the CBFM (i.e., the ratio of the number of low-level basis functions to CBFs) increases if the size of the blocks is increased. Figure 3 shows that the number of CBFs that survives to the SVD filtering does not have a linear behavior with the surface and, therefore, it is always recommended to consider blocks as electrically large as possible to improve the compression.

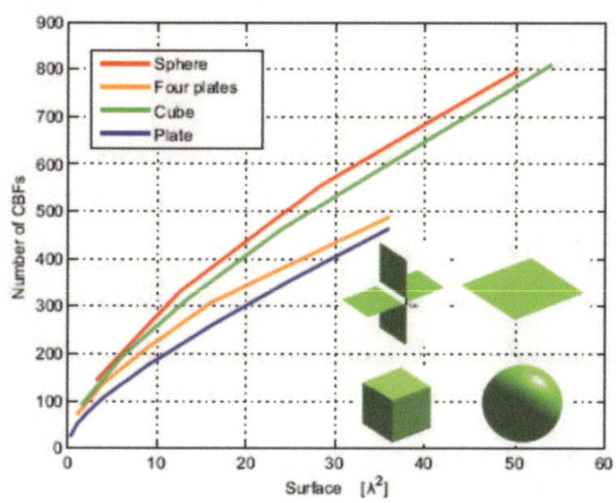

Figure 3. Number of required CBFs for different surfaces as a function of the electrical size.

However, the main drawback of increasing the size of the blocks is that the computation of the CBFs requires more computational resources to solve (9). In order to alleviate this task, a possible alternative is to split the blocks into sublocks and, thus, applied the CBFM to solve (9). Hence, it results into a *recursive scheme* where the CBFs can be expressed in terms of CBFs from a lower level.

The first step to apply the multilevel CBFM (ML-CBFM) is to hierarchically decompose the geometry into blocks until reaching the low-level basis functions. A decomposition example is shown in Figure 4 for a two level decomposition of the NASA almond.

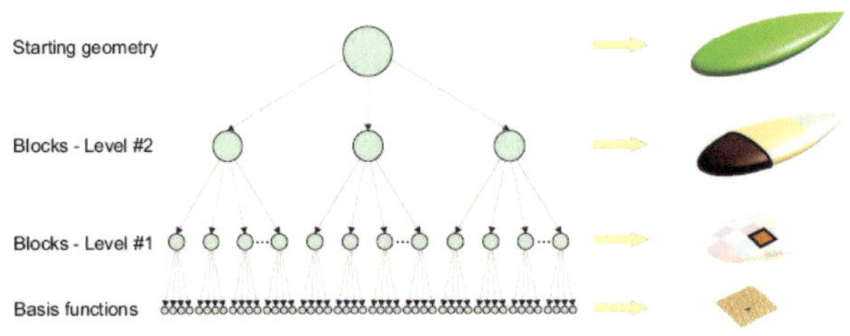

Figure 4. Hierarchical decomposition of the geometry.

In order to mathematically express the multilevel formulation, it is useful to define the notation for a *sequence of blocks* $\{seq\}=\{i_1, i_2, \ldots, i_l\}$ where the values i_k are the indexes of the blocks from the highest level to the lower one. Thus, $\{seq\}=\{2, 5\}$ is used to denote the fifth block, which belongs to the first level, of the second block, which belongs to the second and highest level.

Thus, the n-th CBF inside the block {seq} is given by:

$$\mathcal{F}_{\{seq\},n} = \sum_{b=1}^{B_{\{seq\}}} \sum_{m=1}^{N_{\{seq,b\}}} I^{(n)}_{\{seq,b\},m} \mathcal{F}_{\{seq,b\},m}, \tag{21}$$

where $B_{\{seq\}}$ is the number of block inside the block pointed by $N_{\{seq,b\}}$ is the number of CBFs in the block $I^{(n)}_{\{seq,b\},m}$ are the coefficients for the expansion in terms of the CBFs on the supporting sublocks. The CBFs from the first level, i.e., the level before the low-level basis functions, are defined as in the single-level CBFM in terms of the low-level basis functions:

$$\mathcal{F}_{\{seq\},n} = \sum_{m=1}^{N_{\{seq\}}} I^{(n)}_{\{seq\},m} \mathbf{f}_{\{seq\},m}. \tag{22}$$

The current is as usually expressed as a linear combination of the highest level CBFs:

$$\mathbf{J} = \sum_{b=1}^{B} \sum_{m=1}^{N_{\{b\}}} I_{\{b\},m} \mathcal{F}_{\{b\},m}, \tag{23}$$

where B is the number of blocks and $I_{\{b\},m}$ are the coefficients to be computed by solving the system of equations of the MoM.

The computation of the coefficients of the CBFs for each level is carried out in a similar fashion to the monolevel CBFM. Firstly, the first level blocks are illuminated by multiple plane-waves and the CBFs are obtained on them by computing the currents and applying the SVD filter. Next, the second level blocks are again illuminated by multiple plane-waves and the induced currents (expressed in terms of the first level CBFs) are computed and filtered out. The procedure is repeated until reaching the highest level. It is important to note that the number of plane-waves and the SVD threshold can be different for each level depending on the required accuracy.

Once the CBFs for each block in the highest level have been computed, then it is possible to compute the system of equations matrix containing the reaction terms by applying again a nested scheme. The reaction terms between the CBFs of the $(I - 1)$-th level belonging to the blocks {d} and {e} of the $(I$-1)-th level level can be expressed as:

$$\overline{\overline{Z}}_{\{d\},\{e\}}^{(l-1)}=\begin{bmatrix} \overline{\overline{Z}}_{\{d,1\},\{e,1\}}^{(l-1)} & \overline{\overline{Z}}_{\{d,1\},\{e,2\}}^{(l-1)} & \cdots & \overline{\overline{Z}}_{\{d,1\},\{e,1\}}^{(l-1)} \\ \overline{\overline{Z}}_{\{d,2\},\{e,1\}}^{(l-1)} & \overline{\overline{Z}}_{\{d,2\},\{e,2\}}^{(l-1)} & \cdots & \overline{\overline{Z}}_{\{d,2\},\{e,B_{\{e\}}\}}^{(l-1)} \\ \vdots & \vdots & \ddots & \vdots \\ \overline{\overline{Z}}_{\{d,B_{\{d\}}\},\{e,1\}}^{(l-1)} & \overline{\overline{Z}}_{\{d,1\},\{e,1\}}^{(l-1)} & \cdots & \overline{\overline{Z}}_{\{d,B_{\{d\}}\},\{e,B_{\{e\}}\}}^{(l-1)} \end{bmatrix}, \tag{24}$$

where each submatrix can be computed by considering the coefficients of the CBFs definition and the reaction terms from the CBFs of the previous level:

$$\overline{\overline{Z}}_{\{d\},\{e\}}^{(l)}=\overline{\overline{J}}_{\{d\}}^{T}\overline{\overline{J}}_{\{d\},\{e\}}^{(l-1)}\overline{\overline{J}}_{\{e\}}. \tag{25}$$

Similar steps can be carried out to compute the feeding vector (or matrix in the case of multiple excitations) so that the following system of equations can be solved in order to find the unknown coefficients in (23):

$$\overline{\overline{Z}}^{(L)}\overline{I}^{(L)}=\overline{V}^{(L)}. \tag{26}$$

Once the system of equations has been solved, it is possible to express the CBFs in terms of the CBFs of the next level by means of the following operation:

$$\begin{bmatrix} \overline{I}_{\{seq,1\}}^{(l-1)} \\ \overline{I}_{\{seq,2\}}^{(l-1)} \\ \vdots \\ \overline{I}_{\{seq,B_{\{seq\}}\}}^{(l-1)} \end{bmatrix}=\overline{\overline{J}}_{\{seq\}}\overline{I}_{\{seq\}}^{(l)} \tag{27}$$

this formulation can be applied to all the blocks and repeated until reaching the lowest level so that the solution is expressed in terms of the conventional low-level basis functions.

In the case of considering multiple excitations, the vectors $V^{(L)}$, and the equivalent vectors for each level, becomes matrices as discussed in the single-level CBFM

3. Applications

As it has been shown in the previous sections, the CBFM enables to comprise the MoM matrix. Since the solution of the system of equations is the most time consuming stage for the analysis of electrically large problems with the MoM, a reduction of the number of unknowns will significantly decrease the *total solution time*. Furthermore, the factor limiting the size of the problems that can be solved with the conventional MoM is size of the impedance matrix which must fit in the available RAM memory. Thus, the CBFM compression results on an *extension of the application range* of the MoM. Considering these two facts, the CBFM can be considered a *speed-up technique for the MoM*. In addition, the CBFM is not based on an iterative scheme so it will not suffer from converge problems.

Although the use of the CBFM to compute an iteration-free of relatively large problems may be the most direct application of the MoM, it will be shown in the following sections that the compression of the MoM matrix can be exploited for many other purposes.

3.1 Monostatic Radar Cross Section Computation

The compressed system of equations described by (17) and (26) for the monolevel- and multilevel-CBFM, respectively, can be solved by using conventional direct solutions schemes such as a LU decomposition for problems electrically larger than the ones that can be solved by the MoM. These schemes can handle a relatively large number of RHS with a negligible time increment with respect of the case with a single RHS. As a consequence, many authors (e.g., [33,11]) have benefited from this advantage to compute monostatic radar cross sections (RCS) where there are as many RHS as incident angles.

This application is illustrated for the computation of the monostatic RCS of a 4λ cube. The cube is solved with a single level CBFM by decomposing the geometry into 24 blocks of size $2\lambda \times 2\lambda$ (see Figure 5). The number of plane waves to solve the problem is 400 and the SVD threshold is set to 10^{-3}. The original problem was discretized into 31728 RWG unknowns that were reduced to only 3004 after applying the CBFM.

The results for the $\theta - \theta$ RCS for 91 points in the cut $\phi=0$ are shown in Figure 5 together with the reference results from the MoM. Excellent agreement is found between both of them.

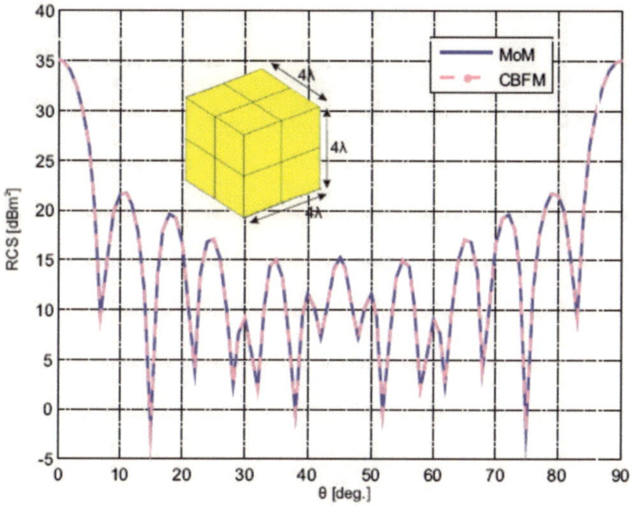

Figure 5. Monostatic RCS for a 4λ cube computed by the MoM and the CBFM.

The CPU times are shown in Table 2. In this case, the ACA has not been used to speed-up the matrix filling time which is the most time consuming stage. The CPU times for the CBFM simulation have been computed on a PC equipped with an Intel® Core™ at 2.4 GHz (only one core was used) and 4GB RAM memory whereas the MoM simulation was run on a computational server in order to fit the impedance matrix on the RAM memory. This server is equipped with 8 Dual Core AMD Opteron™ 880 at 2.4GHz (only one core was used) and 56GB of RAM.

	CBFs generation	Matrix filling time	Solving time	Total time
MoM	-	1468s	14198s	15666s
CBFM	510s	1614s	14s	2138s

Table 2. Computational times for solving a 4λ cube with the MoM and the CBFM.

Although the size of the previous problem is relatively electrically small, it shows how the compression of CBFM enables to solve problems larger than the ones solved by the conventional MoM. The solution of larger problems can be efficiently carried out by combining the multilevel scheme together with an appropriate parallelization scheme [29].

3.2 Fast evaluation of phased arrays

In addition to the monostatic RCS computation, there are other cases where the simultaneous solution of multiple RHS is highly recommended. One important example is the synthesis of phased arrays. In this problem, the optimal combination of voltages at the input of each element of the array has to be found in order to achieved the desired radiation pattern. It is possible to show [1] that there is a matrix, $\overline{\overline{G}}$, that relates the samples of the electromagnetic field on a given set of points (e.g., a circular cut of points in the far-field) with the voltages by the following matrix-vector product:

$$\overline{E} = \overline{\overline{G}}\overline{v} \tag{28}$$

Once this matrix has been obtained, then the computation of the radiated field for a certain choice of voltages can be computed with a very small computational effort. Nonetheless, the computation of this matrix involves solving the geometry of the entire problem (the array and the surrounding elements) for as many RHS as elements [1]. Therefore, the computation of $\overline{\overline{G}}$ can be very time-consuming when the number of elements in the array is high and the structure to be solved is electrically large. As in the previous example, the CBFM can help to mitigate the computational cost due to its ability to solve simultaneously multiple RHS.

In order to illustrate this example, a secondary surveillance radar (SSR) is considered. In this kind of problems, it is very important to have a low level of secondary lobes to avoid false interrogations. The radar under analysis consist on an array of 10×33 elements that is surrounded by the supporting structure, a primary radar, a circular supporting platform and several lighting rods (see Figure 6). The working frequency is 1030 MHz so that the geometry is discretized into 139, 762 RWGs.

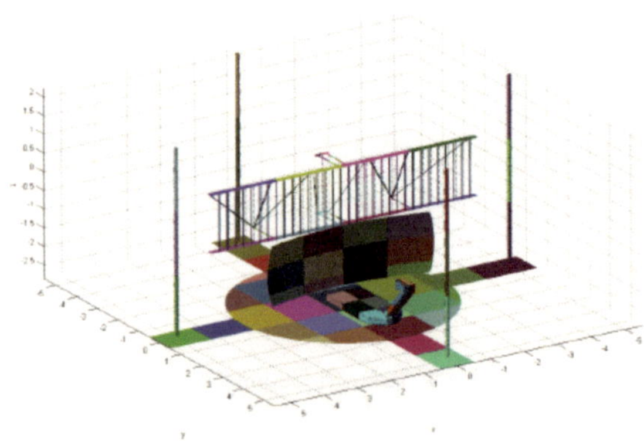

Figure 6. First level decomposition for a secondary surveillance radar.

A two level CBFM is used to efficiently compute the matrix in (28). The first level blocks are shown in Figure 6. After applying an SVD threshold equal to 10^{-3}, the number of first-level CBFs is only 20259.

A second level CBFM is applied on this first-level CBFs by grouping the blocks into five macro blocks. The first one contains the SSR and the supporting structure, and the remaining four blocks contain the rest of the structure. The number of degrees of freedom after this second level is reduced to 13312. In both cases, the number of plane waves to illuminates the first- and second-level blocks is 6720.

In this example, a separable distribution of the feedings is assumed and, therefore, the elevation and azimuth radiation patterns are assumed to be controlled by the feeding distribution on the vertical and horizontal distribution. The vertical distribution is designed as in [25] which considers the isolated structure and computes the vertical feeding to provide a cosecant squared pattern. On the other hand, the azimuth elevation pattern, which must suit the masks depicted in Figure 7, is expected to be affected by the surrounding environment and, in special, by the lighting rods. Thus, the feedings computed in [25] are not expected to be valid and they must be synthesized for the new structure. Hence, the matrix G has been computed to relates the 33 voltages of the horizontal distribution with 181 sample points of the azimuth radiation pattern.

Once the matrix is computed, many synthesis schemes can be efficiently applied. In this book chapter, a genetic algorithm has been used to generated the proper radiation pattern shown in Figure 7 but other optimization techniques could also be used [26]. The evaluation of each generation of the genetic algorithm, which involves computing a hundred of radiation pattern, is accomplished in only 0.22 seconds. In opposite, the radiation pattern of a usual hamming window distribution is also shown revealing that it is out of the mask.

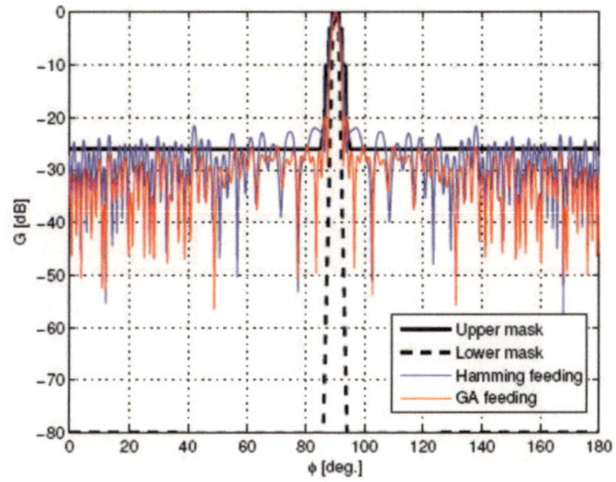

Figure 7. Normalized radiation patterns for a hamming window feeding and a feeding computed by a genetic algorithm.

3.3 Partial modifications for the analysis of on-board antennas

The quick expansion of iterative schemes such as the fast multipole method has enabled us to solve problems electrically larger than in the past. Nonetheless, some enhancements of the MoM cannot be easily applied on these iterative schemes. Since the CBFM is an extended version of the MoM in which the conventional low-level basis functions are replaced by CBFs, then most of MoM enhancements can be easily adapted.

The partial solution of geometries [23,37,21,9] is one of these approaches that can be easily applied to the CBFM. In this book chapter, we will focus on the partial solution based on the Sherman-Morrison formula as described in [24].

In this technique, the structure is splitted into two parts. The first one is assumed to be fixed and it will not be modified. The other part is assumed to be smaller than the first one and it is assumed that it needs to be modified several times during the analysis. Thus, once the fixed geometry is known, it is possible to carry out most of the steps to compute the inverse of the entire impedance matrix so that the analysis of each modification can be resumed from that point saving computational effort. Since the inverse of the entire matrix is involved, then it can only be applied to problems where this matrix fits in the available RAM memory and, hence, the CBFM extends the range of applicability of the method to geometries electrically larger than the structures analyzed by the conventional MoM.

An example of application of these techniques could be the analysis of an antenna on an airplane affected by the turbines rotation. In this problem, the airplane must be solved for different rotation angles of the turbines in order to check the impact of them in the radiation characteristics of it. In this case, the fixed structure would be the airplane without the turbines what comprises most of the total geometry, and the turbines areas are the parts that must be modified.

In this book chapter, the technique is illustrated for a problem similar to the previously described where the analysis of the radiation pattern of an monopole placed on an helicopter affected by the blades rotation is accomplished. The geometry is shown in Figure 8 and the monopole is placed at the beginning of the tail and parallel to it. The problem is solved at 400, MHz so that the geometry is discretized into 96, 851 low-level basis functions. After the application of the CBFM with 400 plane waves and a SVD threshold equal to 10^{-3}, the number of degrees of freedom for the total problem is 9837 with only 830 belonging to the blades.

The analysis of the structure with the CBFM involves 10, 200 s. On the other hand, the required precomputations with the technique described in this section requires 10, 026 s. After that, the analysis of each new blades position can be accomplished in only 865s. Thus, the analysis with this technique becomes quickly more efficient than the use of the conventional CBFM. Times are measured in a machine equipped with four Quad-Core AMD Opteron™ 8378 at 2.4 GHz (only one core was used) and 128 GB of RAM. The radiation patterns for several rotation angles of the main rotor are shown in Figure 8.

Finally, it is important to remark that other approaches can be used for similar purposes. For example, antenna placement is considered in [27] by combining the CBFM with an incomplete Gauss-Jordan elimination.

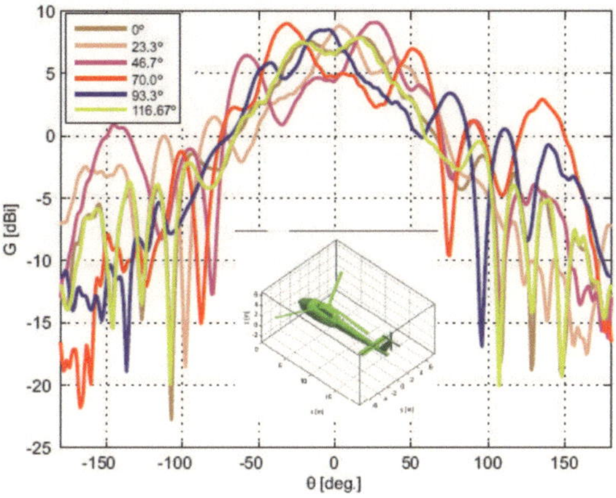

Figure 8. Gain on the cut =90° for different rotation positions in the main rotor. The inset depicts the helicopter with the blades at the zero angle.

4. Conclusions

The characteristic basis function method has been deeply described in this book chapter. New perspectives have been provided to understand the generation of the characteristic basis functions based on plane-waves revealing that it yields a complete set of basis functions if the approximation of external field is accurate enough. Several speeds up have been compared for the case of filling the impedance matrix and the adaptive cross approximation seems to be the most balanced one for a generic analysis.

As a consequence, the CBFM has been proved to be a rigorous method that achieved an iteration-free solution for problems electrically larger than the ones that can be analyzed with the conventional method of moments. In addition, the compression of the matrix that the CBFM carries out is compatible with many of the MoM enhancements. The most direct application is the solution of problems where multiple excitations have to be considered such as monostatic radar cross section computations or characterization of antenna arrays. In addition, techniques for efficiently analysis of small modifications on large structures have been shown to be compatible with the CBFM.

5. Acknowledgments

This work has been supported by the Ministerio de Ciencia e Innovación of Spain /FEDER under projects TEC2011-24492, CSD2008-00068; by the Gobierno del Principado de Asturias (PCTI)/FEDER-FSE under project PC10-06; contract FUO-EM-221-10, and by Cátedra Telefónica-Universidad de Oviedo.

Author details

Jaime Laviada*, Fernando Las-Heras and Marcos R. Pino

*Address all correspondence to: jlaviada@tsc.uniovi.es

Escuela Politécnica de ingeniería de Gijón, Universidad de Oviedo, Electrical Engineering Department, Gijón, España

References

[1] Ayestarán, R.G., Laviada, J., & Las-Heras, F. (2009). Realistic antenna array synthesis in complex environments using a MoM-SVR approach. *J. of Electromagn. Waves and Appl.*, 23, 97-108.

[2] Bebendorf, M. (June 2000). Approximation of boundary element matrices. *Numen. Math*, 86(4), 565-589.

[3] Bebendorf, M., & Rjasanow, S. (March 2003). Adaptive low-rank approximation of collocation matrices. *Computing*, 70(1), 1-24.

[4] Bilotti, F., & Vegni, C. (December 2001). Rigorous and efficient full-wave analysis of trapezoidal patch antennas. *IEEE Trans. Antennas Propagat.*, 49(12), 1773-1776.

[5] Bleszynski, E., Bleszynski, M., & Jaroszewicz, T. (Sep.-Oct. 1996). AIM: adaptive integral method for solving large-scale electromagnetic scattering and radiation problems. *Radio Sci.*, 31(5), 1225-1251.

[6] Cabedo-Fabres, M., Antonino-Daviu, E., Valero-Nogueira, A., & Bataller, M. F. (October 2007). The theory of characteristic modes revisited: A contribution to the design of antennas for modern applications. *IEEE Antennas Propagat. Mag.*, 49(5), 52-68.

[7] Chew, W. C., Jin, J. M., Michielssen, E., & Song, J. (2001). Fast and efficient algorithms in computational electromagnetics. Artech House, Norwood, Massachusetts 02062, USA.

[8] Coifman, R., Rokhlin, V., & Wandzura, S. (June 1993). The fast multipole method for the wave equation: A pedestrian prescription. *IEEE Trans. Antennas Propagat. Mag.*, 53(3), 7-12.

[9] Cormos, D., Loison, R., & Gillard, R. (June 2005). Fast optimization and sensitivity analysis of nonintuitive planar structures. *IEEE Trans. Microwave Theory Tech.*, 53(6), 2019-2025.

[10] Craeye, Christophe. (October 2006). A fast impedance and pattern computation scheme for finite antenna arrays. *IEEE Trans. Antennas Propagat.*, 54(10), 3030-3034.

[11] Delgado, Carlos, Felipe Cátedra, Manuel, & Mittra, Raj. (March 2008). Application of the characteristic basis function method utilizing a class of basis and testing functions defined on NURBS patches. *IEEE Antennas Propagat.*, 56(3), 784-791.

[12] Eshrah, Islam A., & Kishk, Ahmed A. (June 2004). Application of the adaptive basis functions/diagonal moment matrix technique to arrays of dielectric scatterers. *Proc. IEEE AP-S Int. Symp.*, 1, 647-650.

[13] Eshrah, Islam A., & Kishk, Ahmed A. (May 2004). Implementation of the adaptive basis functions/diagonal moment matrix technique to array of identical elements. *Proc. URSI International Symposium on Electromagnetic Theory*, 837-839.

[14] Eshrah, Islam A., & Kishk, Ahmed A. (March 2005). Analysis of linear arrays using the adaptive basis functions/diagonal moment matrix technique. *IEEE Trans. Antennas Propagat.*, 53(3), 1121-1125.

[15] Glisson, A. W., & Wilton, D. R. (September 1980). Simple and efficient numerical methods for problems of electromagnetic radiation and scattering from surfaces. *IEEE Trans. Antennas Propagat.*, AP-28(5), 593-603.

[16] Golub, G. H., & Loan, C. F. van. (1989). Matrix computations. John Hopkins Univ. Press, Baltimore, MD.

[17] Hansen, T. B., & Yaghjian, Arthur D. (1999). Plane-Wave Theory of time domain fields. IEEE press, New York, NY.

[18] Harrington, R. F., & Mautz, J. R. (September 1971). Computation of characteristic modes for conducting bodies. *IEEE Trans. Antennas Propagat.*, AP-19, 629-639.

[19] Harrington, R. F., & Mautz, J. R. (September 1971). Theory of characteristic modes for conducting bodies. *IEEE Trans. Antennas Propagat.*, AP-19, 622-628.

[20] Harrington, R.F. (1993). Field Computation by Moment Method. IEEE press, New York, NY.

[21] Johnson, J. Michael, & Rahmat-Samii, Yahya. (October 1999). Genetic algorithms and method of moments ({GA/MOM}) for the design of integrated antennas. *IEEE Trans. Antennas Propagat.*, 47(10), 1606-1614.

[22] Kalbasi, Khalil, & Demarest, Kenneth R. (May 1993). A multilevel formulation of the method of moments. *IEEE Trans. Antennas Propagat.*, 41(5), 589-599.

[23] Kastner, R. (March 1989). An 'add-on' method for the analysis of scattering from large planar structures. *IEEE Trans. Antennas Propagat.*, 37(3), 353-361.

[24] Kastner, R. (January 1990). Analysis of microstrip antenna structures using the 'add-on' technique. *IEEE Trans. Antennas Propagat.*, 38(1), 114-117.

[25] Las-Heras, F., Galocha, B., & Besada, J. L. (July 1997). Equivalent source modelling and reconstruction for antenna measurement and synthesis. *Proc. IEEE AP-S Int. Symp.*, 1, 156-159, Montreal, Canada.

[26] Laviada, J., Ayestarán, R.G., Pino, M. R., Las-Heras, F., & Mittra, R. (2009). Synthesis of phased arrays in complex environments with the multilevel characteristic basis function method. *Progress In Electromagnetics Research*, 92, 347-360.

[27] Laviada, J., Gutiérrez-Meana, J., Pino, M. R., & Las-Heras, F. (October 2010). A partial solution of MoM matrices based on characteristic basis functions and its application to on-board antennas positioning. *Applied Computational Electromagnetics Society (ACES) Journal*, 25(10), 830-840.

[28] Laviada, J., Pino, M. R., Las-Heras, F., & Mittra, Raj. (October 2009). Solution of electrically large problems with multilevel characteristic basis functions. *IEEE Trans. Antennas Propagat.*, 57(10), 3189-3198.

[29] Laviada, J., Pino, M. R., Mittra, R., & Las-Heras, Fernando. (December 2009). Parallelized multilevel characteristic basis function method for solving electromagnetic scattering problems. *Microwave Optical and Technology Letters*, 51(12), 2963-2969.

[30] Laviada, J., Pino, Marcos R., Las-Heras, F., & Mittra, R. (August 2009). Interpolation scheme for fast calculation of reaction terms in the characteristic basis function method. *Microwave Optical and Technology Letters*, 51(8), 1818-1824.

[31] Losada, V., Boix, R. R., & Horno, M. (April 1999). Resonant modes of circular microstrip patches in multilayered substrates. *IEEE Trans. Antennas Propagat.*, 47, 488-498.

[32] Wei Bing, Lu, Cui, Tie Jun, Qian, Zhi Guo, Yin, Xiao Xing, & Hong, Wei. (November 2004). Accurate analysis of large-scale periodic structures using an efficient sub-entire-domain basis function method. *IEEE Trans. Antennas Propagat.*, 52(11), 3078-3085.

[33] Lucente, Eugenio, Monorchio, Agostino, & Mittra, Raj. (April 2008). An iteration-free MoM approach based on excitation independent characteristic basis functions for solving large multiscale electromagnetic scattering problems. *IEEE Antennas Propagat.*, 56(4), 999-1007.

[34] Maaskant, R., Mittra, R., & Tijhuis, A. (November 2008). Fast analysis of large antenna arrays using the characteristic basis function method and the adaptive cross approximation algorithm. *IEEE Trans. Antennas Propagat.*, 56(11), 3440-3451.

[35] Matekovits, L., Laza, V. A., & Vechhi, G. (September 2007). Analysis of large complex structures with the synthetic-functions approach. *IEEE Trans. Antennas Propagat.*, 55(9), 2509-2521.

[36] MathWorks. (2009). Matlab Getting Started Guide. http://www.mathworks.com.

[37] Misra, P., & Naishadham, K. (December 1996). Order-recursive gaussian elimination (ORGE) and efficient CAD of microwave circuits. *IEEE Trans. Microwave Theory Tech.*, 44(12), 2166-2173.

[38] Ooms, S., & DeZutter, D. (March 1998). A new iterative diakoptics-based multilevel moments method for planar circuits. *IEEE Trans. Microw. Theory Tech.*, MTT-46(3), 280-291.

[39] Prakash, V.V.S., & Mittra, R. (January 2003). Characteristic basis function method: A new technique for efficient solution of method of moments matrix equation. *Microwave Opt. Technol. Lett.*, 36(2), 95-100.

[40] Rao, S. M., Wilton, D. R., & Glisson, A. W. (May 1982). Electromagnetic scattering by surfaces of arbitrary shape. *IEEE Trans. Antennas Propagat.*, AP-30, 409-418.

[41] Rousseau, P. R., & Burkholder, R. J. (October 1995). A hybrid approach for calculating the scattering from obstacles within large, open cavities. *IEEE Trans. Antennas Propagat.*, 43(10), 1068-1075.

[42] Song, J. M., Lu, C. C., & Chew, W. C. (October 1997). Multilevel fast multipole algorithm for electromagnetic scattering by large complex objects. *IEEE Trans. Antennas Propagat.*, 45(10), 1488-1493.

[43] Suter, Eric, & Mosig, Juan R. (August 2000). A subdomain multilevel approach for the efficient MoM analysis of large planar antennas. *Microwave Opt. Technol. Lett.*, 26(4), 270-277.

[44] Vegni, L., Toscano, A., & Popovski, B. (1997). Electromagnetic plane wave scattering by large and finite strip array on dielectric slab. *Ann. Telecommun.*, 52(3-4), 209-218.

[45] Waller, M. L., & Rao, S. M. (October 2002). Application of adaptive basis functions for a diagonal moment matrix solution of arbitrarily shaped three-dimensional conducting body problems. *IEEE Trans. Antennas Propagat.*, 50(10), 1445-1452.

Chapter 5

Analytical Grounds for Modern Theory of Two-Dimensionally Periodic Gratings

L. G. Velychko, Yu. K. Sirenko and E. D. Vinogradova

Additional information is available at the end of the chapter

http://dx.doi.org/10.5772/51007

1. Introduction

Rigorous models of one-dimensionally periodic diffraction gratings made their appearance in the 1970s, when the corresponding theoretical problems had been considered in the context of classical mathematical disciplines such as mathematical physics, computational mathematics, and the theory of differential and integral equations. Periodic structures are currently the objects of undiminishing attention. They are among the most called-for dispersive elements providing efficient polarization, frequency and spatial signal selection. Fresh insights into the physics of wave processes in diffraction gratings are being implemented into radically new devices operating in gigahertz, terahertz, and optical ranges, into new materials with inclusions ranging in size from micro- to nanometers, and into novel circuits for in-situ man-made and natural material measurements.

However, the potentialities of classical two-dimensional models [1-7] are limited. Both theory and applications invite further investigation of three-dimensional, vector models of periodic structures in increasing frequency. In our opinion these models should be based on time-domain (TD) representations and implemented numerically by the mesh methods [8,9]. It follows from the well-known facts: (i) TD-approaches are free from the idealizations inherent in the frequency domain; (ii) they are universal owing to minimal restrictions imposed on geometrical and material parameters of the objects under study; (iii) they allow explicit computational schemes, which do not require inversion of any operators and call for an adequate run time when implementing on present-day computers; (iv) they result in data easy convertible into a standard set of frequency-domain characteristics. To this must be added that in recent years the local and nonlocal exact absorbing conditions (EAC) have been derived and tested [6,7]. They allow one to replace an open initial boundary value problem occurring in the electrodynamic theory of gratings with a closed problem. In addi-

INTECH
open science | open minds

© 2012 Velychko et al.; licensee InTech. This is an open access article distributed under the terms of the Creative Commons Attribution License (http://creativecommons.org/licenses/by/3.0), which permits unrestricted use, distribution, and reproduction in any medium, provided the original work is properly cited.

tion, the efficient fast Fourier transform accelerated finite-difference schemes with EAC for characterizing different resonant structures have been constructed and implemented [10].

It is evident that the computational scheme solving a grating problem must be stable and convergent, computational error must be predictable, while the numerical results are bound to be unambiguously treated in physical terms. To comply with these requirements, it is important to carry out theoretical analysis at each stage of the modeling (formulation of boundary value and initial boundary value problems, determination of the correctness classes for them, study of qualitative characteristics of singularities of analytical continuation for solutions of model boundary value problems into a domain of complex-valued frequencies, etc.).

In the present work, we present a series of analytical results providing the necessary theoretical background to the numerical solution of initial boundary value problems as applied to two-dimensionally periodic structures. Section 1 is an Introduction. In Section 2 we give general information required to formulate a model problem in electrodynamic theory of gratings. Sections 3 and 4 are devoted to correct and efficient truncation of the computational space in the problems describing spatial-temporal electromagnetic wave transformation in two-dimensionally periodic structures. Some important characteristics and properties of transient and steady-state fields in regular parts of the rectangular Floquet channel are discussed in Sections 5 and 7. In Section 6, the method of transformation operators (the TD-analog of the generalized scattering matrix method) is described; by applying this method the computational resources can be optimized when calculating a multi-layered periodic structure or a structure on a thick substrate. In Section 8, elements of spectral theory for two-dimensionally periodic gratings are given in view of its importance to physical analysis of resonant scattering of pulsed and monochromatic waves by open periodic resonators.

2. Fundamental Equations, Domain of Analysis, Initial and Boundary Conditions

Space-time and space-frequency transformations of electromagnetic waves in diffraction gratings, waveguide units, open resonators, radiators, etc. are described by the solutions of initial boundary value problems and boundary value problems for Maxwell's equations. In this chapter, we consider the problems of electromagnetic theory of gratings resulting from the following system of Maxwell's equations for waves propagating in stationary, locally inhomogeneous, isotropic, and frequency dispersive media [9,11]:

$$\mathrm{rot}\vec{H}(g,t)=\eta_0^{-1}\frac{\partial\left[\vec{E}(g,t)+\chi_\varepsilon(g,t)*\vec{E}(g,t)\right]}{\partial t}+\chi_\sigma(g,t)*\vec{E}(g,t)+\vec{j}(g,t), \qquad (1)$$

$$\mathrm{rot}\vec{E}(g,t)=-\eta_0\frac{\partial\left[\vec{H}(g,t)+\chi_\mu(g,t)*\vec{H}(g,t)\right]}{\partial t}, \qquad (2)$$

where

$g=\{x, y, z\}$ is the point in a three-dimensional spaceR^3;

x, y, and z are the Cartesian coordinates;

$\vec{E}(g, t)=\{E_x, E_y, E_z\}$ and $\vec{H}(g, t)=\{H_x, H_y, H_z\}$ are the electric and magnetic field vectors;

$\eta_0=(\mu_0/\varepsilon_0)^{1/2}$ is the intrinsic impedance of free space;

ε_0 and μ_0 are permittivity and permeability of free space;

$\vec{j}(g, t)$ is the extraneous current density vector;

$\chi_\varepsilon(g, t)$, $\chi_\mu(g, t)$, and $\chi_\sigma(g, t)$ are the electric, magnetic, and specific conductivity susceptibilities; $f_1(t) * f_2(t)=\int f_1(t-\tau)f_2(\tau)d\tau$stands for the convolution operation.

We use the SI system of units. From here on we shall use the term "time" for the parameter$t$, which is measured in meters, to mean the product of the natural time and the velocity of light in vacuum.

With no frequency dispersion in the domain$G\subset R^3$, for the points $g\in G$ we have

$\chi_\varepsilon(g, t)=\delta(t)[\varepsilon(g)-1]$, $\chi_\mu(g, t)=\delta(t)[\mu(g)-1]$, $\chi_\sigma(g, t)=\delta(t)\sigma(g)$,

where $\delta(t)$ is the Dirac delta-function;$\varepsilon(g)$, $\mu(g)$, and $\sigma(g)$ are the relative permittivity, relative permeability, and specific conductivity of a locally inhomogeneous medium, respectively. Then equations (1) and (2) take the form:

$$\text{rot}\vec{H}(g, t)=\eta_0^{-1}\varepsilon(g)\frac{\partial\vec{E}(g, t)}{\partial t}+\sigma(g)\vec{E}(g, t)+\vec{j}(g, t), \tag{3}$$

$$\text{rot}\vec{E}(g, t)=-\eta_0\mu(g)\frac{\partial\vec{H}(g, t)}{\partial t}. \tag{4}$$

In vacuum, where $\varepsilon(g)=\mu(g)=1$ and$\sigma(g)=0$, they can be rewritten in the form of the following vector problems [6]:

$$\begin{gathered}\left[\Delta-\text{grad div}-\frac{\partial^2}{\partial t^2}\right]\vec{E}(g, t)=\vec{F}_E(g, t), \quad \frac{\partial}{\partial t}\vec{H}(g, t)=-\eta_0^{-1}rot\,\vec{E}(g, t),\\ \vec{F}_E(g, t)=\eta_0\frac{\partial}{\partial t}\vec{j}(g, t)\end{gathered} \tag{5}$$

or

$$\begin{gathered}\left[\Delta-\frac{\partial^2}{\partial t^2}\right]\vec{H}(g, t)=\vec{F}_H(g, t), \quad \eta_0^{-1}\frac{\partial}{\partial t}\vec{E}(g, t)=\text{rot}\vec{H}(g, t)-\vec{j}(g, t),\\ \vec{F}_H(g, t)=-\text{rot}\,\vec{j}(g, t)\ .\end{gathered} \tag{6}$$

By Δ we denote the Laplace operator. As shown in [6], the operator grad div$\vec{E}$ can be omitted in (5) from the following reasons. By denoting the volume density of induced and external electric charge through $\rho_1(g, t)$ and $\rho_2(g, t)$, we can write grad div$\vec{E} = \varepsilon_0^{-1}$grad$(\rho_1 + \rho_2)$. In homogeneous medium, where ε and σ are positive and non-negative constants, we have $\rho_1(g, t) = \rho_1(g,0)\exp(-t\sigma / \varepsilon)$, and if $\rho_1(g,0)=0$, then $\rho_1(g, t)=0$ for any$t>0$. The remaining term ε_0^{-1}gradρ_2 can be moved to the right-hand side of (5) as a part of the function defining current sources of the electric field.

To formulate the initial boundary value problem for hyperbolic equations (1)-(6) [12], initial conditions at $t=0$ and boundary conditions on the external and internal boundaries of the domain of analysis Q should be added. In 3-D vector or scalar problems, the domain Q is a part of the R^3-space bounded by the surfaces S that are the boundaries of the domains int$S$, filled with a perfect conductor:$Q = R^3 \setminus \overline{\text{int}S}$. In the so-called open problems, the domain of analysis may extend to infinity along one or more spatial coordinates.

The system of boundary conditions for initial boundary value problems is formulated in the following way [11]:

- on the perfectly conducting surface S the tangential component of the electric field vector is zero at all times t

$$E_{tg}(g, t)\,|_{g\in S} = 0 \quad \text{for} \quad t \geq 0; \tag{7}$$

the normal component of the magnetic field vector on S is equal to zero ($H_{nr}(g, t)\,|_{g\in S}=0$), and the function $H_{tg}(g, t)\,|_{g\in S}$ defines the so-called surface currents generated on S by the external electromagnetic field;

- on the surfaces$S^{\varepsilon,\mu,\sigma}$, where material properties of the medium have discontinuities, as well as all over the domain$Q$, the tangential components $E_{tg}(g, t)$ and $H_{tg}(g, t)$ of the electric and magnetic field vectors must be continuous;
- in the vicinity of singular points of the boundaries of$Q$, i.e. the points where the tangents and normals are undetermined, the field energy density must be spatially integrable;
- if the domain Q is unbounded and the field $\{\vec{E}(g, t), \vec{H}(g, t)\}$ is generated by the sources having bounded supports in Q then for any finite time interval $(0,T)$ one can construct a closed virtual boundary $M \subset Q$ sufficiently removed from the sources such that

$$\{\vec{E}(g, t), \vec{H}(g, t)\}\,|_{g\in M, t\in(0,T)} = 0. \tag{8}$$

The initial state of the system is determined by the initial conditions at$t=0$. The reference states $\vec{E}(g,0)$ and $\vec{H}(g,0)$ in the system (1), (2) or the system (3), (4) are the same as $\vec{E}(g,0)$ and $[\partial\vec{E}(g, t)/\partial t]\,|_{t=0}$ ($\vec{H}(g,0)$and$[\partial\vec{H}(g, t)/\partial t]\,|_{t=0}$) in the differential forms of the second order (in the terms of t), to which (1), (2) or (3), (4) are transformed if the vector $\vec{H}$ (vector $\vec{E}$) is eliminated (see, for example, system (5), (6)). Thus, (5) should be complemented with the initial conditions

$$\vec{E}(g,0)=\vec{\varphi}(g),\ \frac{\partial}{\partial t}\vec{E}(g,\ t)\Big|_{t=0}=\vec{\psi}(g),\ g\in\bar{Q}. \tag{9}$$

The functions $\vec{\varphi}(g)$, $\vec{\psi}(g)$, and $\vec{F}(g,\ t)$ (called the instantaneous and current source functions) usually have limited support in the closure of the domain Q. It is the practice to divide current sources into hard and soft [9]: soft sources do not have material supports and thus they are not able to scatter electromagnetic waves. Instantaneous sources are obtained from the pulsed wave $\vec{U}^i(g,\ t)$ exciting an electrodynamic structure: $\vec{\varphi}(g)=\vec{U}^i(g,0)$ and $\vec{\psi}(g)=[\partial\vec{U}^i(g,\ t)/\partial t]\big|_{t=0}$. The pulsed signal $\vec{U}^i(g,\ t)$ itself should satisfy the corresponding wave equation and the causality principle. It is also important to demand that the pulsed signal has not yet reached the scattering boundaries by the moment $t=0$.

The latter is obviously impossible if infinite structures (for example, gratings) are illuminated by plane pulsed waves that propagate in the direction other than the normal to certain infinite boundary. Such waves are able to run through a part of the scatterer's surface by any moment of time. As a result a mathematically correct modeling of the process becomes impossible: the input data required for the initial boundary value problem to be set are defined, as a matter of fact, by the solution of this problem.

3. Time Domain: Initial Boundary Value Problems

The vector problem describing the transient states of the field nearby the gratings whose geometry is presented in Figure 1 can be written in the form

$$\begin{cases}\operatorname{rot}\vec{H}(g,t)=\eta_0^{-1}\dfrac{\partial\left[\vec{E}(g,t)+\chi_\varepsilon(g,t)*\vec{E}(g,t)\right]}{\partial t}+\chi_\sigma(g,t)*\vec{E}(g,t)+\vec{j}(g,t),\\ \operatorname{rot}\vec{E}(g,t)=-\eta_0\dfrac{\partial\left[\vec{H}(g,t)+\chi_\mu(g,t)*\vec{H}(g,t)\right]}{\partial t},\quad g=\{x,y,z\}\in\mathbf{Q},\quad t>0\\ \vec{E}(g,0)=\vec{\varphi}_E(g),\quad \vec{H}(g,0)=\vec{\varphi}_H(g),\quad g\in\bar{\mathbf{Q}}\\ E_{tg}(g,t)\big|_{g\in\mathbf{S}}=0,\quad H_{nr}(g,t)\big|_{g\in\mathbf{S}}=0,\quad t\geq 0\ .\end{cases} \tag{10}$$

Here, $\bar{Q}$ is the closure of Q, $\chi_{\varepsilon,\mu,\sigma}(g,\ t)$ are piecewise continuous functions and the surfaces S are assumed to be sufficiently smooth. From this point on it will be also assumed that the continuity conditions for tangential components of the field vectors are satisfied, if required. The domain of analysis $Q=R^3\setminus\overline{\operatorname{int}S}$ occupies a great deal of the R^3-space. The problem formulated for that domain can be resolved analytically or numerically only in two following cases.

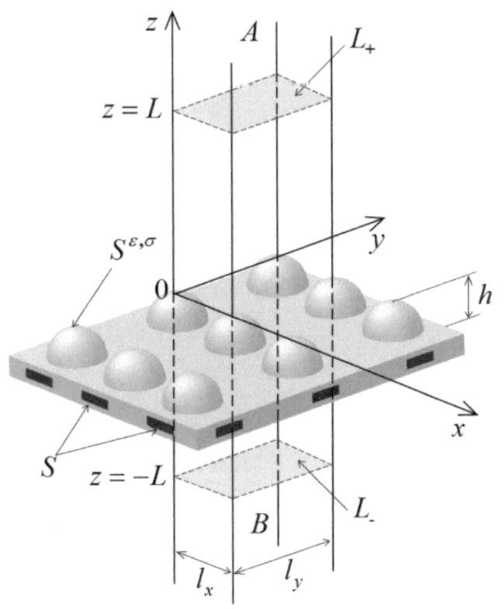

Figure 1. Geometry of a two-dimensionally periodic grating.

- The problem (10) degenerates into a conventional Cauchy problem ($\overline{\mathrm{int}S} = \varnothing$, the medium is homogeneous and nondispersive, while the supports of the functions $\vec{F}(g, t)$, $\vec{\varphi}(g)$, and $\vec{\psi}(g)$ are bounded). With some inessential restrictions for the source functions, the classical and generalized solution of the Cauchy problem does exist; it is unique and is described by the well-known Poisson formula [12].
- The functions $\vec{F}(g, t)$, $\vec{\varphi}(g)$, and $\vec{\psi}(g)$ have the same displacement symmetry as the periodic structure. In this case, the domain of analysis can be reduced to $Q^N = \{g \in Q : 0<x<l_x; 0<y<l_y\}$, by adding to problem (10) periodicity conditions [7] on lateral surfaces of the rectangular Floquet channel $R = \{g \in R^3 : 0<x<l_x; 0<y<l_y\}$.

The domain of analysis can also be reduced to Q^N in a more general case. The objects of analysis are in this case not quite physical (complex-valued sources and waves). However, by simple mathematical transformations, all the results can be presented in the customary, physically correct form. There are several reasons (to one of them we have referred at the end of Section 3) why the modeling of physically realizable processes in the electromagnetic theory of gratings should start with the initial boundary value problems for the images $f^N(g, t, \Phi_x, \Phi_y)$ of the functions $f(g, t)$ describing the actual sources:

$$\begin{aligned} f(g, t) &= \int_{-\infty}^{\infty}\int_{-\infty}^{\infty} \tilde{f}(z, t, \Phi_x, \Phi_y) \exp\left(2\pi i \Phi_x \frac{x}{l_x}\right) \exp\left(2\pi i \Phi_y \frac{y}{l_y}\right) d\Phi_x d\Phi_y \\ &= \int_{-\infty}^{\infty}\int_{-\infty}^{\infty} f^N(g, t, \Phi_x, \Phi_y) d\Phi_x d\Phi_y \end{aligned} \tag{11}$$

From (11) it follows that

$$f^N\left\{\frac{\partial f^N}{\partial x}\right\}(x+l_x,\ y,\ z,\ t,\ \Phi_x,\ \Phi_y)=e^{2\pi i\Phi_x}f^N\left\{\frac{\partial f^N}{\partial x}\right\}(x,\ y,\ z,\ t,\ \Phi_x,\ \Phi_y),$$

$$f^N\left\{\frac{\partial f^N}{\partial y}\right\}(x,\ y+l_y,\ z,\ t,\ \Phi_x,\ \Phi_y)=e^{2\pi i\Phi_y}f^N\left\{\frac{\partial f^N}{\partial y}\right\}(x,\ y,\ z,\ t,\ \Phi_x,\ \Phi_y)$$

or, in other symbols,

$$D[f^N](x+l_x,\ y)=e^{2\pi i\Phi_x}D[f^N](x,\ y), D[f^N](x,\ y+l_y)=e^{2\pi i\Phi_y}D[f^N](x,\ y).$$

The use of the foregoing conditions truncates the domain of analysis to the domain Q^N, which is a part of the Floquet channel R, and allows us to rewrite problem (10) in the form

$$\vec{E}(g,\ t)=\int_{-\infty}^{\infty}\int_{-\infty}^{\infty}\vec{E}^N(g,\ t,\ \Phi_x,\ \Phi_y)d\Phi_x d\Phi_y,\ \vec{H}(g,\ t)=\int_{-\infty}^{\infty}\int_{-\infty}^{\infty}\vec{H}^N(g,\ t,\ \Phi_x,\ \Phi_y)d\Phi_x d\Phi_y \qquad (12)$$

and

$$\begin{cases} \operatorname{rot}\vec{H}^N(g,t)=\eta_0^{-1}\dfrac{\partial\left[\vec{E}^N(g,t)+\chi_\varepsilon(g,t)*\vec{E}^N(g,t)\right]}{\partial t}+\chi_\sigma(g,t)*\vec{E}^N(g,t)+\vec{j}^N(g,t), \\ \operatorname{rot}\vec{E}^N(g,t)=-\eta_0\dfrac{\partial\left[\vec{H}^N(g,t)+\chi_\mu(g,t)*\vec{H}^N(g,t)\right]}{\partial t}, \quad g\in Q^N, \quad t>0 \\ \vec{E}^N(g,0)=\vec{\varphi}_E^N(g), \quad \vec{H}^N(g,0)=\vec{\varphi}_H^N(g), \quad g\in\bar{Q}^N \\ D\left[\vec{E}^N\left(\vec{H}^N\right)\right](l_x,y)=e^{2\pi i\Phi_x}D\left[\vec{E}^N\left(\vec{H}^N\right)\right](0,y), \quad 0\le y\le l_y \\ D\left[\vec{E}^N\left(\vec{H}^N\right)\right](x,l_y)=e^{2\pi i\Phi_y}D\left[\vec{E}^N\left(\vec{H}^N\right)\right](x,0), \quad 0\le x\le l_x \\ E_{tg}^N(g,t)\Big|_{g\in S}=0, \quad H_{nr}^N(g,t)\Big|_{g\in S}=0, \quad t\ge 0. \end{cases} \qquad (13)$$

It is known [6-8] that initial boundary value problems for the above discussed equations can be formulated such that they are uniquely solvable in the Sobolev space $W_2^1(Q^T)$, where $Q^T=Q\times(0,T)$ and $0\le t\le T$. On this basis we suppose in the subsequent discussion that the problem (13) for all $t\in[0,T]$ has also a generalized solution from the space $W_2^1(Q^{N,T})$ and that the uniqueness theorem is true in this space. Here symbol $\times$ stands for the operation of direct product of two sets, $(0,T)$ and $[0,T]$ are open and closed intervals, $W_m^n(G)$ is the set of all elements $\vec{f}(g)$ from the space $L_m(G)$ whose generalized derivatives up to the order n inclusive also belong to $L_m(G)$. $L_m(G)$ is the space of the functions $\vec{f}(g)=\{f_x,\ f_y,\ f_z\}$ (for $g\in G$) such that the functions $|f_{...}(g)|^m$ are integrable on the domain G.

4. Exact Absorbing Conditions for the Rectangular Floquet Channel

In this section, we present analytical results relative to the truncation of the computational space in open 3-D initial boundary value problems of the electromagnetic theory of gratings. In Section 3, by passing on to some special transforms of the functions describing physically realizable sources, the problem for infinite gratings have been reduced to that formulated in the rectangular Floquet channel R or, in other words, in the rectangular waveguide with quasi-periodic boundary conditions. Now we perform further reduction of the domain Q^N to the region $Q_L^N = \{g \in Q^N : |z| < L\}$ (all the sources and inhomogeneities of the Floquet channel R are supposedly located in this domain). For this purpose the exact absorbing conditions [6,7,10,13,14] for the artificial boundaries $L_{\pm}$ $(z = \pm L)$ of the domain Q_L^N will be constructed such that their inclusion into (13) does not change the correctness class of the problem and its solution $\vec{E}^N(g, t), \vec{H}^N(g, t)$.

From here on we omit the superscripts N in (13). By applying the technique similar to that described in [13,14], represent the solution $\vec{E}(g, t)$ of (13) in the closure of the domains $A = \{g \in R : z > L\}$ and $B = \{g \in R : z < -L\}$ in the following form:

$$\vec{E}(g, t) = \sum_{n,m=-\infty}^{\infty} \vec{u}_{nm}^{\pm}(z, t)\mu_{nm}(x, y), \ \{x, y\} \in \bar{R}_z, \ t \geq 0, \tag{14}$$

where the superscript '+' corresponds to $z \geq L$ and '−' to $z \leq -L$ and the following notation is used:

$R_z = (0 < x < l_x) \times (0 < y < l_y)$;

$\{\mu_{nm}(x, y)\}$ $(n, m = 0, \pm 1, \pm 2, ...)$ is the complete in $L_2(R_z)$ orthonormal system of the functions $\mu_{nm}(x, y) = (l_x l_y)^{-1/2} \exp(i\alpha_n x) \exp(i\beta_m y)$;

$\alpha_n = 2\pi(\Phi_x + n)/l_x$, $\beta_m = 2\pi(\Phi_y + m)/l_y$, and $\lambda_{nm}^2 = \alpha_n^2 + \beta_m^2$.

The space-time amplitudes $\vec{u}_{nm}^{\pm}(z, t)$ satisfy the equations

$$\begin{cases} \left[-\dfrac{\partial^2}{\partial t^2} + \dfrac{\partial^2}{\partial z^2} - \lambda_{nm}^2 \right] \vec{u}_{nm}^{\pm}(z, t) = 0, \ t > 0 \\ \vec{u}_{nm}^{\pm}(z, 0) = 0, \quad \dfrac{\partial}{\partial t} \vec{u}_{nm}^{\pm}(z, t) \Big|_{t=0} = 0 \end{cases}, \quad \begin{Bmatrix} z \geq L \\ z \leq -L \end{Bmatrix}. \tag{15}$$

Equations (14) and (15) are obtained by separating variables in the homogeneous boundary value problems for the equation $[\Delta - \partial^2/\partial t^2]\vec{E}(g, t) = \vec{0}$ (see formula (5)) and taking into account that in the domains A and B we have grad div $\vec{E}(g, t) = 0$ and $\vec{F}_E(g, t) = 0$. It is also assumed that the field generated by the current and instantaneous sources located in Q_L has not yet reached the boundaries $L_{\pm}$ by the moment of time $t = 0$.

The solutions $\vec{u}_{nm}^{\pm}(z, t)$ of the vector problems (15), as well as in the case of scalar problems [13,14], can be written as

$$\vec{u}_{nm}^{\pm}(\pm L, t) = \mp \int_0^t J_0[\lambda_{nm}(t-\tau)]\vec{u}_{nm}^{\pm\,\prime}(\pm L, \tau)d\tau, \; t \geq 0 \quad . \tag{16}$$

The above formula represents nonlocal EAC for the space-time amplitudes of the field $\vec{E}(g, t)$ in the cross-sections $z = \pm L$ of the Floquet channel$R$. The exact nonlocal and local absorbing conditions for the field $\vec{E}(g, t)$ on the artificial boundaries $L_{\pm}$ follow immediately from (16) and (14):

$$\vec{E}(x, y, \pm L, t) = \mp \sum_{n,m=-\infty}^{\infty} \left\{ \int_0^t J_0[\lambda_{nm}(t-\tau)] \left[\int_0^{l_x}\int_0^{l_y} \frac{\partial \vec{E}(\tilde{x}, \tilde{y}, z, \tau)}{\partial z}\bigg|_{z=\pm L} \mu_{nm}^*(\tilde{x}, \tilde{y})d\tilde{x}d\tilde{y} \right] d\tau \right\} \mu_{nm}(x, y), \quad \{x, y\} \in \bar{R}_z, \quad t \geq 0 \tag{17}$$

and

$$\begin{gathered}
\vec{E}(x,y,\pm L,t) = \frac{2}{\pi}\int_0^{\pi/2} \frac{\partial \vec{W}_E(x,y,t,\varphi)}{\partial t} d\varphi, \quad \{x,y\} \in \bar{R}_z, \quad t \geq 0 \\
\begin{cases}
\left[\frac{\partial^2}{\partial t^2} - \sin^2\varphi\left(\frac{\partial^2}{\partial x^2} + \frac{\partial^2}{\partial y^2}\right)\right]\vec{W}_E(x,y,t,\varphi) = \mp \frac{\partial \vec{E}(g,t)}{\partial z}\bigg|_{z=\pm L}, & \{x,y\} \in R_z, \quad t > 0 \\
\vec{W}_E(x,y,t,\varphi)\big|_{t=0} = \frac{\partial \vec{W}_E(x,y,t,\varphi)}{\partial t}\bigg|_{t=0} = 0, & \{x,y\} \in \bar{R}_z \\
D\left[\vec{W}_E\right](l_x,y) = e^{2\pi i \Phi_x} D\left[\vec{W}_E\right](0,y), & 0 \leq y \leq l_y \\
D\left[\vec{W}_E\right](x,l_y) = e^{2\pi i \Phi_y} D\left[\vec{W}_E\right](x,0), & 0 \leq x \leq l_x, \quad t \geq 0
\end{cases}
\end{gathered} \tag{18}$$

Here, $\vec{u}_{nm}^{\pm\,\prime}(\pm L, \tau) = \partial \vec{u}_{nm}^{\pm}(z, \tau)/\partial z\,|_{z=\pm L}$, $J_0(t)$is the zero-order Bessel function, the superscript '$*$' stands for the complex conjugation operation, $\vec{W}_E(x, y, t, \varphi)$is some auxiliary function, where the numerical parameter φ lies in the range$0 \leq \varphi \leq \pi/2$.

It is obvious that the magnetic field vector $\vec{H}(g, t)$ of the pulsed waves $\vec{U}(g, t) = \{\vec{E}(g, t), \vec{H}(g, t)\}$outgoing towards the domains A and B satisfies similar boundary conditions on$L_{\pm}$. The boundary conditions for $\vec{E}(g, t)$ and $\vec{H}(g, t)$ (nonlocal or local) taken together reduce the computational space for the problem (13) to the domain Q_L (a part of the Floquet channel$R$) that contains all the sources and obstacles.

Now suppose that in addition to the sources $\vec{j}(g,t)$, $\vec{\varphi}_E(g)$, and $\vec{\varphi}_H(g)$, there exist sources $\vec{j}^A(g,t)$, $\vec{\varphi}_E^A(g)$, and $\vec{\varphi}_H^A(g)$ located in A and generating some pulsed wave $\vec{U}^i(g,t)=\{\vec{E}^i(g,t),\vec{H}^i(g,t)\}$ being incident on the boundary L_+ at times $t>0$. The field $\vec{U}^i(g,t)$ is assumed to be nonzero only in the domain A. Since the boundary conditions (17), (18) remain valid for any pulsed wave outgoing through $L_\pm$ towards $z=\pm\infty$ [13,14], then the total field $\{\vec{E}(g,t),\vec{H}(g,t)\}$ is the solution of the initial boundary value problem (13) in the domain Q_L with the boundary conditions (17) or (18) on L_- and the following conditions on the artificial boundary L_+:

$$\vec{E}^s(x,y,L,t)$$
$$=-\sum_{n,m=-\infty}^{\infty}\left\{\int_0^t J_0\left[\lambda_{nm}(t-\tau)\right]\left[\int_0^{l_x}\int_0^{l_y}\left.\frac{\partial\vec{E}^s(\tilde{x},\tilde{y},z,\tau)}{\partial z}\right|_{z=L}\mu_{nm}^*(\tilde{x},\tilde{y})d\tilde{x}d\tilde{y}\right]d\tau\right\}\mu_{nm}(x,y), \tag{19}$$
$$\{x,y\}\in\bar{R}_z,\quad t\geq 0$$

or

$$\begin{gathered}\vec{E}^s(x,y,L,t)=\frac{2}{\pi}\int_0^{\pi/2}\frac{\partial\vec{W}_E(x,y,t,\varphi)}{\partial t}d\varphi,\quad \{x,y\}\in\bar{R}_z,\quad t\geq 0\\ \begin{cases}\left[\dfrac{\partial^2}{\partial t^2}-\sin^2\varphi\left(\dfrac{\partial^2}{\partial x^2}+\dfrac{\partial^2}{\partial y^2}\right)\right]\vec{W}_E(x,y,t,\varphi)=-\left.\dfrac{\partial\vec{E}^s(g,t)}{\partial z}\right|_{z=L}, & \{x,y\}\in R_z,\quad t>0\\ \left.\vec{W}_E(x,y,t,\varphi)\right|_{t=0}=\left.\dfrac{\partial\vec{W}_E(x,y,t,\varphi)}{\partial t}\right|_{t=0}=0, & \{x,y\}\in\bar{R}_z\\ D\left[\vec{W}_E\right](l_x,y)=e^{2\pi i\Phi_x}D\left[\vec{W}_E\right](0,y), & 0\leq y\leq l_y\\ D\left[\vec{W}_E\right](x,l_y)=e^{2\pi i\Phi_y}D\left[\vec{W}_E\right](x,0), & 0\leq x\leq l_x,\quad t\geq 0\ .\end{cases}\end{gathered} \tag{20}$$

Here $\vec{U}^s(g,t)=\{\vec{E}^s(g,t),\vec{H}^s(g,t)\}=\vec{U}(g,t)-\vec{U}^i(g,t)$ $(g\in A, t>0)$ is the pulsed wave outgoing towards $z=+\infty$. It is generated by the incident wave $\vec{U}^i(g,t)$ ('reflection' from the virtual boundary L_+) and the sources $\vec{j}(g,t)$, $\vec{\varphi}_E(g)$, and $\vec{\varphi}_H(g)$.

5. Some Important Characteristics of Transient Fields in the Rectangular Floquet Channel

For numerical implementation of the computational schemes involving boundary conditions like (19) or (20), the function $\vec{U}^i(g,t)$ for $t\in[0,T]$ and its normal derivative with respect to the boundary L_+ are to be known. To obtain the required data for the wave $\vec{U}^i(g,t)$ generated by a given set of sources $\vec{j}^A(g,t)$, $\vec{\varphi}_E^A(g)$, and $\vec{\varphi}_H^A(g)$, the following initial boundary value problem for a regular hollow Floquet channel R are to be solved:

$$
\begin{cases}
\left[-\frac{\partial^2}{\partial t^2}+\Delta\right]\begin{Bmatrix}\vec{E}^i\\ \vec{H}^i\end{Bmatrix}=\begin{Bmatrix}\eta_0\partial\vec{j}^{\mathbf{A}}/\partial t+\varepsilon_0^{-1}\operatorname{grad}\rho_2^{\mathbf{A}}\\ -\operatorname{rot}\vec{j}^{\mathbf{A}}\end{Bmatrix}=\begin{Bmatrix}\vec{F}_E^{\mathbf{A}}\\ \vec{F}_H^{\mathbf{A}}\end{Bmatrix}, & g=\{x,y,z\}\in R,\quad t>0\\
\begin{Bmatrix}\vec{E}^i(g,0)\\ \vec{H}^i(g,0)\end{Bmatrix}=\begin{Bmatrix}\vec{\varphi}_E^{\mathbf{A}}\\ \vec{\varphi}_H^{\mathbf{A}}\end{Bmatrix},\quad \begin{Bmatrix}\partial\vec{E}^i(g,t)/\partial t\big|_{t=0}=\eta_0\operatorname{rot}\vec{H}^i(g,0)\\ \partial\vec{H}^i(g,t)/\partial t\big|_{t=0}=-\eta_0^{-1}\operatorname{rot}\vec{E}^i(g,0)\end{Bmatrix}=\begin{Bmatrix}\vec{\psi}_E^{\mathbf{A}}\\ \vec{\psi}_H^{\mathbf{A}}\end{Bmatrix}, & g\in\bar{R}\\
D\left[\vec{E}^i\left(\vec{H}^i\right)\right](l_x,y)=e^{2\pi i\Phi_x}D\left[\vec{E}^i\left(\vec{H}^i\right)\right](0,y), & 0\le y\le l_y\\
D\left[\vec{E}^i\left(\vec{H}^i\right)\right](x,l_y)=e^{2\pi i\Phi_y}D\left[\vec{E}^i\left(\vec{H}^i\right)\right](x,0), & 0\le x\le l_x,\quad t\ge 0\ .
\end{cases}
\tag{21}
$$

The function $\rho_2^A(g, t)$ here determines the volume density of foreign electric charge.

First we determine the longitudinal components E_z^i and H_z^i of the field $\{\vec{E}^i, \vec{H}^i\}$ at all points $g$of the domain R for all times$t>0$. Let us consider the scalar initial boundary value problems following from (21):

$$
\begin{cases}
\left[-\frac{\partial^2}{\partial t^2}+\Delta\right]\begin{Bmatrix}E_z^i\\ H_z^i\end{Bmatrix}=\begin{Bmatrix}F_{z,E}^{\mathbf{A}}\\ F_{z,H}^{\mathbf{A}}\end{Bmatrix}, & g\in R,\quad t>0\\
\begin{Bmatrix}E_z^i(g,0)\\ H_z^i(g,0)\end{Bmatrix}=\begin{Bmatrix}\varphi_{z,E}^{\mathbf{A}}\\ \varphi_{z,H}^{\mathbf{A}}\end{Bmatrix},\quad \begin{Bmatrix}\partial E_z^i(g,t)/\partial t\big|_{t=0}\\ \partial H_z^i(g,t)/\partial t\big|_{t=0}\end{Bmatrix}=\begin{Bmatrix}\psi_{z,E}^{\mathbf{A}}\\ \psi_{z,H}^{\mathbf{A}}\end{Bmatrix}, & g\in\bar{R}\\
D\left[E_z^i\left(H_z^i\right)\right](l_x,y)=e^{2\pi i\Phi_x}D\left[E_z^i\left(H_z^i\right)\right](0,y), & 0\le y\le l_y\\
D\left[E_z^i\left(H_z^i\right)\right](x,l_y)=e^{2\pi i\Phi_y}D\left[E_z^i\left(H_z^i\right)\right](x,0), & 0\le x\le l_x,\quad t\ge 0\ .
\end{cases}
\tag{22}
$$

By separating of the transverse variables x and y in (22) represent the solution of the problem as

$$
\begin{Bmatrix}E_z^i(g,t)\\ H_z^i(g,t)\end{Bmatrix}=\sum_{n,m=-\infty}^{\infty}\begin{Bmatrix}v_{nm(z,E)}(z,t)\\ v_{nm(z,H)}(z,t)\end{Bmatrix}\mu_{nm}(x,y)
\tag{23}
$$

To determine the scalar functions $v_{nm(z,E)}(z, t)$ and$v_{nm(z,H)}(z, t)$, we have to invert the following Cauchy problems for the one-dimensional Klein-Gordon equations:

$$
\begin{cases}
\left[-\frac{\partial^2}{\partial t^2}+\frac{\partial^2}{\partial z^2}-\lambda_{nm}\right]\begin{Bmatrix}v_{nm(z,E)}(z,t)\\ v_{nm(z,H)}(z,t)\end{Bmatrix}=\begin{Bmatrix}F_{nm(z,E)}^{A}\\ F_{nm(z,H)}^{A}\end{Bmatrix}, & t>0,\quad -\infty<z<\infty\\
\begin{Bmatrix}v_{nm(z,E)}(z,0)\\ v_{nm(z,H)}(z,0)\end{Bmatrix}=\begin{Bmatrix}\varphi_{nm(z,E)}^{A}\\ \varphi_{nm(z,H)}^{A}\end{Bmatrix},\quad \frac{\partial}{\partial t}\begin{Bmatrix}v_{nm(z,E)}(z,t)\\ v_{nm(z,H)}(z,t)\end{Bmatrix}\Bigg|_{t=0}=\begin{Bmatrix}\psi_{nm(z,E)}^{A}\\ \psi_{nm(z,E)}^{A}\end{Bmatrix}, & \\
\qquad -\infty<z<\infty, \qquad n,m=0,\pm1,\pm2,\ldots\ . &
\end{cases}
\tag{24}
$$

Here $F^{A}_{nm(z,E)}, \varphi^{A}_{nm(z,E)}, \psi^{A}_{nm(z,E)}$ and $F^{A}_{nm(z,H)}, \varphi^{A}_{nm(z,H)}, \psi^{A}_{nm(z,H)}$ are the amplitudes of the Fourier transforms of the functions $F^{A}_{z,E}, \varphi^{A}_{z,E}, \psi^{A}_{z,E}$ and $F^{A}_{z,H}, \varphi^{A}_{z,H}, \psi^{A}_{z,H}$ in the basic set $\{\mu_{nm}(x, y)\}$.

Let us continue analytically the functions $v_{nm(z,E)}(z, t)$, $v_{nm(z,H)}(z, t)$ and $F^{A}_{nm(z,E)}, F^{A}_{nm(z,H)}$ by zero on the semi-axis $t0$ and pass on to the generalized formulation of the Cauchy problem (24) [12]:

$$
\begin{gathered}
B(\lambda_{nm})\begin{bmatrix} v_{nm(z,E)}(z,t) \\ v_{nm(z,H)}(z,t) \end{bmatrix} \equiv \left[-\frac{\partial^2}{\partial t^2}+\frac{\partial^2}{\partial z^2}-\lambda^2_{nm}\right]\begin{Bmatrix} v_{nm(z,E)}(z,t) \\ v_{nm(z,H)}(z,t) \end{Bmatrix} \\
=\begin{Bmatrix} F^{A}_{nm(z,E)} \\ F^{A}_{nm(z,H)} \end{Bmatrix} - \delta^{(1)}(t)\begin{Bmatrix} \varphi^{A}_{nm(z,E)} \\ \varphi^{A}_{nm(z,H)} \end{Bmatrix} - \delta(t)\begin{Bmatrix} \psi^{A}_{nm(z,E)} \\ \psi^{A}_{nm(z,H)} \end{Bmatrix} = \begin{Bmatrix} f_{nm(z,E)} \\ f_{nm(z,H)} \end{Bmatrix}, \\
-\infty<z<\infty, \quad -\infty<t<\infty, \quad n, m=1, \pm 2, \pm 3, \ldots ,
\end{gathered} \tag{25}
$$

where $\delta(t)$ and $\delta^{(m)}(t)$ are the Dirac delta-function and its derivative of the order m. Taking into account the properties of the fundamental solution $G(z, t, \lambda) = -(1/2)\chi(t-|z|)J_0\left(\lambda\sqrt{t^2-z^2}\right)$ of the operator $B(\lambda)$ [6,13,14] ($\chi(t)$ is the Heaviside step function), the solutions $v_{nm(z,E)}(z, t)$ and $v_{nm(z,H)}(z, t)$ of equations (25) can be written as

$$
\begin{gathered}
\begin{Bmatrix} v_{nm(z,E)}(z,t) \\ v_{nm(z,H)}(z,t) \end{Bmatrix} = G(z, t, \lambda_{nm}) * \begin{Bmatrix} f_{nm(z,E)} \\ f_{nm(z,H)} \end{Bmatrix} = \\
-\frac{1}{2}\int_{-\infty}^{t-|z-\omega|}\int_{-\infty}^{\infty} J_0\left(\lambda_{nm}\sqrt{(t-\tau)^2-(z-\omega)^2}\right)\left(\begin{Bmatrix} F^{A}_{nm(z,E)} \\ F^{A}_{nm(z,H)} \end{Bmatrix} - \delta^{(1)}(t)\begin{Bmatrix} \varphi^{A}_{nm(z,E)} \\ \varphi^{A}_{nm(z,H)} \end{Bmatrix} - \delta(t)\begin{Bmatrix} \psi^{A}_{nm(z,E)} \\ \psi^{A}_{nm(z,H)} \end{Bmatrix}\right) d\omega d\tau, \\
-\infty<z<\infty, \quad t\geq 0, \quad n, m=1, \pm 2, \pm 3, \ldots .
\end{gathered} \tag{26}
$$

Relations (23) and (26) completely determine the longitudinal components of the field $\{\vec{E}^{i}, \vec{H}^{i}\}$.

Outside the bounded domain enclosing all the sources, in the domain $G\subset R$, where the waves generated by these sources propagate freely, the following relations [6,14] are valid:

$$
\begin{cases}
\vec{E}^{i} = \left(\dfrac{\partial^2 U^E}{\partial x \partial z} - \dfrac{\partial^2 U^H}{\partial y \partial t}\right)\vec{x} + \left(\dfrac{\partial^2 U^E}{\partial y \partial z} + \dfrac{\partial^2 U^H}{\partial x \partial t}\right)\vec{y} + \left(\dfrac{\partial^2 U^E}{\partial z^2} - \dfrac{\partial^2 U^E}{\partial t^2}\right)\vec{z} \\
\eta_0\vec{H}^{i} = \left(\dfrac{\partial^2 U^E}{\partial y \partial t} + \dfrac{\partial^2 U^H}{\partial x \partial z}\right)\vec{x} + \left(-\dfrac{\partial^2 U^E}{\partial x \partial t} + \dfrac{\partial^2 U^H}{\partial y \partial z}\right)\vec{y} + \left(\dfrac{\partial^2 U^H}{\partial z^2} - \dfrac{\partial^2 U^H}{\partial t^2}\right)\vec{z}
\end{cases} \tag{27}
$$

in which

$$U^{E,H}(g,t)=\sum_{n,m=-\infty}^{\infty} u_{nm}^{E,H}(z,t)\mu_{nm}(x,y) \tag{28}$$

are the scalar Borgnis functions such that$[\Delta-\partial^2/\partial t^2][\partial U^{E,H}(g,t)/\partial t]=0$. Equations (23), (26)-(28) determine the field $\{\vec{E}^i, \vec{H}^i\}$ at all points g of the domain G for all times$t>0$. Really, since at the time point $t=0$ the domain G is undisturbed, then we have $[\Delta-\partial^2/\partial t^2]U^{E,H}=0$ $(g\in G, t>0)$. Hence, in view of (27), (28), it follows:

$$E_z=\frac{\partial^2 U^E}{\partial z^2}-\frac{\partial^2 U^E}{\partial t^2}=-\left(\frac{\partial^2 U^E}{\partial x^2}+\frac{\partial^2 U^E}{\partial y^2}\right)=\sum_{n,m=-\infty}^{\infty}\lambda_{nm}^2 u_{nm}^E\mu_{nm},$$

$$\eta_0 H_z=\frac{\partial^2 U^H}{\partial z^2}-\frac{\partial^2 U^H}{\partial t^2}=-\left(\frac{\partial^2 U^H}{\partial x^2}+\frac{\partial^2 U^H}{\partial y^2}\right)=\sum_{n,m=-\infty}^{\infty}\lambda_{nm}^2 u_{nm}^H\mu_{nm}$$

and (see representation (23))

$$u_{nm}^E(z,t)=(\lambda_{nm})^{-2}v_{nm(z,E)}(z,t),\ u_n^H(z,t)=\eta_0(\lambda_{nm})^{-2}v_{nm(z,H)}(z,t). \tag{29}$$

Hence the functions $U^{E,H}(g,t)$ as well as the transverse components of the field $\{\vec{E}^i, \vec{H}^i\}$ are determined.

The foregoing suggests the following important conclusion: the fields generated in the reflection zone (the domain$A$) and transmission zone (the domain$B$) of a periodic structure are uniquely determined by their longitudinal (directed along z-axis) components and can be represented in the following form (see also formulas (14) and (23)). For the incident wave we have

$$\begin{Bmatrix} E_z^i(g,t) \\ H_z^i(g,t) \end{Bmatrix}=\sum_{n,m=-\infty}^{\infty}\begin{Bmatrix} v_{nm(z,E)}(z,t) \\ v_{nm(z,H)}(z,t) \end{Bmatrix}\mu_{nm}(x,y),\ g\in\bar{A},\ t\geq 0, \tag{30}$$

for the reflected wave $\vec{U}^s(g,t)$(which coincides with the total field $\vec{U}(g,t)$ if$\vec{U}^i(g,t)\equiv 0$) we have

$$\begin{Bmatrix} E_z^s(g,t)\text{or}E_z(g,t) \\ H_z^s(g,t)\text{or}H_z(g,t) \end{Bmatrix}=\sum_{n,m=-\infty}^{\infty}\begin{Bmatrix} \overset{+}{u}_{nm(z,E)}(z,t) \\ \overset{+}{u}_{nm(z,H)}(z,t) \end{Bmatrix}\mu_{nm}(x,y),\ g\in\bar{A},\ t\geq 0 \tag{31}$$

and for the transmitted wave (coinciding in the domain B with the total field$\vec{U}(g,t)$) we can write

$$\begin{Bmatrix} E_z(g, t) \\ H_z(g, t) \end{Bmatrix} = \sum_{n,m=-\infty}^{\infty} \begin{Bmatrix} u_{nm(z,E)}^{-}(z, t) \\ u_{nm(z,H)}^{-}(z, t) \end{Bmatrix} \mu_{nm}(x, y), \; g \in \bar{B}, \; t \geq 0. \tag{32}$$

In applied problems, the most widespread are situations where a periodic structure is excited by one of the partial components of TE-wave (with $E_z^i(g, t)=0$) or TM-wave (with $H_z^i(g, t)=0$) [7]. Consider, for example, a partial wave of order pq. Then we have

$$\vec{U}^i(g, t)=\vec{U}^i_{pq(H)}(g, t): H_z^i(g, t)=v_{pq(z,H)}(z, t)\mu_{pq}(x, y)$$

or

$$\vec{U}^i(g, t)=\vec{U}^i_{pq(E)}(g, t): E_z^i(g, t)=v_{pq(z,E)}(z, t)\mu_{pq}(x, y).$$

The excitation of this kind is implemented in our models in the following way. The time function $v_{pq(z,H)}(L, t)$ or $v_{pq(z,E)}(L, t)$ is defined on the boundary L_+. This function determines the width of the pulse $\vec{U}^i(g, t)$, namely, the frequency range $[K_1, K_2]$ such that for all frequencies k from this range ($k=2\pi / \lambda$, λ is the wavelength in free space) the value

$$\gamma = \frac{|\tilde{v}_{pq(z,H\,\mathrm{or}\,E)}(L, k)|}{\max\limits_{k \in [K_1;K_2]} |\tilde{v}_{pq(z,H\,\mathrm{or}\,E)}(L, k)|}$$

where $\tilde{v}_{pq(z,H\,\mathrm{or}\,E)}(L, k)$ is the spectral amplitude of the pulse $v_{pq(z,H\,\mathrm{or}\,E)}(L, t)$, exceeds some given value $\gamma=\gamma_0$. All spectral characteristics $\tilde{f}(k)$ are obtainable from the temporal characteristics $f(t)$ by applying the Laplace transform

$$\tilde{f}(k)=\int_0^{\infty} f(t)e^{ikt}dt \leftrightarrow f(t)=\frac{1}{2\pi}\int_{i\alpha-\infty}^{i\alpha+\infty} \tilde{f}(k)e^{-ikt}dk, \; 0 \leq \alpha \leq \mathrm{Im}k. \tag{33}$$

For numerical implementation of the boundary conditions (19) and (20) and for calculating space-time amplitudes of the transverse components of the wave $\vec{U}^i(g, t)$ in the cross-section $z=L$ of the Floquet channel (formulas (27) and (29)), the function $(v_{pq(z,H\,\mathrm{or}\,E)})'(L, t)$ are to be determined. To do this, we apply the following relation [7,14]:

$$\vec{v}_{pq(H\,\mathrm{or}\,E)}(L, t)=\int_0^t J_0[\lambda_{pq}(t-\tau)](\vec{v}_{pq(H\,\mathrm{or}\,E)})'(L, t)d\tau, \; t \geq 0. \tag{34}$$

which is valid for all the amplitudes of the pulsed wave $\vec{U}^i(g, t)$ outgoing towards $z=-\infty$ and does not violate the causality principle.

http://dx.doi.org/10.5772/51007

6. Transformation Operator Method

6.1. Evolutionary basis of a signal and transformation operators

Let us place an arbitrary periodic structure of finite thickness between two homogeneous dielectric half-spaces $z_1=z-L>0$ (with $\varepsilon=\varepsilon_1$) and $z_2=-z-L>0$ (with $\varepsilon=\varepsilon_2$). Let also a local coordinate system $g_j=\{x_j, y_j, z_j\}$ be associated with each of these half-spaces (Figure 2). Assume that the distant sources located in the domain A of the upper half-space generate a primary wave $\vec{U}_1^i(g, t)=\{\vec{E}_1^i(g, t), \vec{H}_1^i(g, t)\}$ being incident on the artificial boundary L_+ (on the plane $z_1=0$) as viewed from $z_1=\infty$.

Denote by $\vec{U}_j^s(g, t)=\{\vec{E}_j^s(g, t), \vec{H}_j^s(g, t)\}$ the fields resulting from scattering of the primary wave $\vec{U}_1^i(g, t)$ in the domains A (where the total field is $\vec{U}(g, t)=\{\vec{E}(g, t), \vec{H}(g, t)\}=\vec{U}_1^s(g, t)+\vec{U}_1^i(g, t)$) and B (where $\vec{U}(g, t)=\vec{U}_2^s(g, t)$). In Section 5, we have shown that the fields under consideration are uniquely determined by their longitudinal components, which can be given, for example, as:

$$\begin{Bmatrix} E_z^i(g, t) \\ H_z^i(g, t) \end{Bmatrix} = \sum_{n,m=-\infty}^{\infty} \begin{Bmatrix} v_{nm(1,E)}(z_1, t) \\ v_{nm(1,H)}(z_1, t) \end{Bmatrix} \mu_{nm}(x, y),\ z_1\geq 0,\ t\geq 0 \tag{35}$$

$$\begin{Bmatrix} E_z^s(g, t) \\ H_z^s(g, t) \end{Bmatrix} = \sum_{n,m=-\infty}^{\infty} \begin{Bmatrix} u_{nm(j,E)}(z_j, t) \\ u_{nm(j,H)}(z_j, t) \end{Bmatrix} \mu_{nm}(x, y),\ z_j\geq 0,\ t\geq 0,\ j=1,2 \tag{36}$$

(see also formulas (30)-(32)). Here, as before, $\{\mu_{nm}(x, y)\}_{n,m=-\infty}^{\infty}$ is the complete (in $L_2(R_z)$) orthonormal system of transverse eigenfunctions of the Floquet channel R (see Section 4), while the space-time amplitudes $u_{nm(j,E)}(z_j, t)$ and $u_{nm(j,H)}(z_j, t)$ are determined by the solutions of the following problems (see also problem (15)) for the one-dimensional Klein-Gordon equations:

$$\begin{cases} \left[-\varepsilon_j \frac{\partial^2}{\partial t^2} + \frac{\partial^2}{\partial z_j^2} - \lambda_{nm}^2\right] u_{nm(j,E\text{ or }H)}(z_j, t)=0, & t>0 \\ u_{nm(j,E\text{ or }H)}(z_j,0)=0, & \frac{\partial}{\partial t} u_{nm(j,E\text{ or }H)}(z_j, t)\Big|_{t=0}=0 \end{cases},\quad j=1,2,\quad n, m=0, \pm 1, \pm 2, \ldots . \tag{37}$$

Compose from the functions $v_{nm(1,E)}(z_1, t)$, $v_{nm(1,H)}(z_1, t)$, $u_{nm(j,E)}(z_j, t)$, $u_{nm(j,H)}(z_j, t)$ and the eigenvalues λ_{nm} $(n, m=0, \pm 1, \pm 2, \ldots)$ the sets $v_{(1)}(z_1, t)=\{v_{p(1)}(z_1, t)\}_{p=-\infty}^{\infty}$, $u_{(j)}(z_j, t)=\{u_{p(j)}(z_j, t)\}_{p=-\infty}^{\infty}$, and $\{\lambda_p\}_{p=-\infty}^{\infty}$ such that their members are defined according to the rules depicted in Figure 3. The sets $v_{(1)}(z_1, t)$ and $u_{(j)}(z_j, t)$ are said to be evolutionary bases of signals $\vec{U}_1^i(g, t)$ and $\vec{U}_j^s(g, t)$. They describe completely and unambiguously transformation of the corresponding nonsine waves in the regular Floquet channels A and B filled with dielectric.

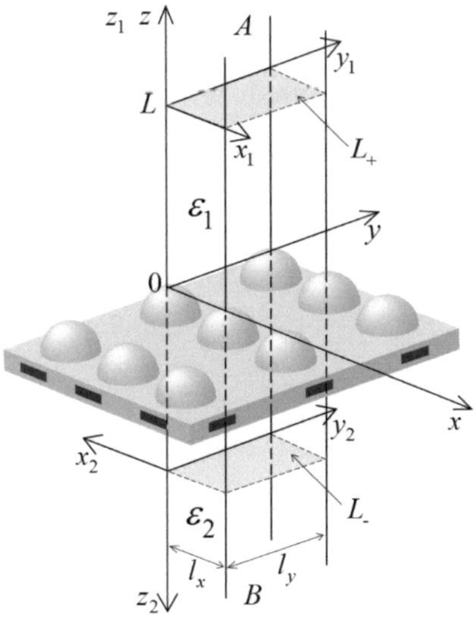

Figure 2. A two-dimensionally periodic grating between two dielectric half-spaces as element of a multi-layered structure.

Let us introduce by the relations

$$u_{p(j)}'(0,t) \equiv \frac{\partial}{\partial z_j} u_{p(j)}(z_j, t) \Big|_{z_j=0} = \int_0^t \sum_{q=-\infty}^{\infty} [S_{pq}^{AA}(t-\tau)\delta_j^1 + S_{pq}^{BA}(t-\tau)\delta_j^2] v_{q(1)}(0,\tau)d\tau, \tag{38}$$

$t \geq 0,\ p=0, \pm 1, \pm 2, \ldots,\ j=1,2$

$$u_{(j)}'(0,t) = \{u_{p(j)}'(0,t)\}_p = [S^{AA}\delta_j^1 + S^{BA}\delta_j^2][v_{(1)}(0,\tau)],\ t \geq 0, j=1,2 \tag{39}$$

the boundary (on the boundaries $z_j=0$) transformation operators S^{AA} and S^{BA} of the evolutionary basis $v_{(1)}(z_1, t)$ of the wave $\vec{U}_1^i(g, t)$ incoming from the domain A. Here δ_m^n stands for the Kronecker delta, the operators' elements S_{nm}^{XY} specify the space-time energy transformation from the domain Y into the domain X and from the mode of order m into the mode of order n.

It is evident that the operators S^{AA} and S^{BA} working in the space of evolutionary bases are intrinsic characteristics of the periodic structure placed between two dielectric half-spaces. They totalize an impact of the structure on elementary excitations composing any incident signal $\vec{U}_1^i(g, t)$. Thus for $v_{q(1)}(0,t)=\delta_q^r \delta(t-\eta)$, where r is an integer and $\eta>0$, we have $u_{p(1)}'(0,t)=S_{pr}^{AA}(t-\eta)$ and $u_{p(2)}'(0,t)=S_{pr}^{BA}(t-\eta)$. We use this example with an abstract non-physical signal by methodological reasons in order to associate the transformation operators' components $S_{pr}^{AA}(t-\tau)$ and $S_{pr}^{BA}(t-\tau)$ with an 'elementary excitation'.

http://dx.doi.org/10.5772/51007

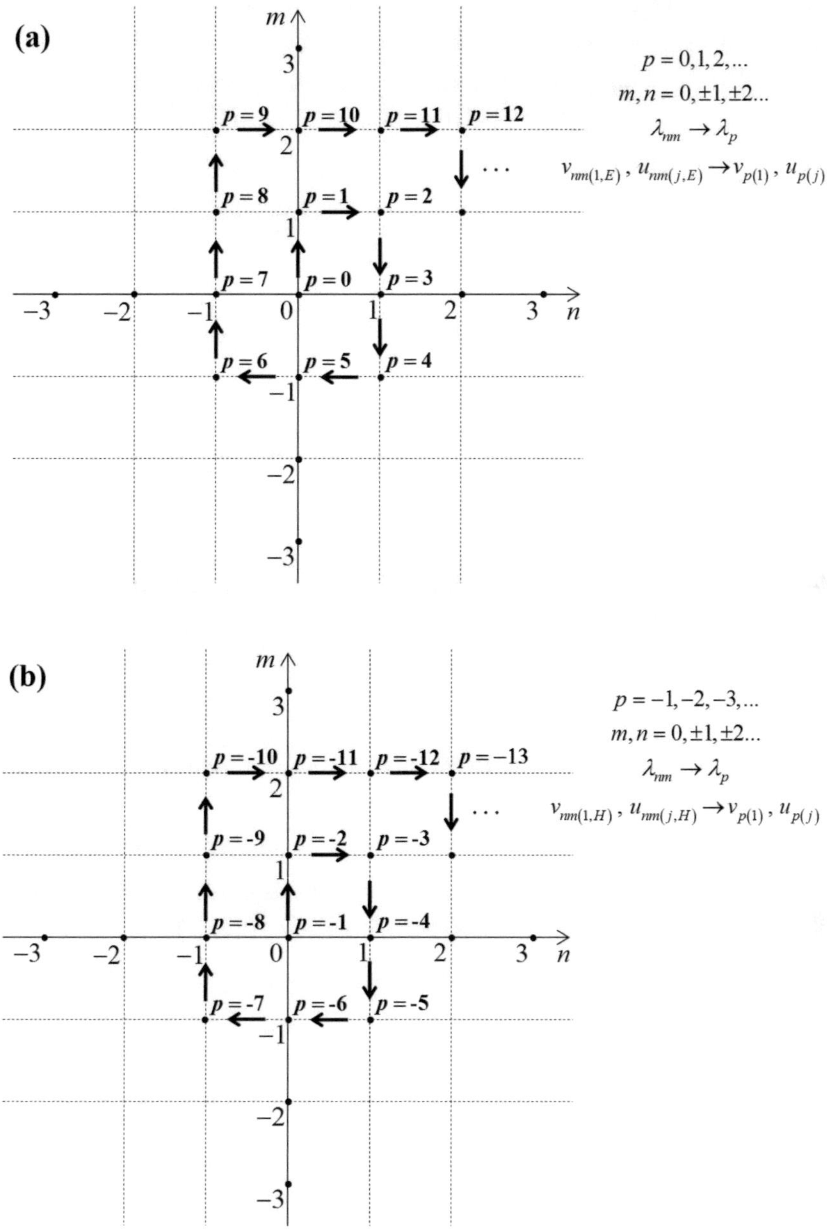

Figure 3. Construction of sets of the values $v_{p(1)}$, $u_{p(j)}$, and λ_p ($p=0,\pm 1,\pm 2,...$) from sets of the values $v_{nm(1,E)}$, $u_{nm(j,E)}$, $v_{nm(1,H)}$, $u_{nm(j,H)}$, and λ_{nm} ($m,\ n=0,\pm 1,\pm 2,...$): (a) $p=0,1,2,...$; (b) $p=-1,-2,-3,...$.

The operators S^{AA} and S^{BA} determine all the features of transient states on the upper and bottom boundaries of the layer enclosing the periodic structure. Secondary waves outgoing from these boundaries propagate freely in the regular Floquet channels A and B therewith undergoing deformations (see, for example, [6]). The space-time amplitudes $u_{p(j)}(z_j, t)$ of the partial components of these waves (the elements of the evolutionary bases of the signals $\vec{U}_j^s(g, t)$) vary differently for different values of p and j. These variations on any finite sections of the Floquet channels A и B are described by the diagonal transporting operators $Z_{0\to z_1}^A$ and $Z_{0\to z_2}^B$ acting according the rule:

$$u_{(j)}(z_j, t)=\{u_{p(j)}(z_j, t)\}=\left[Z_{0\to z_1}^A\delta_j^1 + Z_{0\to z_2}^B\delta_j^2\right]\left[u_{(j)}'(0,\tau)\right],\ j=1,2. \tag{40}$$

The structure of the operators given by (40) can be detailed by the formula

$$u_{p(j)}(z_j, t)=-\frac{1}{\varepsilon_j}\int_0^t J_0\left[\lambda_p\sqrt{\frac{(t-\tau)^2}{\varepsilon_j}-z_j^2}\right]\chi\left(\frac{t-\tau}{\sqrt{\varepsilon_j}}-z_j\right)u_{p(j)}'(0,\tau)d\tau, \tag{41}$$
$$t\geq 0,\ z_j\geq 0,\ p=0,\pm 1,\pm 2,\ldots,\ j=1,2,$$

which reflects general properties of solutions of homogeneous problems (37), i.e. the solutions that satisfy zero initial conditions and are free from the components propagating in the direction of decreasing z_j. The derivation technique for (41) is discussed at length in [6,13,14].

6.2. Equations of the operator method in the problems for multilayer periodic structures

The operators S^{AA} and S^{BA} completely define properties of the periodic structure excited from the channel A. By analogy with (38) we can determine transformation operators S^{BB} and S^{AB} for evolutionary basis $v_{(2)}(z_2, t)=\{v_{p(2)}(z_2, t)\}_{p=-\infty}^{\infty}$ of the wave $\vec{U}_2^i(g, t)=\{\vec{E}_2^i(g, t), \vec{H}_2^i(g, t)\}$ incident onto the boundary $z_2=0$ from the channel B:

$$u_{p(j)}'(0,t)=\int_0^t \sum_{m=-\infty}^{\infty}\left[S_{pq}^{AB}(t-\tau)\delta_j^1 + S_{pq}^{BB}(t-\tau)\delta_j^2\right]v_{q(2)}(0,\tau)d\tau, \tag{42}$$
$$t\geq 0,\ p=0,\pm 1,\pm 2,\ldots,\ j=1,2.$$

Let us construct the algorithm for calculating scattering characteristics of a multilayer structure consisting of two-dimensionally periodic gratings, for which the operators S^{AA}, S^{BA}, S_{pq}^{AB}, and S_{pq}^{BB} are known. Consider a double-layer structure, whose geometry is given in Figure 4. Two semi-transparent periodic gratings I and II are separated by a dielectric layer of finite thickness M (here $\varepsilon=\varepsilon_2(\text{I})=\varepsilon_1(\text{II})$) and placed between the upper and the bottom dielectric half-spaces with the permittivity $\varepsilon_1(\text{I})$ and $\varepsilon_2(\text{II})$, respectively. Let also a pulsed wave like (35) be incident onto the boundary $z_1(\text{I})=0$ from the Floquet channel A.

Retaining previously accepted notation (the evident changes are conditioned by the presence of two different gratings I and II), represent the solution of the corresponding initial boundary value problem in the regular domains$A$, B, and C in a symbolic form

$$U(A)=\sum_{p=-\infty}^{\infty}\left[v_{p(1)}(z_1(\mathrm{I}),\ t)+u_{p(1)}(z_1(\mathrm{I}),\ t)\right]\mu_p(x,\ y),$$

$$U(B)=\sum_{p=-\infty}^{\infty}\left[u_{p(2)}(z_2(\mathrm{I}),\ t)+u_{p(1)}(z_1(\mathrm{II}),\ t)\right]\mu_p(x,\ y),$$

$$U(C)=\sum_{p=-\infty}^{\infty}u_{p(2)}(z_2(\mathrm{II}),\ t)\mu_p(x,\ y).$$

The first terms in the square brackets correspond to the waves propagating towards the domain$C$, while the second ones correspond to the waves propagating towards the domain A (Figure 4). The set $\{\mu_p(x,\ y)\}_{p=-\infty}^{\infty}$ is formed from the functions$\mu_{nm}(x,\ y)$, $(n,\ m=0, \pm 1, \pm 2,...)$, while the set $\{\lambda_p\}_{p=-\infty}^{\infty}$ is composed from the valuesλ_{nm}, $(n,\ m=0, \pm 1, \pm 2,...)$ (Figure 3).

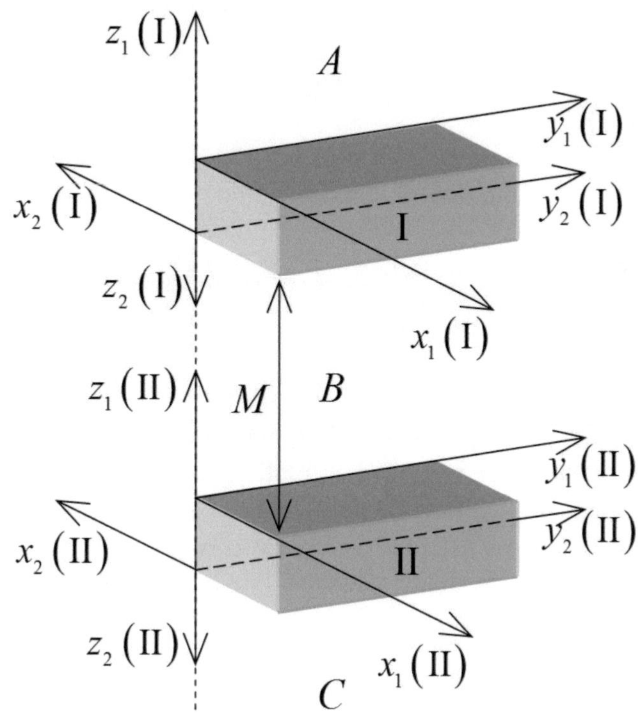

Figure 4. Schematic drawing of a double-layered structure.

By denoting

$$u_{(j)}'(\mathrm{I}) \equiv \frac{\partial}{\partial z_j(\mathrm{I})} u_{(j)}(z_j(\mathrm{I}), t) \Big|_{z_j(\mathrm{I})=0'} \qquad u_{(j)}(\mathrm{I}) = \{u_{p(j)}(z_j(\mathrm{I}), t)\} \,|_{z_j(\mathrm{I})=0'}$$

according to formulas (38)-(42), we construct the following system of operator equations:

$$\begin{cases} u_{(1)}'(\mathrm{I}) = S^{AA}(\mathrm{I})\left[v_{(1)}(\mathrm{I})\right] + S^{AB}(\mathrm{I}) Z^B_{z_1(\mathrm{II})=0\to M}\left[u_{(1)}'(\mathrm{II})\right] \\ u_{(2)}'(\mathrm{I}) = S^{BA}(\mathrm{I})\left[v_{(1)}(\mathrm{I})\right] + S^{BB}(\mathrm{I}) Z^B_{z_1(\mathrm{II})=0\to M}\left[u_{(1)}'(\mathrm{II})\right] \\ u_{(1)}'(\mathrm{II}) = S^{BB}(\mathrm{II}) Z^B_{z_2(\mathrm{I})=0\to M}\left[u_{(2)}'(\mathrm{I})\right] \\ u_{(2)}'(\mathrm{II}) = S^{CB}(\mathrm{II}) Z^B_{z_2(\mathrm{I})=0\to M}\left[u_{(2)}'(\mathrm{I})\right]. \end{cases} \tag{43}$$

Equations (43) clearly represent step-by-step response of the complex structure on the excitation by the signal $\vec{U}_1^i(g, t)$ with the evolutionary basis $v_{(1)}(z_1(\mathrm{I}), t) = \{v_{p(1)}(z_1(\mathrm{I}), t)\}_{p=-\infty}^{\infty}$ (or simply$v_{(1)}(\mathrm{I})$). Thus, for example, the first equation can be interpreted as follows. A signal $u_{(1)}(\mathrm{I})$ (the secondary field in$A$) is a sum of two signals, where the first signal is a result of the reflection of the incident signal $v_{(1)}(\mathrm{I})$ by the gratingI, while another one is determined by the signal $u_{(1)}(\mathrm{II})$being deformed during propagation in the channel $B$and interaction with the gratingI.

By method of elimination the system (43) is reduced to the operator equation of the second kind

$$u_{(2)}'(\mathrm{I}) = S^{BA}(\mathrm{I})[v_{(1)}(\mathrm{I})] + S^{BB}(\mathrm{I}) Z^B_{z_1(\mathrm{II})=0\to M} S^{BB}(\mathrm{II}) Z^B_{z_2(\mathrm{I})=0\to M}[u_{(2)}'(\mathrm{I})] \tag{44}$$

and some formulas for calculating the electromagnetic field components in all regions of the two-layered structure. The observation time t for the unknown function $u_{(2)}'(\mathrm{I})$ from the left-hand side of equation (44) strictly greater of any moment of time τ for the function $u_{(2)}'(\mathrm{I})$ in the right-hand side of the equation (owing to finiteness of wave velocity). Therefore equation (44) can be inverted explicitly in the framework of standard algorithm of step-by-step progression through time layers. Upon realization of this scheme and calculation of the boundary operators by (38), (42), the two-layered structure can be used as 'elementary' unit of more complex structures.

Turning back to (38)-(42), we see that the operators entering these equations act differently that their analogues in the frequency domain, where the boundary operators relate a pair 'field → field'. Reasoning from the structure of the transport operators $Z^A_{0\to z_1}$ and $Z^B_{0\to z_2}$ (formulas (40) and (41)), we relate a pair 'field → directional derivative with respect to the propagation direction' to increase numerical efficiency of the corresponding computational algorithms.

7. Some Important Properties of Steady-State Fields in the Rectangular Floquet Channel

7.1. Excitation by a TM-wave

Let a grating (Figure 1) be excited form the domain A by a pulsed TM-wave $\vec{U}^i(g,t)=\vec{U}^i_{pq(E)}(g,t): E_z^i(g,t)=v_{pq(z,E)}(z,t)\mu_{pq}(x,y)$ and the region Q_L is free from the sources $\vec{j}(g,t)$, $\vec{\varphi}_E(g)$, and $\vec{\varphi}_H(g)$. The field generated in the domains A and B is determined completely by their longitudinal components. They can be represented in the form of (31), (32). Define steady-state fields $\{\tilde{\vec{E}}(g,k), \tilde{\vec{H}}(g,k)\}$ (see formula (33) with $\mathrm{Im}k=0$) corresponding to the pulsed fields $\{\vec{E}^i, \vec{H}^i\}$, $\{\vec{E}^s, \vec{H}^s\}$ in A and the pulsed field $\{\vec{E}, \vec{H}\}$ in B, by their z-components:

$$\begin{Bmatrix}\tilde{E}_z^i(g,k)\\ \tilde{H}_z^i(g,k)\end{Bmatrix}=\begin{Bmatrix}\tilde{v}_{pq(z,E)}(k)\\ 0\end{Bmatrix}e^{-i\Gamma_{pq}(z-L)}\mu_{pq}(x,y),\ g\in\bar{A} \tag{45}$$

$$\begin{Bmatrix}\tilde{E}_z^s(g,k)\\ \tilde{H}_z^s(g,k)\end{Bmatrix}=\sum_{n,m=-\infty}^{\infty}\begin{Bmatrix}\tilde{u}^+_{nm(z,E)}(k)\\ \tilde{u}^+_{nm(z,H)}(k)\end{Bmatrix}e^{i\Gamma_{nm}(z-L)}\mu_{nm}(x,y),\ g\in\bar{A} \tag{46}$$

$$\begin{Bmatrix}\tilde{E}_z(g,k)\\ \tilde{H}_z(g,k)\end{Bmatrix}=\sum_{n,m=-\infty}^{\infty}\begin{Bmatrix}\tilde{u}^-_{nm(z,E)}(k)\\ \tilde{u}^-_{nm(z,H)}(k)\end{Bmatrix}e^{-i\Gamma_{nm}(z+L)}\mu_{nm}(x,y),\ g\in\bar{B} \tag{47}$$

where the following notation is used: $\tilde{v}_{pq(z,E)}(k)\leftrightarrow v_{pq(z,E)}(L,t)$, $\tilde{u}^{\pm}_{nm(z,E\,\mathrm{or}\,H)}(k)\leftrightarrow u^{\pm}_{nm(z,E\,\mathrm{or}\,H)}(\pm L,t)$, $\Gamma_{nm}=(k^2-\lambda^2_{nm})^{1/2}$, $\mathrm{Re}\Gamma_{nm}\mathrm{Re}k\geq 0$, $\mathrm{Im}\Gamma_{nm}\geq 0$[7].

The amplitudes $\tilde{u}^{\pm}_{nm(z,E\,\mathrm{or}\,H)}(k)$ form the system of the so-called scattering coefficients of the grating, namely, the reflection coefficients

$$R^{nm(H)}_{pq(E)}=\frac{\tilde{u}^+_{nm(z,H)}(k)}{\tilde{v}_{pq(z,E)}(k)},\quad R^{nm(E)}_{pq(E)}=\frac{\tilde{u}^+_{nm(z,E)}(k)}{\tilde{v}_{pq(z,E)}(k)},\quad n,m=0,\pm1,\pm2,... \tag{48}$$

specifying efficiency of transformation of pq-th harmonic of a monochromatic TM-wave into of order nm-th harmonics of the scattered field $\{\tilde{\vec{E}}^s, \tilde{\vec{H}}^s\}$ in the reflection zone, and the transmission coefficients

$$T^{nm(H)}_{pq(E)}=\frac{\tilde{u}^-_{nm(z,H)}(k)}{\tilde{v}_{pq(z,E)}(k)},\quad T^{nm(E)}_{pq(E)}=\frac{\tilde{u}^-_{nm(z,E)}(k)}{\tilde{v}_{pq(z,E)}(k)},\quad n,m=0,\pm1,\pm2,... \tag{49}$$

determining the efficiency of excitation of the transmitted harmonics in the domain B.

These coefficients are related by the energy balance equations

$$\sum_{n,m=-\infty}^{\infty} \frac{1}{\lambda_{nm}^2}\left[\left(\left|R_{pq(E)}^{nm(E)}\right|^2+\left|T_{pq(E)}^{nm(E)}\right|^2\right)\pm\eta_0^2\left(\left|R_{pq(E)}^{nm(H)}\right|^2+\left|T_{pq(E)}^{nm(H)}\right|^2\right)\right]\begin{Bmatrix}\mathrm{Re}\Gamma_{nm}\\ \mathrm{Im}\Gamma_{nm}\end{Bmatrix}$$
$$=\frac{1}{\lambda_{pq}^2}\begin{Bmatrix}\mathrm{Re}\Gamma_{pq}+2\mathrm{Im}\Gamma_{pq}\mathrm{Im}R_{pq(E)}^{pq(E)}\\ \mathrm{Im}\Gamma_{pq}-2\mathrm{Re}\Gamma_{pq}\mathrm{Im}R_{pq(E)}^{pq(E)}\end{Bmatrix}\mp\frac{1}{\varepsilon_0}\begin{Bmatrix}W_1\\ W_2\end{Bmatrix},\ p,\ q=0,\pm1,\pm2,\ldots, \tag{50}$$

$$W_1=\frac{\varepsilon_0\eta_0}{k}\int_{Q_L}\sigma(g,k)\left|\vec{E}(g,k)\right|^2 dg,$$
$$W_2=\int_{Q_L}\left[\mu_0\mu(g,k)\left|\vec{H}(g,k)\right|^2-\varepsilon_0\varepsilon(g,k)\left|\vec{E}(g,k)\right|^2\right]dg \tag{51}$$

They follow from the complex power theorem (Poynting theorem) in the integral form [11]

$$\oint_{S_L}\left(\left[\vec{E}\times\vec{H}^*\right]\cdot\vec{ds}\right)=\int_{Q_L}\mathrm{div}\left[\vec{E}\times\vec{H}^*\right]dg=ik\eta_0\int_{Q_L}\mu\left|\vec{H}\right|^2dg-\frac{ik}{\eta_0}\int_{Q_L}\varepsilon\left|\vec{E}\right|^2dg-\int_{Q_L}\sigma\left|\vec{E}\right|^2dg \tag{52}$$

where $\varepsilon(g,k)-1=\tilde{\chi}_\varepsilon(g,k)\leftrightarrow\chi_\varepsilon(g,t)$, $\mu(g,k)-1=\tilde{\chi}_\mu(g,k)\leftrightarrow\chi_\mu(g,t)$, $\sigma(g,k)=\tilde{\chi}_\sigma(g,k)\leftrightarrow\chi_\sigma(g,t)$, $\vec{ds}$ is the vector element of the surface S_L bounding the domain Q_L. Equations (50)-(52) have been derived starting from the following boundary value problem for a diffraction grating illuminated by a plane TM-wave $\vec{U}_{pq(E)}^{i}(g,k):\tilde{E}_z^i(g,k)=\exp\left[-i\Gamma_{pq}(z-L)\right]\mu_{pq}(x,y)$:

$$\begin{cases}\eta_0\mathrm{rot}\tilde{\vec{H}}(g,k)=-ik\bar{\varepsilon}(g,k)\tilde{\vec{E}}(g,k),\\ \mathrm{rot}\tilde{\vec{E}}(g,k)=ik\eta_0\mu(g,k)\tilde{\vec{H}}(g,k), \quad g\in Q_L\\ D\left[\tilde{\vec{E}}\left(\tilde{\vec{H}}\right)\right](l_x,y)=e^{2\pi i\Phi_x}D\left[\tilde{\vec{E}}\left(\tilde{\vec{H}}\right)\right](0,y), \quad 0\le y\le l_y, \quad |z|<L\\ D\left[\tilde{\vec{E}}\left(\tilde{\vec{H}}\right)\right](x,l_y)=e^{2\pi i\Phi_y}D\left[\tilde{\vec{E}}\left(\tilde{\vec{H}}\right)\right](x,0), \quad 0\le x\le l_x, \quad |z|<L\\ \tilde{E}_{tg}(g,k)\big|_{g\in S}=0, \quad \tilde{H}_{nr}(g,k)\big|_{g\in S}=0\end{cases} \tag{53}$$

$$\begin{Bmatrix}\tilde{E}_z(g,k)\\ \tilde{H}_z(g,k)\end{Bmatrix}=\begin{Bmatrix}1\\ 0\end{Bmatrix}e^{-i\Gamma_{pq}(z-L)}\mu_{pq}(x,y)+\sum_{n,m=-\infty}^{\infty}\begin{Bmatrix}R_{pq(E)}^{nm(E)}(k)\\ R_{pq(E)}^{nm(H)}(k)\end{Bmatrix}e^{i\Gamma_{nm}(z-L)}\mu_{nm}(x,y),\ g\in\bar{A},$$
$$\begin{Bmatrix}\tilde{E}_z(g,k)\\ \tilde{H}_z(g,k)\end{Bmatrix}=\sum_{n,m=-\infty}^{\infty}\begin{Bmatrix}T_{pq(E)}^{nm(E)}(k)\\ T_{pq(E)}^{nm(H)}(k)\end{Bmatrix}e^{-i\Gamma_{nm}(z+L)}\mu_{nm}(x,y),\ g\in\bar{B}. \tag{54}$$

When deriving (50), (51) we have also used the equations relating z-components of the eigenmode of the Floquet channel

$$\vec{U}(g,\, k):\;\; \tilde{E}_z(g,\, k)=Ae^{\pm i\Gamma z}\mu(x,\, y) \quad \text{and} \quad \tilde{H}_z(g,\, k)=Be^{\pm i\Gamma z}\mu(x,\, y) \tag{55}$$

(subscripts nm are omitted) with its longitudinal components:

$$\begin{aligned}\tilde{E}_x&=-\frac{\beta k\eta_0}{\lambda^2}\tilde{H}_z\mp\frac{\alpha\Gamma}{\lambda^2}\tilde{E}_z,\; \tilde{E}_y=\frac{\alpha k\eta_0}{\lambda^2}\tilde{H}_z\mp\frac{\beta\Gamma}{\lambda^2}\tilde{E}_z,\\ \tilde{H}_x&=\mp\frac{\alpha\Gamma}{\lambda^2}\tilde{H}_z+\frac{\beta k}{\eta_0\lambda^2}\tilde{E}_z,\; \tilde{H}_y=\mp\frac{\beta\Gamma}{\lambda^2}\tilde{H}_z-\frac{\alpha k}{\eta_0\lambda^2}\tilde{E}_z.\end{aligned} \tag{56}$$

Here, $\bar{\varepsilon}(g,\, k)=\varepsilon(g,\, k)+i\eta_0\sigma(g,\, k)/k$, $\mu(x,\, y)=(l_x l_y)^{-1/2}\exp(i\alpha x)\exp(i\beta y)$, $\Gamma=\sqrt{k^2-\lambda^2}$, $\lambda^2=\alpha^2+\beta^2$.

According to the Lorentz lemma [11], the fields $\{\vec{E}^{(1)},\, \vec{H}^{(1)}\}$ and $\{\vec{E}^{(2)},\, \vec{H}^{(2)}\}$ resulting from the interaction of a grating with two plane TM-waves

$\vec{U}^{i(1)}_{pq(E)}(g,\, k):\tilde{E}^{i(1)}_z(g,\, k)=\exp[-i\Gamma_{pq}(\Phi_x,\, \Phi_y)(z-L\,)]\mu_{pq}(x,\, y,\, \Phi_x,\, \Phi_y)$ and

$\vec{U}^{i(2)}_{-r,-s(E)}(g,\, k):\tilde{E}^{i(2)}_z(g,\, k)=\exp[-i\Gamma_{-r,-s}(-\Phi_x,\, -\Phi_y)(z-L\,)]\mu_{-r,-s}(x,\, y,\, -\Phi_x,\, -\Phi_y)$,

satisfy the following equation

$$\oint_{S_L}\left(\left([\vec{E}^{(1)}\times\vec{H}^{(2)}]-[\vec{E}^{(2)}\times\vec{H}^{(1)}]\right)\cdot d\vec{s}\right)=0. \tag{57}$$

From (57), using (54) and (56), we obtain

$$\frac{R^{rs(E)}_{pq(E)}(\Phi_x,\, \Phi_y)\lambda^2_{p,q}(\Phi_x,\, \Phi_y)}{\Gamma_{pq}(\Phi_x,\, \Phi_y)}=\frac{R^{-p,-q(E)}_{-r,-s(E)}(-\Phi_x,\, -\Phi_y)\lambda^2_{-r,-s}(-\Phi_x,\, -\Phi_y)}{\Gamma_{-r,-s}(-\Phi_x,\, -\Phi_y)}, \quad p,\, q,\, r,\, s=0,\pm 1,\pm 2,\ldots \tag{58}$$

– the reciprocity relations, which are of considerable importance in the physical analysis of wave scattering by periodic structures as well as when testing numerical algorithms for boundary problems (53), (54).

Assume now that the first wave $\vec{U}^{i(1)}_{pq(E)}(g,\, k)$: $:\tilde{E}^{i(1)}_z(g,\, k)=\exp[-i\Gamma_{pq}(\Phi_x,\, \Phi_y)(z-L\,)]\mu_{pq}(x,\, y,\, \Phi_x,\, \Phi_y)=\vec{U}^{i(1)}_{pq(E)}(g,\, k,\, A)$ be incident on the grating from the domain$A$, as in the case considered above, while another wave $\vec{U}^{i(2)}_{-r,-s(E)}(g,\, k):\tilde{E}^{i(2)}_z(g,\, k,\, B)=\exp[i\Gamma_{-r,-s}(-\Phi_x,\, -\Phi_y)(z+L\,)]\mu_{-r,-s}(x,\, y,\, -\Phi_x,\, -\Phi_y)$ is incident from$B$. Both of these waves satisfy equation (57), whence we have

$$\frac{T^{rs(E)}_{pq(E)}(\Phi_x, \Phi_y, A)\lambda^2_{p,q}(\Phi_x, \Phi_y)}{\Gamma_{pq}(\Phi_x, \Phi_y)} = \frac{T^{-p,-q(E)}_{-r,-s(E)}(-\Phi_x, -\Phi_y, B)\lambda^2_{-r,-s}(-\Phi_x, -\Phi_y)}{\Gamma_{-r,-s}(-\Phi_x, -\Phi_y)}, \quad p, q, r, s = 0, \pm 1, \pm 2, \ldots \tag{59}$$

7.2. Excitation by a *TE*-wave

Let a grating be excited form the domain A by a pulsed TE-wave $\vec{U}^i(g, t) = \vec{U}^i_{pq(H)}(g, t): H^i_z(g, t) = v_{pq(z,H)}(z, t)\mu_{pq}(x, y)$ and the region Q_L is free from the sources $\vec{j}(g, t)$, $\vec{\varphi}_E(g)$, and$\vec{\varphi}_H(g)$. The field generated in the domains A and B is determined completely by their longitudinal components. They can be represented in the form of (31), (32). Define steady-state fields $\{\tilde{\vec{E}}(g, k), \tilde{\vec{H}}(g, k)\}$ corresponding to the pulsed fields$\{\vec{E}^i, \vec{H}^i\}$, $\{\vec{E}^s, \vec{H}^s\}$in A and the pulsed field $\{\vec{E}, \vec{H}\}$ in$B$, by their z-components as was done for the TM-case (see equations (45)-(47)). Introduce the scattering coefficients$R^{nm(E)}_{pq(H)}$, $R^{nm(H)}_{pq(H)}$, $T^{nm(E)}_{pq(H)}$, and $T^{nm(H)}_{pq(H)}$ by the relations like (48). These coefficients can be determined from the problems

$$\begin{cases} \eta_0 \mathrm{rot}\tilde{\vec{H}}(g,k) = -ik\bar{\varepsilon}(g,k)\tilde{\vec{E}}(g,k), \\ \mathrm{rot}\tilde{\vec{E}}(g,k) = ik\eta_0\mu(g,k)\tilde{\vec{H}}(g,k), \quad g \in Q_L \\ D\left[\tilde{\vec{E}}\left(\tilde{\vec{H}}\right)\right](l_x, y) = e^{2\pi i\Phi_x} D\left[\tilde{\vec{E}}\left(\tilde{\vec{H}}\right)\right](0, y), \quad 0 \le y \le l_y, \quad |z| < L \\ D\left[\tilde{\vec{E}}\left(\tilde{\vec{H}}\right)\right](x, l_y) = e^{2\pi i\Phi_y} D\left[\tilde{\vec{E}}\left(\tilde{\vec{H}}\right)\right](x, 0), \quad 0 \le x \le l_x, \quad |z| < L \\ \tilde{E}_{tg}(g,k)\Big|_{g\in S} = 0, \quad \tilde{H}_{nr}(g,k)\Big|_{g\in S} = 0, \end{cases} \tag{60}$$

$$\begin{aligned} \begin{Bmatrix} \tilde{E}_z(g, k) \\ \tilde{H}_z(g, k) \end{Bmatrix} &= \begin{Bmatrix} 0 \\ 1 \end{Bmatrix} e^{-i\Gamma_{pq}(z-L)}\mu_{pq}(x, y) + \sum_{n,m=-\infty}^{\infty} \begin{Bmatrix} R^{nm(E)}_{pq(H)}(k) \\ R^{nm(H)}_{pq(H)}(k) \end{Bmatrix} e^{i\Gamma_{nm}(z-L)}\mu_{nm}(x, y), \ g \in \bar{A}, \\ \begin{Bmatrix} \tilde{E}_z(g, k) \\ \tilde{H}_z(g, k) \end{Bmatrix} &= \sum_{n,m=-\infty}^{\infty} \begin{Bmatrix} T^{nm(E)}_{pq(H)}(k) \\ T^{nm(H)}_{pq(H)}(k) \end{Bmatrix} e^{-i\Gamma_{nm}(z+L)}\mu_{nm}(x, y), \ g \in \bar{B} \end{aligned} \tag{61}$$

and satisfy the following relations, which are corollaries from the Poynting theorem and the Lorentz lemma:

$$\begin{aligned} &\sum_{n,m=-\infty}^{\infty} \frac{1}{\lambda^2_{nm}}\left[\left(|R^{nm(H)}_{pq(H)}|^2 + |T^{nm(H)}_{pq(H)}|^2\right) \pm \frac{1}{\eta_0^2}\left(|R^{nm(E)}_{pq(H)}|^2 + |T^{nm(E)}_{pq(H)}|^2\right)\right] \begin{Bmatrix} \mathrm{Re}\Gamma_{nm} \\ \mathrm{Im}\Gamma_{nm} \end{Bmatrix} = \\ &= \frac{1}{\lambda^2_{pq}} \begin{Bmatrix} \mathrm{Re}\Gamma_{pq} + 2\mathrm{Im}\Gamma_{pq}\mathrm{Im}R^{pq(H)}_{pq(H)} \\ \mathrm{Im}\Gamma_{pq} - 2\mathrm{Re}\Gamma_{pq}\mathrm{Im}R^{pq(H)}_{pq(H)} \end{Bmatrix} - \frac{1}{\mu_0}\begin{Bmatrix} W_1 \\ W_2 \end{Bmatrix}, \ p, q = 0, \pm 1, \pm 2, \ldots \end{aligned} \tag{62}$$

http://dx.doi.org/10.5772/51007

and

$$\frac{R_{pq(H)}^{rs(H)}(\Phi_x, \Phi_y)\lambda_{p,q}^2(\Phi_x, \Phi_y)}{\Gamma_{pq}(\Phi_x, \Phi_y)} = \frac{R_{-r,-s(H)}^{-p,-q(H)}(-\Phi_x, -\Phi_y)\lambda_{-r,-s}^2(-\Phi_x, -\Phi_y)}{\Gamma_{-r,-s}(-\Phi_x, -\Phi_y)}, \quad p, q, r, s = 0, \pm 1, \pm 2, ... \tag{63}$$

$$\frac{T_{pq(H)}^{rs(H)}(\Phi_x, \Phi_y, A)\lambda_{p,q}^2(\Phi_x, \Phi_y)}{\Gamma_{pq}(\Phi_x, \Phi_y)} = \frac{T_{-r,-s(H)}^{-p,-q(H)}(-\Phi_x, -\Phi_y, B)\lambda_{-r,-s}^2(-\Phi_x, -\Phi_y)}{\Gamma_{-r,-s}(-\Phi_x, -\Phi_y)}, \quad p, q, r, s = 0, \pm 1, \pm 2, ... \tag{64}$$

7.3. General properties of the grating's secondary field

Let now k be a real positive frequency parameter, and let an arbitrary semi-transparent grating (Figure 1) be excited from the domain A by a homogeneous TM- or TE-wave

$$\vec{U}_{pq(E \text{ or } H)}^i(g, k): \{\tilde{E}_z^i(g, k) \text{ or } \tilde{H}_z^i(g, k)\} = e^{-i\Gamma_{pq}(z-L)}\mu_{pq}(x, y), \qquad p, q: \text{Im}\Gamma_{pq} = 0. \tag{65}$$

The terms of infinite series in (54) and (61) are z-components of nm-th harmonics of the scattered field for the domains A and$B$. The complex amplitudes $R_{pq(E\text{or}H)}^{nm(E\text{or}H)}$ and $T_{pq(E\text{or}H)}^{nm(E\text{or}H)}$ are the functions of$k$, Φ_x, Φ_y, as well as of the geometry and material parameters of the grating. Every harmonic for which $\text{Im}\Gamma_{nm}=0$ and $\text{Re}\Gamma_{nm}>0$ is a homogeneous plane wave propagating away from the grating along the vector$\vec{k}_{nm}$:$k_x=\alpha_n$, $k_y=\beta_m$, $k_z=\Gamma_{nm}$(in$A$; Figure 5) or $k_z=-\Gamma_{nm}$ (in$B$). The frequencies $k$such that $\Gamma_{nm}(k)=0$ ($k=k_{nm}^{\pm}=\pm|\lambda_{nm}|$) are known as threshold frequency or sliding points [1-6]. At those points, a spatial harmonic of order nm with $\text{Im}\Gamma_{nm}>0$ are transformed into a propagating homogeneous pane wave.

It is obvious that the propagation directions $\vec{k}_{nm}$ of homogeneous harmonics of the secondary field depends on their ordernm, on the values of k and on the directing vector of the incident wave$\vec{k}_{pq}^i$:$k_x^i=\alpha_p$, $k_y^i=\beta_q$,$k_z^i=-\Gamma_{pq}$. According to (50) and (62), we can write the following formulas for the values, which determine the 'energy content' of harmonics, or in other words, the relative part of the energy directed by the structure into the relevant spatial radiation channel:

$$\begin{aligned}(WR)_{pq}^{nm} &= \left(|R_{pq(E)}^{nm(E)}|^2 + \eta_0^2|R_{pq(E)}^{nm(H)}|^2\right)\frac{\text{Re}\Gamma_{nm}}{\lambda_{nm}^2}\frac{\lambda_{pq}^2}{\Gamma_{pq}} = (WR)_{pq(E)}^{nm(E)} + (WR)_{pq(E)}^{nm(H)}, \\ (WT)_{pq}^{np} &= \left(|T_{pq(E)}^{nm(E)}|^2 + \eta_0^2|T_{pq(E)}^{nm(H)}|^2\right)\frac{\text{Re}\Gamma_{nm}}{\lambda_{nm}^2}\frac{\lambda_{pq}^2}{\Gamma_{pq}} = (WT)_{pq(E)}^{np(E)} + (WT)_{pq(E)}^{np(H)}\end{aligned} \tag{66}$$

(for TM-case) and

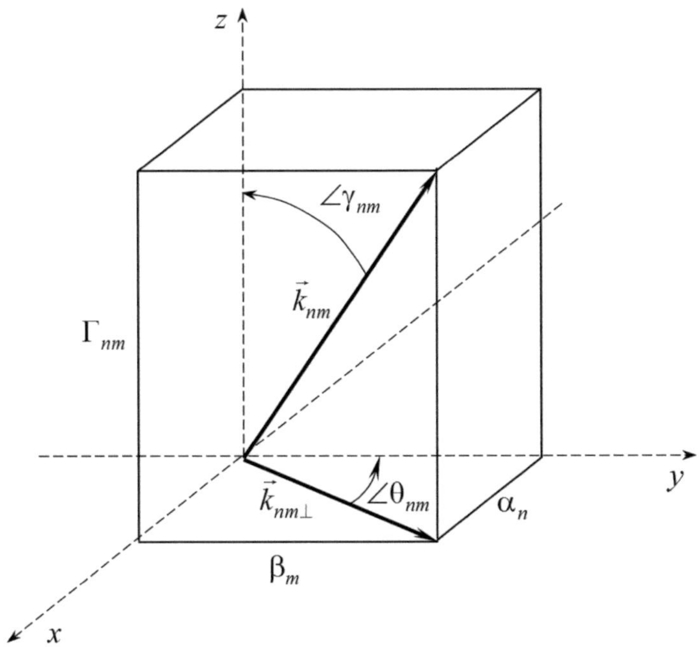

Figure 5. On determination of propagation directions for spatial harmonics of the field formed by a two-dimensionally periodic structure.

$$\begin{aligned}(WR)_{pq}^{nm} &= \left(\left| R_{pq(H)}^{nm(H)} \right|^2 + \frac{1}{\eta_0^2} \left| R_{pq(H)}^{nm(E)} \right|^2 \right) \frac{\mathrm{Re}\Gamma_{nm}}{\lambda_{nm}^2} \frac{\lambda_{pq}^2}{\Gamma_{pq}} = (WR)_{pq(H)}^{nm(H)} + (WR)_{pq(H)}^{nm(E)}, \\ (WT)_{pq}^{np} &= \left(\left| T_{pq(H)}^{nm(H)} \right|^2 + \frac{1}{\eta_0^2} \left| T_{pq(H)}^{nm(E)} \right|^2 \right) \frac{\mathrm{Re}\Gamma_{nm}}{\lambda_{nm}^2} \frac{\lambda_{pq}^2}{\Gamma_{pq}} = (WT)_{pq(H)}^{nm(H)} + (WT)_{pq(H)}^{nm(E)}\end{aligned} \tag{67}$$

(for TE-case). The channel corresponding to the nm-th harmonic will be named 'open' if $\mathrm{Im}\Gamma_{nm}=0$. The regime with a single open channel ($nm=pq$) will be called the single-mode regime.

Since $|\vec{k}_{pq}^{i}| = |\vec{k}_{nm}| = k$, the nm-th harmonic of the secondary field in the reflection zone propagates in opposition to the incident wave only if $\alpha_n = -\alpha_p$ and $\beta_m = -\beta_q$ or, in other notation, if

$$n = -2\Phi_x - p \quad \text{and} \quad m = -2\Phi_y - q \tag{68}$$

Generation of the nonspecularly reflected mode of this kind is termed the auto-collimation.

The amplitudes $R_{pq(E\,\mathrm{or}H)}^{nm(E\,\mathrm{or}H)}$ or $T_{pq(E\,\mathrm{or}H)}^{nm(E\,\mathrm{or}H)}$ are not all of significance for the physical analysis. In the far-field zone, the secondary field is formed only by the propagating harmonics of the orders nm such that$\mathrm{Re}\Gamma_{nm} \geq 0$. However, the radiation field in the immediate proximity of the grating requires a consideration of the contribution of damped harmonics ($nm:\mathrm{Im}\Gamma_{nm}>0$). Moreover, in some situations (resonance mode) this contribution is the dominating one [6].

7.4. The simplest corollaries of the reciprocity relations and the energy conservation law

Let us formulate several corollaries of the relations (50), (58), (59), and (62)-(64) basing on the results presented in [3] and [7] for one-dimensionally periodic gratings and assuming that $\varepsilon(g,\,k)\geq 0$, $\mu(g,\,k)\geq 0$, and$\sigma(g,\,k)\geq 0$.

- The upper lines in (50) and (62) represent the energy conservation law for propagating waves. If$\mathrm{Im}\Gamma_{pq}=0$, the energy of the scattered field is clearly related with the energy of the incident wave. The energy of the wave $\vec{U}^{\,i}_{pq(E\,\mathrm{or}H)}(g,\,k)$ is partially absorbed by the grating (only if$W_1 \neq 0$), and the remaining part is distributed between spatial TM- and TE-harmonics propagating in the domains A and B (the wave is reradiating into the directions $z=\pm\infty$). If a plane inhomogeneous wave be incident on a grating ($\mathrm{Im}\Gamma_{pq}>0$), the total energy is defined by the imaginary part of reflection coefficient$R_{pq(E\,\mathrm{or}H)}^{pq(E\,\mathrm{or}H)}$, which in this case is nonnegative.
- The relations in the bottom lines in (50), (62) limit the values of $\sum_{n,m=-\infty}^{\infty}|R_{pq(E)}^{nm(E)}|^2\lambda_{nm}^{-2}\mathrm{Im}\Gamma_{nm}$, $\sum_{n,m=-\infty}^{\infty}|T_{pq(E)}^{nm(E)}|^2\lambda_{nm}^{-2}\mathrm{Im}\Gamma_{nm}$, etc. and determine thereby the class of infinite sequences

$$l_2=\left\{a=\{a_{nm}\}_{nm=-\infty}^{\infty}:\ \sum_{nm=-\infty}^{\infty}\frac{|a_{nm}|^2}{\sqrt{n^2+m^2}}\infty\right\} \tag{69}$$

or energetic space, to which amplitudes of the scattered harmonics$R_{pq(E)}^{nm(E)}$, $T_{pq(E)}^{nm(E)}$, etc. belong.

- It follows from (58), (59), (63), and (64) that for all semi-transparent and reflecting gratings we can write

$$\begin{aligned}(WR)_{00(E\,\mathrm{or}H)}^{00(E\,\mathrm{or}H)}(\Phi_x,\,\Phi_y)&=(WR)_{00(E\,\mathrm{or}H)}^{00(E\,\mathrm{or}H)}(-\Phi_x,\,-\Phi_y),\\ (WT)_{00(E\,\mathrm{or}H)}^{00(E\,\mathrm{or}H)}(\Phi_x,\,\Phi_y,\,A)&=(WT)_{00(E\,\mathrm{or}H)}^{00(E\,\mathrm{or}H)}(-\Phi_x,\,-\Phi_y,\,B).\end{aligned} \tag{70}$$

The first equation in (70) proves that the efficiency of transformation of the TM- or TE-wave into the specular reflected wave of the same polarization remains unchanged if the grating is rotated in the plane $x0y$ about z-axis through$180°$. The efficiency of transformation into the principal transmitted wave of the same polarizations does not also vary with the grating rotation about the axis lying in the plane $x0y$ and being normal to the vector $\vec{k}_{00}$ (Figure 5).

- When $r=s=p=q=0$ we derive from (58), (59), (63), and (64) that

$$R_{00(E\,\mathrm{or}H)}^{00(E\,\mathrm{or}H)}(\Phi_x, \Phi_y)=R_{00(E\,\mathrm{or}H)}^{00(E\,\mathrm{or}H)}(-\Phi_x, -\Phi_y),$$
$$T_{00(E\,\mathrm{or}H)}^{00(E\,\mathrm{or}H)}(\Phi_x, \Phi_y, A)=T_{00(E\,\mathrm{or}H)}^{00(E\,\mathrm{or}H)}(-\Phi_x, -\Phi_y, B). \tag{71}$$

That means that even if a semi-transparent or reflecting grating is non symmetric with respect to the any planes, the reflection and transmission coefficients entering (71) do not depend on the proper changes in the angles of incidence of the primary wave.

- Relations (50), (58) allow the following regularities to be formulated for ideal ($\sigma(g, k)\equiv 0$) asymmetrical reflecting gratings. Let the parameters$k$, Φ_x, and Φ_y be such that $\mathrm{Re}\Gamma_{00}(\Phi_x, \Phi_y)>0$ and $\mathrm{Re}\Gamma_{nm}(\Phi_x, \Phi_y)=0$ for$n$, $m\neq 0$. If the incident wave is an inhomogeneous plane wave$\vec{U}^{\,i}_{\pm p,\pm q(E)}(g, k, \pm\Phi_x, \pm\Phi_y)$, then

$$\left(\left|R_{\pm p,\pm q(E)}^{00(E)}(\pm\Phi_x, \pm\Phi_y)\right|^2+\eta_0^2\left|R_{\pm p,\pm q(E)}^{00(H)}(\pm\Phi_x, \pm\Phi_y)\right|^2\right)\frac{\mathrm{Re}\Gamma_{00}(\pm\Phi_x, \pm\Phi_y)}{\lambda_{00}^2(\pm\Phi_x, \pm\Phi_y)}=$$
$$=2\mathrm{Im}R_{\pm p,\pm q(E)}^{\pm p,\pm q(E)}(\pm\Phi_x, \pm\Phi_y)\frac{\mathrm{Im}\Gamma_{\pm p,\pm q}(\pm\Phi_x, \pm\Phi_y)}{\lambda_{\pm p,\pm q}^2(\pm\Phi_x, \pm\Phi_y)}. \tag{72}$$

Since$R_{pq(E)}^{pq(E)}(\Phi_x, \Phi_y)=R_{-p,-q(E)}^{-p,-q(E)}(-\Phi_x, -\Phi_y)$, we derive from (72)

$$\left|R_{p,q(E)}^{00(E)}(\Phi_x, \Phi_y)\right|^2+\eta_0^2\left|R_{p,q(E)}^{00(H)}(\Phi_x, \Phi_y)\right|^2=$$
$$=\left|R_{-p,-q(E)}^{00(E)}(-\Phi_x, -\Phi_y)\right|^2+\eta_0^2\left|R_{-p,-q(E)}^{00(H)}(-\Phi_x, -\Phi_y)\right|^2. \tag{73}$$

It is easy to realize a physical meaning of the equation (73) and of similar relation for TE-case, which may be of interest for diffraction electronics. If a grating is excited by a damped harmonic, the efficiency of transformation into the unique propagating harmonic of spatial spectrum is unaffected by the structure rotation in the plane $x0y$ about z-axis through$180°$. The above-stated corollaries have considerable utility in testing numerical results and making easier their physical interpretation. The use of these corollaries may considerably reduce amount of calculations.

8. Elements of Spectral Theory for Two-Dimensionally Periodic Gratings

The spectral theory of gratings studies singularities of analytical continuation of solutions of boundary value problems formulated in the frequency domain (see, for example, problems (53), (54) and (60), (61)) into the domain of complex-valued (nonphysical) values of real parameters (like frequency, propagation constants, etc.) and the role of these singularities in resonant and anomalous modes in monochromatic and pulsed wave scattering. The fundamental results of this theory for one-dimensionally periodic gratings are presented in [4,6,7]. We present some elements of the spectral theory for two-dimensionally periodic structures, which follow immediately form the results obtained in the previous sections. The frequency k acts as a spectral parameter; a two-dimensionally periodic grating is considered as an open periodic resonator.

8.1. Canonical Green function

Let a solution $\tilde{G}_0(g,\ p,\ k)$ of the scalar problem

$$\begin{cases} \left[\Delta_g+k^2\right]\left[\tilde{G}_0(g,\ p,\ k)\right]=\delta(g-p),\ g=\{x_g,\ y_g,\ z_g\}\in R,\ p=\{x_p,\ y_p,\ z_p\}\in Q_L \\ D\left[\tilde{G}_0\right](l_x,\ y_g)=e^{2\pi i\Phi_x}D\left[\tilde{G}_0\right](0,y_g),\ 0\le y_g\le l_y,\ |z_g|\le L \\ D\left[\tilde{G}_0\right](x_g,\ l_y)=e^{2\pi i\Phi_y}D\left[\tilde{G}_0\right](x_g,0),\ 0\le x_g\le l_x,\ |z_g|\le L \\ \tilde{G}_0(g,\ p,\ k)=\sum_{n,m=-\infty}^{\infty}\begin{Bmatrix}A_{nm}(p,\ k)\\ B_{nm}(p,\ k)\end{Bmatrix}e^{\pm i\Gamma_{nm}(z_g\mp L)}\mu_{nm}(x_g,\ y_g),\ g\in\begin{Bmatrix}\bar{A}\\ \bar{B}\end{Bmatrix} \end{cases} \tag{74}$$

is named the canonical Green function for 2-D periodic gratings. In the case of the elementary periodic structure with the absence of any material scatterers, the problems of this kind but with arbitrary right-hand parts of the Helmholtz equation are formulated for the monochromatic waves generated by quasi-periodic current sources located in the region $|z|<L$.

Let us construct $\tilde{G}_0(g,\ p,\ k)$ as a superposition of free-space Green functions:

$$\tilde{G}_0(g,\ p,\ k)=-\frac{1}{4\pi}\sum_{n,m=-\infty}^{\infty}\frac{\exp\left[ik\,|g-p_{nm}|\right]}{|g-p_{nm}|}e^{2\pi in\Phi_x}e^{2\pi im\Phi_y},\ p_{nm}=\{x_p+nl_x,\ y_p+ml_y,\ z_p\}. \tag{75}$$

By using in (75) the Poisson summation formula [15] and the tabulated integrals [16]

$$\int_{-\infty}^{\infty}\frac{\exp\left(ip\sqrt{x^2+a^2}\right)}{\sqrt{x^2+a^2}}e^{ibx}dx=\pi iH_0^{(1)}\left(a\sqrt{|p^2-b^2|}\right)\ \text{and}\ \int_{-\infty}^{\infty}H_0^{(1)}\left(p\sqrt{x^2+a^2}\right)e^{ibx}dx=2\frac{\exp\left(ia\sqrt{p^2-b^2}\right)}{\sqrt{p^2-b^2}},$$

where $H_0^{(1)}(x)$ is the Hankel function of the first kind, we obtain

$$\widetilde{G}_0(g,\ p,\ k)=-\frac{i}{2l_x l_y}\sum_{n,m=-\infty}^{\infty} e^{i\left[\alpha_n(x_g-x_p)+\beta_m(y_g-y_p)\right]}\frac{\exp\left[i\mid z_g-z_p\mid\Gamma_{nm}\right]}{\Gamma_{nm}} \tag{76}$$

The surface K of analytic continuation of the canonical Green function (76) into the domain of complex-valued k is an infinite-sheeted Riemann surface consisting of the complex planes $k\in C$ with cuts along the lines$(\mathrm{Re}k)^2-(\mathrm{Im}k)^2-\lambda_{nm}^2=0,\ n,\ m=0,\pm 1,\pm 2,...,\ \mathrm{Im}k\le 0$(Figure 6). The first (physical) sheet C_k of the surface K is uniquely determined by the radiation conditions for $\widetilde{G}_0(g,\ p,\ k)$ in the domains A and$B$, i.e. by the choice of $\mathrm{Re}\Gamma_{nm}\mathrm{Re}k\ge 0$ and $\mathrm{Im}\Gamma_{nm}\ge 0$ on the axis$\mathrm{Im}k=0$. On this sheet, in the domain$0\le\arg k<\pi$, we have$\mathrm{Im}\Gamma_{nm}>0$, while $\mathrm{Re}\Gamma_{nm}\ge 0$ for $0\arg k\le\pi/2$ and $\mathrm{Re}\Gamma_{nm}\le 0$ for$\pi/2\le\arg k\pi$. In the domain $3\pi/2\le\arg k<2\pi$ for finite number of functions $\Gamma_{nm}(k)$ (with n and m such that$(\mathrm{Re}k)^2-(\mathrm{Im}k)^2-\lambda_{nm}^2>0$), the inequalities $\mathrm{Im}\Gamma_{nm}<0$ and $\mathrm{Re}\Gamma_{nm}>0$ hold; for the rest of these functions we have $\mathrm{Im}\Gamma_{nm}>0$ and$\mathrm{Re}\Gamma_{nm}\le 0$. In the domain$\pi<\arg k\le 3\pi/2$, the situation is similar only the signs of $\mathrm{Re}\Gamma_{nm}$ are opposite. On the subsequent sheets (each of them with its own pair$\{k;\Gamma_{nm}(k)\}$), the signs (root branches) of $\Gamma_{nm}(k)$ are opposite to those they have on the first sheet for a finite number of n and$m$. The cuts (solid lines in Fig. 6) originate from the real algebraic branch points$k_{nm}^{\pm}=\pm\mid\lambda_{nm}\mid$.

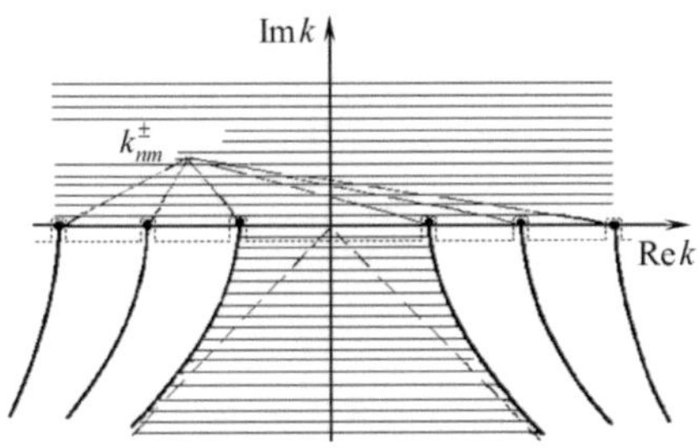

Figure 6. Natural domain of variation of the spectral parameter$k$: the first sheet of the surface$K$.

In the vicinity of some fixed point $K\in K$ the function $\widetilde{G}_0(g,\ p,\ k)$ can be expanded into a Loran series in terms of the local variable [17]

$$\kappa=\begin{cases}k-K;K\notin\{k_{nm}^{\pm}\}\\ \sqrt{k-K};K\in\{k_{nm}^{\pm}\}\end{cases}.$$

Therefore, this function is meromorphic on the surface$K$. Calculating the residuals $\underset{k=\bar{k}}{\mathrm{Res}}\widetilde{G}_0(g,\ p,\ k)$ at the simple poles$\bar{k}\in\{k_{nm}^{\pm}\}$, we obtain nontrivial solutions of homogeneous $(\vec{\widetilde{U}}^{i}(g,\ k)\equiv 0)$ canonical $(\bar{\varepsilon}(g,\ k)\equiv 1,\ \mu(g,\ k)\equiv 1,\overline{\mathrm{int}S}=\varnothing)$ problems (53), (54) and (60), (61):

$$\vec{\widetilde{E}}\left(g,\ k_{nm}^{\pm}\right)=\{\widetilde{E}_x,\ \widetilde{E}_y,\ \widetilde{E}_z\};\quad \widetilde{E}_{x,\ y\ \text{or}\ z}$$
$$=a_{x,\ y\ \text{or}\ z}\exp\left[i(\alpha_n x+\beta_m y)\right]\quad\text{and}\quad \vec{\widetilde{H}}\left(g,\ k_{nm}^{\pm}\right)=\left(ik_{nm}^{\pm}\eta_0\right)^{-1}\mathrm{rot}\vec{\widetilde{E}}\left(g,\ k_{nm}^{\pm}\right) \tag{77}$$

where $a_{x,\, y \text{ or } z}$ are the arbitrary constants. These solutions determine free oscillations in the space stratified by the following conditions:

$$D[\vec{E}(\vec{H})](x+l_x,\, y)=e^{2\pi i\Phi_x}D[\vec{E}(\vec{H})](x,\, y),\; D[\vec{E}(\vec{H})](x,\, y+l_y)=e^{2\pi i\Phi_y}D[\vec{E}(\vec{H})](x,\, y). \qquad (78)$$

8.2. Spectrum qualitative characteristics

Let a set Ω_k of the points $\{\bar{k}_j\}_j \in K$ such that for all $k \in \{\bar{k}_j\}_j$ the homogeneous (spectral) problem

$$\begin{cases} \eta_0 \text{rot}\tilde{\vec{H}}(g,k) = -ik\bar{\varepsilon}(g,k)\tilde{\vec{E}}(g,k), \\ \text{rot}\tilde{\vec{E}}(g,k) = ik\eta_0\mu(g,k)\tilde{\vec{H}}(g,k), \quad g \in Q_L \\ D\left[\tilde{\vec{E}}\left(\tilde{\vec{H}}\right)\right](l_x,y) = e^{2\pi i\Phi_x} D\left[\tilde{\vec{E}}\left(\tilde{\vec{H}}\right)\right](0,y), \quad 0 \le y \le l_y, \quad |z| < L \\ D\left[\tilde{\vec{E}}\left(\tilde{\vec{H}}\right)\right](x,l_y) = e^{2\pi i\Phi_y} D\left[\tilde{\vec{E}}\left(\tilde{\vec{H}}\right)\right](x,0), \quad 0 \le x \le l_x, \quad |z| < L \\ \tilde{E}_{tg}(g,k)\Big|_{g\in S} = 0, \quad \tilde{H}_{nr}(g,k)\Big|_{g\in S} = 0 \quad , \end{cases} \qquad (79)$$

$$\begin{aligned} \begin{Bmatrix} \tilde{E}_z(g,\, k) \\ \tilde{H}_z(g,\, k) \end{Bmatrix} &= \sum_{n,m=-\infty}^{\infty} \begin{Bmatrix} A_{nm(E)}(k) \\ A_{nm(H)}(k) \end{Bmatrix} e^{i\Gamma_{nm}(z-L)} \mu_{nm}(x,\, y),\; g \in \bar{A} \\ \begin{Bmatrix} \tilde{E}_z(g,\, k) \\ \tilde{H}_z(g,\, k) \end{Bmatrix} &= \sum_{n,m=-\infty}^{\infty} \begin{Bmatrix} B_{nm(E)}(k) \\ B_{nm(H)}(k) \end{Bmatrix} e^{-i\Gamma_{nm}(z+L)} \mu_{nm}(x,\, y),\; g \in \bar{B} \end{aligned} \qquad (80)$$

has a nontrivial (not necessarily unique) solution $\vec{U}(g,\, \bar{k}_j)=\{\vec{E}(g,\, \bar{k}_j),\, \vec{H}(g,\, \bar{k}_j)\}$be called the point spectrum of the grating. It is obvious that these solutions characterize the so-called free oscillations, whose field pattern, structure of their spatial harmonics and behavior of these harmonics for large $|z|$ and t are determined by the value of $\bar{k}_j=\text{Re}\bar{k}_j+i\text{Im}\bar{k}_j$ and by a position of the point $\bar{k}_j$ (the eigen frequency associated with a free oscillation$\vec{U}(g,\, \bar{k}_j)$) on the surface K [4,6,7]. By continuing analytically the problems (53), (54) and (60), (61) together with their solutions $\vec{U}(g,\, k)=\{\vec{E}(g,\, k),\, \vec{H}(g,\, k)\}$ into the domain K of the complex-valued k, we detect poles of the function $\vec{U}(g,\, k)$at the points$k=\bar{k}_j$. In the vicinity of these poles, the desired solutions can be represented by the Loran series in terms of the local on K variable κ [17]. The analytical findings of this kind may form the basis for detailed study of physical features of resonant wave scattering by one-dimensionally and two-dimensionally periodic structures [4,6,7,18,19].

Derive now the conditions that constrain existence of nontrivial solutions of the problem (79), (80). These conditions can be considered as uniqueness theorems for the problems (53), (54) and (60), (61) formulated for different domains of the surface$K$. Notice that the study of the uniqueness allows one to estimate roughly a domain where elements of the set Ω_k are localized and simplify substantially the subsequent numerical solution of spectral problems owing to reduction of a search zone of the eigen frequencies. The uniqueness theorems serve also as a basis for application of the 'meromorphic' Fredholm theorem [20] when constructing well grounded algorithms for solving diffraction problems as well as when studying qualitative characteristics of gratings' spectra [4,7].

Assume that grating scattering elements are nondispersive ($\varepsilon(g, k)=\varepsilon(g)$, $\mu(g, k)=\mu(g)$, and $\sigma(g, k)=\sigma(g)$). In this case, the analytical continuation of the spectral problem (79), (80) into the domain of complex-valued k are simplified considerably. From the complex power theorem in the integral form formulated for the nontrivial solutions $\vec{U}(g, \bar{k}_j)$ like

$$\begin{aligned}\oint_{S_L}([\vec{E}\times\vec{H}^*]\cdot d\vec{s}) &= \int_{Q_L}\operatorname{div}[\vec{E}\times\vec{H}^*]dg \\ &= ik\eta_0\int_{Q_L}\mu|\vec{H}|^2dg - \frac{ik^*}{\eta_0}\int_{Q_L}\varepsilon|\vec{E}|^2dg - \int_{Q_L}\sigma|\vec{E}|^2dg\end{aligned} \qquad (81)$$

the following relations result:

$$\begin{aligned}&\sum_{n,m=-\infty}^{\infty}\frac{1}{\lambda_{nm}^2}\begin{Bmatrix}(\mathrm{Re}\Gamma_{nm}\mathrm{Re}k+\mathrm{Im}\Gamma_{nm}\mathrm{Im}k)\\(\mathrm{Im}\Gamma_{nm}\mathrm{Re}k-\mathrm{Re}\Gamma_{nm}\mathrm{Im}k)\end{Bmatrix}[(|A_{nm(E)}|^2+|B_{nm(E)}|^2)\\ &\pm\eta_0^2(|A_{nm(H)}|^2+|B_{nm(H)}|^2)]=\frac{1}{\varepsilon_0}\begin{Bmatrix}-\mathrm{Im}k(V_3+V_2)-V_1\\\mathrm{Re}k(V_3-V_2)\end{Bmatrix}\end{aligned} \qquad (82)$$

Notation: $k=\bar{k}_j$, $\vec{E}=\vec{E}(g, \bar{k}_j)$, $\Gamma_{nm}=\Gamma_{nm}(\bar{k}_j)$, $A_{nm(E)}=A_{nm(E)}(\bar{k}_j)$, etc., and

$$V_1=\varepsilon_0\eta_0\int_{Q_L}\sigma|\vec{E}|^2dg,\ V_2=\int_{Q_L}\varepsilon_0\varepsilon|\vec{E}|^2dg, V_3=\int_{Q_L}\mu_0\mu|\vec{H}|^2dg.$$

No free oscillations exist whose amplitudes do not satisfy equations (82). From this general statement, several important consequences follow. Below some of them are formulated for gratings with $\varepsilon(g)>0$, $\mu(g)>0$, and $\sigma(g)\geq 0$.

- There are no free oscillations whose eigen frequencies $\bar{k}_j$ are located on the upper half-plane (Imk0) of the first sheet of the surface K. This can be verified by taking into account the upper relation in (82), the function $\Gamma_{nm}(k)$ on C_k, and the inequalities $V_1\geq 0$, $V_2>0$, $V_3>0$.
- If $\sigma(g)\equiv 0$ (the grating is non-absorptive), no free oscillations exist whose eigen frequencies $\bar{k}_j$ are located on the bottom half-plane (Im$k<0$) of the sheet C_k between the cuts corresponding to the least absolute values of $k_{nm}^{\pm}$. In Figure 6, this region of the first sheet of K and the above-mentioned domain are shaded by horizontal lines.

- If $\sigma(g)>0$ on some set of zero-measure points$g\in Q_L$, then there are no elements $\bar{k}_j$ of grating's point spectrum Ω_k that are located on the real axis of the planeC_k.

Investigation of the entire spectrum of a grating, i.e. a set of the points$k\in K$, for which the diffraction problems given by (53), (54) and (60), (61) are not uniquely solvable, is a complicated challenge. Therefore below we do no more than indicate basic stages for obtaining well grounded results. The first stage is associated with regularization of the boundary value problem describing excitation of a metal-dielectric grating by the currents $\tilde{\vec{J}}(g, k)\leftrightarrow \vec{J}(g, t)$ located in the domainQ_L :

$$\begin{cases} \eta_0 \mathrm{rot}\tilde{\vec{H}}(g,k) = -ik\bar{\varepsilon}(g,k)\tilde{\vec{E}}(g,k) + \tilde{\vec{J}}(g,k), \\ \mathrm{rot}\tilde{\vec{E}}(g,k) = ik\eta_0\mu(g,k)\tilde{\vec{H}}(g,k), \quad g\in Q_L \\ D\left[\tilde{\vec{E}}\left(\tilde{\vec{H}}\right)\right](l_x,y) = e^{2\pi i\Phi_x} D\left[\tilde{\vec{E}}\left(\tilde{\vec{H}}\right)\right](0,y), \quad 0\le y\le l_y, \quad |z|<L \\ D\left[\tilde{\vec{E}}\left(\tilde{\vec{H}}\right)\right](x,l_y) = e^{2\pi i\Phi_y} D\left[\tilde{\vec{E}}\left(\tilde{\vec{H}}\right)\right](x,0), \quad 0\le x\le l_x, \quad |z|<L \\ \tilde{E}_{tg}(g,k)\Big|_{g\in S} = 0, \quad \tilde{H}_{nr}(g,k)\Big|_{g\in S} = 0 \ , \end{cases} \tag{83}$$

$$\begin{aligned} \begin{Bmatrix} \tilde{E}_z(g, k) \\ \tilde{H}_z(g, k) \end{Bmatrix} &= \sum_{n,m=-\infty}^{\infty} \begin{Bmatrix} A_{nm(E)}(k) \\ A_{nm(H)}(k) \end{Bmatrix} e^{i\Gamma_{nm}(z-L)} \mu_{nm}(x, y), \ g\in\bar{A}, \\ \begin{Bmatrix} \tilde{E}_z(g, k) \\ \tilde{H}_z(g, k) \end{Bmatrix} &= \sum_{n,m=-\infty}^{\infty} \begin{Bmatrix} B_{nm(E)}(k) \\ B_{nm(H)}(k) \end{Bmatrix} e^{-i\Gamma_{nm}(z+L)} \mu_{nm}(x, y), \ g\in\bar{B}. \end{aligned} \tag{84}$$

By regularization is meant (see, for example, [7]) a reduction of the boundary value electrodynamic problem to the equivalent operator equation of the second kind

$$[E + B(\tilde{G}_0, S, \bar{\varepsilon}, \mu, k)]X = Y, \ EX = X \tag{85}$$

with a compact (in some space W of vector fields) finite-meromorphic (in local on Kvariablesκ) operator-function $B(\tilde{G}_0, S, \bar{\varepsilon}, \mu, k)$ [20,21]. If the problem given by (83), (84) is considered separately for metal gratings ($\mathrm{int}\overline{S}\neq\varnothing$and S are sufficiently smooth surfaces; $\bar{\varepsilon}(g, k)=\mu(g, k)\equiv 1$) and dielectric gratings ($\mathrm{int}\overline{S}=\varnothing$, $\bar{\varepsilon}(g, k)=\varepsilon(g)$and $\mu(g, k)=\mu(g)$ are sufficiently smooth functions), then its regularization can be performed by applying the potential theory methods [4,7,22].

In the second stage, the following statements should be proved: (i) the resolvent $[E + B(k)]^{-1}$ ($k\in K$) of the problem in (85) is a finite-meromorphic operator-function; (ii) its poles are located at the points $k=\bar{k}_j$ ($j=1,2,3,...$); (iii) the entire spectrum coincides with its point spectrumΩ_k; (iv) $\Omega_k$is nothing more than a countable set without finite accumulation points. All these statements are corollaries of the 'meromorphic' Fredholm theorem [4,20,21] and the uniqueness theorem proved previously.

By inverting homogeneous operator equation (85), we can construct a numerical solution of the spectral problem given by (79), (80) [4,6], in other words, calculate the complex-valued eigen frequencies $\bar{k}_j$ and associated eigen waves $\vec{U}(g,\bar{k}_j)=\{\vec{E}(g,\bar{k}_j),\vec{H}(g,\bar{k}_j)\}$ or free oscillations of an open two-dimensionnaly periodic resonator. Commonly, this operation is reduced to an approximate solution of the characteristic equation like:

$$\det[C(k)]=0. \tag{86}$$

Here $C(k)$ is some infinite matrix-function; the compactness of the operator $B(k)$ ensures (i) existence of the determinant $\det[C(k)]$ and (ii) the possibility to approximate the solutions $\bar{k}$ of equation (86) by the solutions $\bar{k}(N)$ of the equation $\det[C(k,N)]$ with the matrix $C(k,N)$ reduced to dimension$N\times N$.

Let $\bar{k}$ be a root of characteristic equation (86) that do not coincide with any pole of the operator-function$B(k)$. The multiplicity of this root determines the multiplicity of the eigen value $\bar{k}$ of homogeneous operator equation (85), i.e. the value $M=M(1)+M(2)+...+M(Q)$ [21]. Here, $Q$is the number of linearly-independent eigen functions$\vec{U}^{(q)}(g,\bar{k})$; $q=1,2,...,Q$(the number of free oscillations) corresponding to the eigen value (eigenfrequency)$\bar{k}$, while $M(q)-1$ is the number of the associated functions$\vec{U}_{(m)}^{(q)}(g,\bar{k})$;$m=1,2,...,M(q)-1$. The order of pole of the resolvent $[E+B(k)]^{-1}$ (and of the Green function $\tilde{G}(g,p,k)$ of the problem in (83), (84)) for $k=\bar{k}$is determined by a maximal value of$M(q)$.

9. Conclusion

The analytical results presented in the chapter are of much interest in the development of rigorous theory of two-dimensionally periodic gratings as well as in numerical solution of the associated initial boundary value problems. We derived exact absorbing boundary conditions truncating the unbounded computational space of the initial boundary value problem for two-dimensionally periodic structures to a bounded part of the Floquet channel. Some important features of transient and steady-state fields in rectangular parts of the Floquet channel were discussed. The technique for calculating electrodynamic characteristics of multi-layered structure consisting of two-dimensionally periodic gratings was developed by introducing the transformation operators similar to generalized scattering matrices in the frequency domain. In the last section, the elements of spectral theory for two-dimensionally periodic gratings were discussed.

Author details

L. G. Velychko[1], Yu. K. Sirenko[1] and E. D. Vinogradova[2*]

*Address all correspondence to: elena.vynogradova@mq.edu.au

1 Usikov Institute of Radiophysics and Electronics, National Academy of Sciences of Ukraine, Kharkov, Ukraine

2 Macquarie University, Department of Mathematics, North Ryde, Australia

References

[1] Shestopalov, V. P., Litvinenko, L. N., Masalov, S. A., & Sologub, V. G. (1973). Wave Diffraction by Gratings. Kharkov: Kharkov State Univ. Press.

[2] Petit, R. editor (1980). Electromagnetic Theory of Gratings. New York: Springer.

[3] Shestopalov, V. P., Kirilenko, A. A., Masalov, S. A., & Sirenko, Y. K. (1986). Resonance Wave Scattering. Vol. 1. Diffraction Gratings. Kiev: Naukova Dumka; (in Russian).

[4] Shestopalov, V. P., & Sirenko, Y. K. (1989). Dynamic Theory of Gratings. Kiev: Naukova Dumka; (in Russian).

[5] Neviere, M., & Popov, E. (2003). Light Propagation in Periodic Media: Differential Theory and Design. New York: Marcel Dekker.

[6] Sirenko, Y. K., Strom, S., & Yashina, N. P. (2007). Modeling and Analysis of Transient Processes in Open Resonant Structures. New Methods and Techniques. New York: Springer.

[7] Sirenko, Y. K., & Strom, S. editors (2010). Modern Theory of Gratings. Resonant Scattering: Analysis Techniques and Phenomena. New York: Springer.

[8] Ladyzhenskaya, O. A. (1985). The Boundary Value Problems of Mathematical Physics. New York: Springer-Verlag.

[9] Taflove, A., & Hagness, S. C. (2000). Computational Electrodynamics: the Finite-Difference Time-Domain Method. Boston: Artech House.

[10] Sirenko, K., Pazynin, V., Sirenko, Y., & Bagci, H. (2011). An FFT-Accelerated FDTD Scheme with Exact Absorbing Conditions for Characterizing Axially Symmetric Resonant Structures. *Progress in Electromagnetics Research*, 111, 331-64.

[11] Rothwell, E. J., & Cloud, M. J. (2001). Electromagnetics. New York: CRC Press.

[12] Vladimirov, V. S. (1971). Equations of Mathematical Physics. New York: Dekker.

[13] Sirenko, K. Y., & Sirenko, Y. K. (2005). Exact'Absorbing' Conditions in the Initial Boundary-Value Problems of the Theory of Open Waveguide Resonators. *Computational Mathematics and Mathematical Physics*, 45, 490-506.

[14] Kravchenko, V. F., Sirenko, Y. K., & Sirenko, K. Y. (2011). Electromagnetic Wave Transformation and Radiation by the Open Resonant Structures. Modelling and Analysis of Transient and Steady-State Processes. Moscow: FizMathLit; (in Russian).

[15] Titchmarsh, E. (1948). Introduction to the Theory of Fourier Integrals. Oxford: Clarendon Press.

[16] Gradshteyn, I. S., & Ryzhik, I. M. (1994). Table of Integrals, Series, and Products. New York: Academic.

[17] von Hurwitz, A., & von Courant, R. (1964). Allgemeine Funktionentheorie und Elliptische Funktionen: Geometrische Funktionentheorie. Berlin: Springer-Verlag; (in German).

[18] Sirenko, Y. K., Velychko, L. G., & Erden, F. (2004). Time-Domain and Frequency-Domain Methods Combined in the Study of Open Resonance Structures of Complex Geometry. *Progress in Electromagnetics Research*, 44, 57-79.

[19] Velychko, L. G., Sirenko, Y. K., & Shafalyuk, O. S. (2006). Time-Domain Analysis of Open Resonators. Analytical Grounds. *Progress in Electromagnetics Research*, 61, 1-26.

[20] Reed, M., & Simon, B. (1978). Methods of Modern Mathematical Physics. IV: Analysis of Operators. New York: Academic Press.

[21] Hokhberg, I. Z., & Seagul, Y. I. (1971). Operator Generalization of the Theorem about Logarithmic Residue and the Rouche Theorem. *Matematicheskii Sbornik*, 84, 607-629, (in Russian).

[22] Colton, D., & Kress, R. (1983). Integral Equation Methods in Scattering Theory. New York: Wiley-Interscience.

Chapter 6

The Sources Reconstruction Method for Antenna Diagnostics and Imaging Applications

Yuri Álvarez, Fernando Las-Heras and Cebrián García

Additional information is available at the end of the chapter

http://dx.doi.org/10.5772/50744

1. Introduction

The advances on communications systems require more challenging antenna designs, increasing the complexity of design, manufacturing, and testing stages. Different sources of error may affect the antenna final prototype, degrading its performance, and then requiring a re-design step to identify and correct these manufacturing errors. In general, this re-design step requires invasive techniques based on trial-error procedure, whose technical and economical costs are directly related to the antenna complexity. For this reason, the development of fast and accurate non-invasive methods for antenna faults detection (antenna diagnostics) has been of great interest in the last decade.

This chapter describes the Sources Reconstruction Method (SRM), an integral equation-based inverse technique, and its applications on antenna diagnostics, near-field to far-field transformation, and the recent advances on imaging applications. The method provides an electromagnetic model of the problem-under-study based on an equivalent currents distribution, which can be used to evaluate the field radiated at any point of the space. The equivalent model can be used for antenna diagnostics applications in the field of antenna measurements, as well as for geometry reconstruction purposes in the case of scattering acquisitions.

1.1. Antenna Diagnostics

Antenna diagnostics comprises these techniques based on the recovery of the extreme near-field of the Antenna-Under-Test (AUT) from the field radiated by it. Extreme near-field is defined as the field in the proximity of the AUT (i.e. a distance < 1 wavelength (λ), or on the AUT aperture plane). The field radiated by the AUT can be acquired in the near or far field region (Yaghian, 1986).

INTECH
open science | open minds

© 2012 Álvarez et al.; licensee InTech. This is an open access article distributed under the terms of the Creative Commons Attribution License (http://creativecommons.org/licenses/by/3.0), which permits unrestricted use, distribution, and reproduction in any medium, provided the original work is properly cited.

The extreme near-field is strongly related to the physical currents (in the case of metallic surfaces) or aperture field distribution of the AUT. In consequence, an accurate retrieval of this field provides a useful mapping of the fields and currents distribution on the manufactured antenna, which can be compared to the initial design to identify which parts of the antenna have faults or differences with respect to the designed one.

1.1.1. Backpropagation techniques

It is well-known that the AUT radiation pattern can be calculated as the Fourier Transform of the fields on the AUT aperture plane (Yaghian, 1986). In consequence, this calculation can be reverted in order to recover the AUT aperture fields from the radiation pattern. The simplest case relates a planar field acquisition domain with the AUT aperture plane. For example, in (Kaplan et al., 1979; Rahmat-Samii, 1984) the fields acquired on a plane are backpropagated to the aperture plane by using the inverse Fast Fourier transform (iFFT). Polar and bi-polar measurement setups are also suitable for this kind of technique (Gatti & Rahmat-Samii, 1988; Yaccarino et al., 1994), which has been extended to phaseless measurements (Yaccarino & Rahmat-Samii, 1999; Razavi & Rahmat-Samii, 2007). Equivalent current techniques have been also developed (Petre & Sarkar, 1992), showing similar performance to the iFFT-based methods.

Next step was the aperture field calculation from the field acquired in spherical domains, which is one of the most extended antenna measurement systems (Yaghian, 1986). A formulation relating spherical wave modes with the AUT plane wave spectrum has been developed in (Cappellin et al., 2008). Also equivalent current methods have been extended to this measurement setup, providing the same accuracy as wave mode based techniques (Álvarez et al., 2008c).

It is out of doubt that fastest diagnostics results are given by iFFT-based methods. So for many years, the computational cost of equivalent current techniques, derived from a kind of inverse Method-of-Moments (MoM), was prohibitive for practical antenna diagnostics. However, the increase of antenna design complexity requires the extreme near-field to be recovered not only in the AUT aperture plane, but also in a surface fitting the AUT geometry. The need of this requirement, together with the improvement of calculation resources fostered the development of equivalent current techniques.

1.1.2. Three-dimensional techniques

Wave mode-based techniques are limited to canonical geometries (mainly planar, cylindrical, and spherical), for both reconstruction domain (e.g. extreme near-field) and field acquisition domain. Equivalent current methods for antenna diagnostics are able to overcome this limitation, so that arbitrary geometry domains can be defined. The only limitation is that the extreme near-field surface must enclose all the radiating elements.

Equivalent current methods provide an accurate electromagnetic model of the AUT, that has been mainly used for antenna diagnostics, as for example, detection of faulting elements in antenna arrays (Álvarez et al., 2009b), reflector antenna distortions (Álvarez et al., 2008c; Las-

http://dx.doi.org/10.5772/50744

Heras et al., 2009), leakage identification (Foged et al., 2011a), and parasitic currents in antenna elements (Araque & Vecchi, 2009). In addition to the equivalent current techniques, it has been proposed the use of electric and magnetic infinitesimal dipoles that model the AUT electromagnetic behaviour (Sijher & Kishk, 2005; Mikki & Kishk, 2007). The orientation and excitation of these infinitesimal dipoles is determined from the acquired near-field distribution.

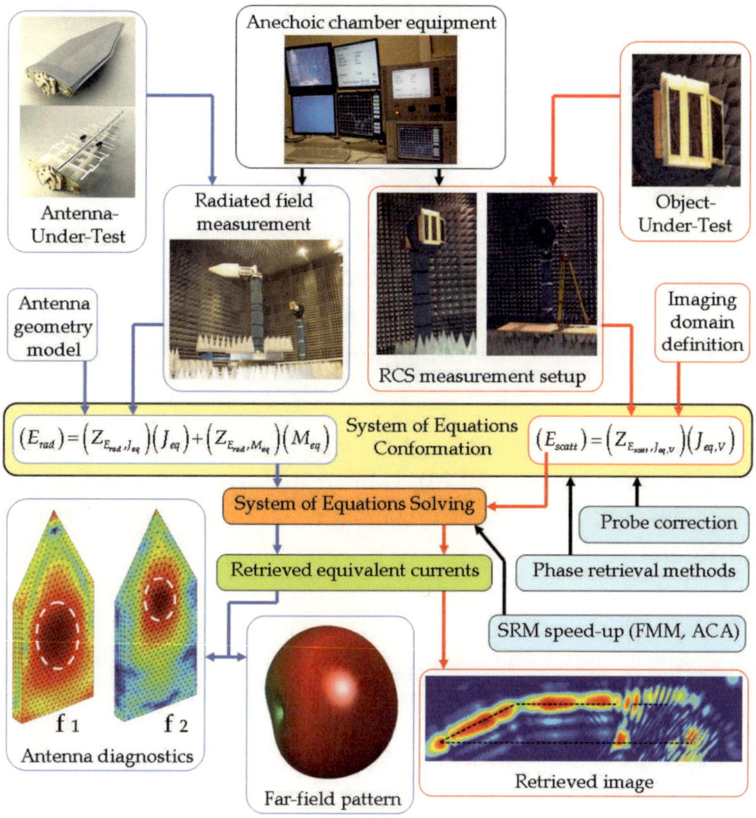

Figure 1. General scheme showing the equivalent current (or sources reconstruction) method for antenna diagnostics and geometry reconstruction.

Apart from antenna diagnostics, these techniques have potential applications in other antenna measurement related topics: near-field to far-field transformation (that is, antenna radiation pattern calculation) (Las-Heras et al., 2006; Álvarez et al., 2008a; Álvarez et al., 2007; Álvarez et al., 2012b), analysis of the AUT interaction with scatterers (Las-Heras, 2001; García et al., 2011), filtering of parasitic elements (Araque & Vecchi, 2009; Cano et al., 2010) and electromagnetic emissions analysis (Laviada et al., 2010; Hernando et al., 2008; Persson & Gustafsson, 2005; Eibert & Schmidt, 2009). Due to its wide field of applications, some antenna measurement companies have even released their own antenna diagnostics software, as for example (Foged et al., 2011b) and (Cappellin et al., 2011).

Equivalent current methods have been extended to cope with phaseless measurements (Las-Heras & Sarkar, 2002; Álvarez et al., 2010c; Álvarez et al., 2012b) which are of special interest on sub-mm and THz frequency bands, where phase measurements require expensive equipment (Hellicar et al., 2009; Hislop & Hellicar, 2009).

1.2. Inverse scattering methods based on equivalent Currents

Equivalent currents provide an equivalent model of the AUT that can be also extrapolated to scattering problems. In this case, the scatterer is illuminated by an incident field, inducing a set of currents on the object. Thus, if the incident and scattered fields are known, it would be possible to recover an equivalent currents distribution related to the scatterer geometry. For metallic objects the idea is simple: the equivalent currents intensity will be related to the regions of the space where the physical induced currents are flowing.

Several inverse scattering techniques are based upon this idea (Habashy et al., 1990; Caorsi et al., 1990; Qin & Ciric, 1993; Van den Berg, Kleimann, 1997; Lin & Kiang, 1996; Álvarez et al., 2010b). The main limitation is the need of volumetric equivalent currents domains, which significantly increases the calculation time. Computational acceleration schemes can be applied to reduce it.

To sum up this section, general schemes of the equivalent current techniques, also referred as Sources Reconstruction Method (SRM), is presented in Fig. 1, together with its applications to antenna diagnostics and characterization, near-field to far-field transformation, and profile reconstruction.

2. The Sources Reconstruction Method

The concept of the Sources Reconstruction Method (SRM) is illustrated in Fig. 2. On the left, the direct radiation problem is plotted: the field radiated by the AUT (E_{rad}) is collected on the acquisition domain S_{obs}. Now, the SRM goal is to recover the extreme near-field on a surface S' enclosing the AUT. The field samples (E_{rad}) are *backpropagated* from S_{obs} to S' (Fig. 2, right) obtaining an electromagnetic equivalent currents distribution, J_{eq}, M_{eq}. These equivalent currents radiate the same electromagnetic fields as the AUT outside the domain enclosed by S'; in other words, the currents provide an *equivalent* electromagnetic model of the AUT.

2.1. The electromagnetic equivalence principle

The SRM is based on the electromagnetic Equivalence Principle (Harrington, 2001; Stratton, 1941; Balanis, 1997): consider the fields (E_n, H_n) and currents (J_n, M_n) defined in Fig. 3 (a) where two media are present: an inner medium (n= 1) with constitutive parameters ε_1, μ_1, σ_1, defined in the volume V', and bounded by S'; and the outer medium (n=2), with constitutive parameters ε_2, μ_2, σ_2.

http://dx.doi.org/10.5772/50744

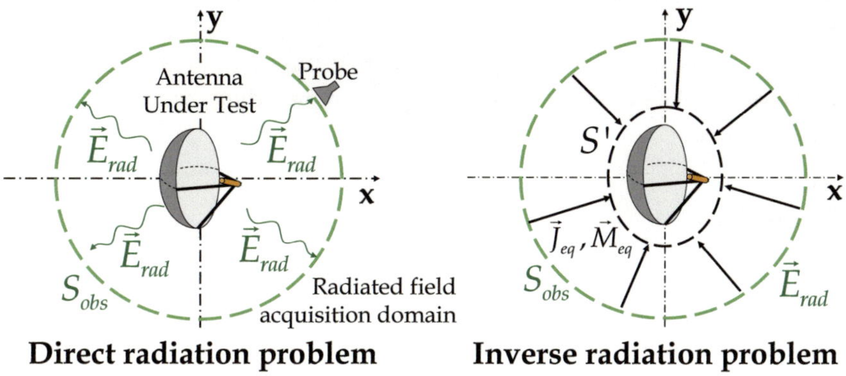

Figure 2. Direct and inverse radiation problem.

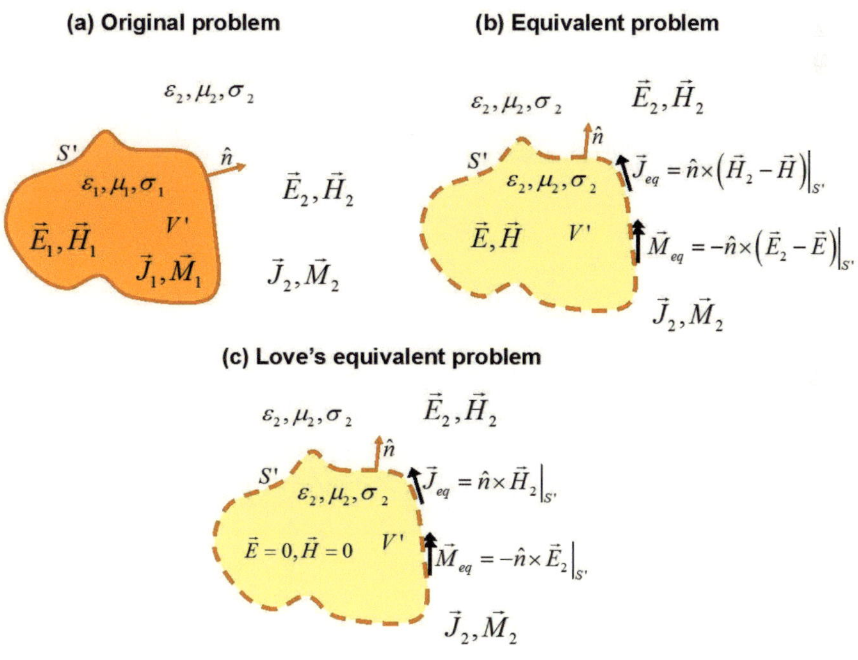

Figure 3. Electromagnetic Equivalence Principle.

In Fig. 3 (b), a set of equivalent electric (J_{eq}) and magnetic currents (M_{eq}) is defined on the boundary between the two media (S'). These currents are defined as the gap between the tangential components of the fields in media 1,2. Note that no assumption about the fields and currents inside S' is considered, just imposing the currents to radiate the same fields outside S′ as in Fig. 3 (a). In consequence, J_{eq} and M_{eq} characterize the electromagnetic behaviour *outside* S′ of the fields (E_1, H_1) and currents (J_1, M_1) enclosed by S′.

Several SRM implementations (Foged et al., 2011a; Araque & Vecchi, 2009; Cappellin et al., 2011; Araque & Vecchi, 2010) make use of the Love's Equivalence Principle: in this case, the fields inside S′ are forced to be zero (Fig. 3 (c)), so that the equivalent currents on S′ are directly the field tangential components. In the case of antenna diagnostics, they correspond to the extreme near-fields.

2.2. Integral Equations

The Sources Reconstruction Method (SRM) is based on the integral equations relating the electromagnetic field ($\boldsymbol{E}_{\mathbf{rad}}$) observed in S_{obs} with their sources. In the case of the SRM, the sources are the equivalent currents $\boldsymbol{J}_{\mathbf{eq}}$ and $\boldsymbol{M}_{\mathbf{eq}}$ defined on S′ (Fig. 2 (b)).

The integral equations relating the electric field with an equivalent electric $\boldsymbol{J}_{\mathbf{eq}}(\mathbf{r}')$ and magnetic $\boldsymbol{M}_{\mathbf{eq}}(\mathbf{r}')$ currents distribution defined on a surface S' are given by Eqs. (1) and (2) respectively (Harrington, 2001; Balanis, 1997), so that the total electric field is the sum of the electric field due to electric currents $\boldsymbol{E}_{\mathbf{rad,Jeq}}(\mathbf{r})$ and the electric field due to magnetic currents $\boldsymbol{E}_{\mathbf{rad,Meq}}(\mathbf{r})$.

$$\vec{E}_{rad,J_{eq}}(\vec{r}) = -\frac{j\eta_0}{4\pi k_0}\int_{S'}\left\{k_0^2\frac{e^{-jk_0R(\vec{r};\vec{r}')}}{R(\vec{r};\vec{r}')}\vec{J}_{eq}(\vec{r}')+\nabla\left(\nabla\cdot\left(\frac{e^{-jk_0R(\vec{r};\vec{r}')}}{R(\vec{r};\vec{r}')}\vec{J}_{eq}(\vec{r}')\right)\right)\right\}dS' \tag{1}$$

$$\vec{E}_{rad,M_{eq}}(\vec{r}) = -\frac{1}{4\pi}\int_{S'}\nabla\times\left(\frac{e^{-jk_0R(\vec{r};\vec{r}')}}{R(\vec{r};\vec{r}')}\vec{M}_{eq}(\vec{r}')\right)dS' \tag{2}$$

where k_0 is the wavenumber, η_0 the intrinsic impedance, $\boldsymbol{r}$ the position vector of the points where the field is acquired (surface S_{obs}), and $\boldsymbol{r}'$ the position vector of the points where the currents are reconstructed, which belong to the S' domain. $R(\boldsymbol{r};\boldsymbol{r}')$ is defined as $|\boldsymbol{r} - \boldsymbol{r}'|$.

2.3. Numerical solution

Integral equations (1), (2) have to be discretized for a numerical solution. The field observation domain (surface S_{obs}) is sampled in N points. The equivalent currents domain (S') is discretized in M planar facets (e.g. triangular facets or Rao-Wilton-Glisson basis functions (Rao et al., 1982) when modelling arbitrary 3D geometries).

Equivalent currents vectors are expressed as a linear combination of two perpendicular components defined on each facet $\Delta S'_m$. These components are referred to an orthogonal coordinate system defined on each facet using the unitary vectors $\hat{u}_m$, $\hat{v}_m$, $\hat{n}_m$ (see Fig. 4).

The definition of the components is given by Eq. (3), assuming that the equivalent currents are constant on each facet. This approach is, in general, valid, provided that $\Delta S'_m \ll (\lambda)^2$.

http://dx.doi.org/10.5772/50744

$$\vec{J}_{eq,m} = J_{eq,u_m}\hat{u}_m + J_{eq,v_m}\hat{v}_m$$
$$\vec{M}_{eq,m} = M_{eq,u_m}\hat{u}_m + M_{eq,v_m}\hat{v}_m \tag{3}$$

The discretized integral equations as a function of indexes m, n, are described by (4), (5). All the constants and variables are known except the complex coefficients $J_{eq,um}$, $J_{eq,vm}$, $M_{eq,um}$, $M_{eq,vm}$.

$$\vec{E}_{rad,J_{eq}}(\vec{r}_n) = -\frac{j\eta_0}{4\pi k_0}\sum_{m=1}^{M}\left\{\begin{array}{l}\left(k_0^2\dfrac{e^{-jk_0R(\vec{r}_n;\vec{r}_m')}}{R(\vec{r}_n;\vec{r}_m')} + \nabla\left(\nabla\cdot\left(\dfrac{e^{-jk_0R(\vec{r}_n;\vec{r}_m')}}{R(\vec{r}_n;\vec{r}_m')}\right)\right)\right)\cdot \\ \left(J_{eq,u_m}\hat{u}_m + J_{eq,v_m}\hat{v}_m\right)\Delta S_m'\end{array}\right\} \tag{4}$$

$$\vec{E}_{rad,M_{eq}}(\vec{r}_n) = -\frac{1}{4\pi}\sum_{m=1}^{M}\left\{\nabla\times\left(\frac{e^{-jk_0R(\vec{r}_n;\vec{r}_m')}}{R(\vec{r}_n;\vec{r}_m')}\left(M_{eq,u_m}\hat{u}_m + M_{eq,v_m}\hat{v}_m\right)\right)\Delta S_m'\right\} \tag{5}$$

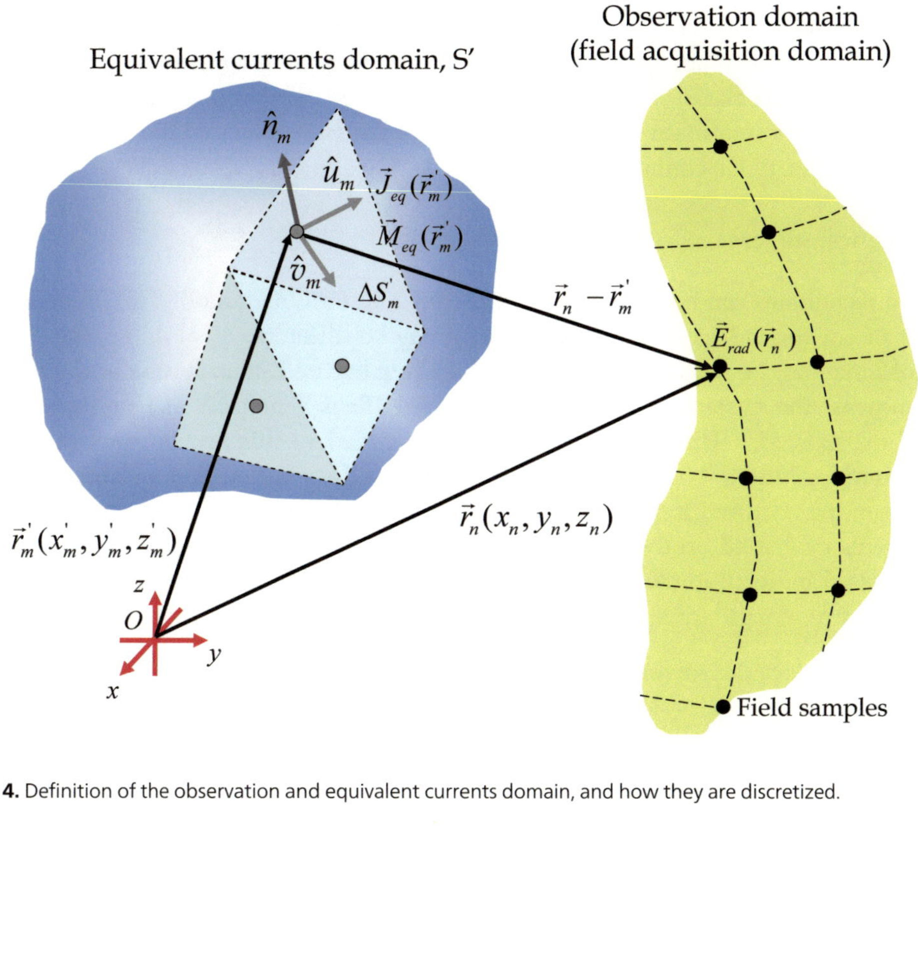

Figure 4. Definition of the observation and equivalent currents domain, and how they are discretized.

The discretized equations can be rewritten in a more compact form, resulting in the matrix system (6). The unknown coefficients of the equivalent currents correspond to the right column vector of (6), the field samples correspond to the independent column vector, and the rest of variables and coefficients of Eqs. (4) and (5) are grouped in a matrix called Z-matrix.

$$\begin{pmatrix} E_{rad}^{t_1} \\ E_{rad}^{t_2} \end{pmatrix} = \begin{pmatrix} Z_{E_{rad}^{t_1};J_{eq,u}} & Z_{E_{rad}^{t_1};J_{eq,v}} & Z_{E_{rad}^{t_1};M_{eq,u}} & Z_{E_{rad}^{t_1};M_{eq,v}} \\ Z_{E_{rad}^{t_2};J_{eq,u}} & Z_{E_{rad}^{t_2};J_{eq,v}} & Z_{E_{rad}^{t_2};M_{eq,u}} & Z_{E_{rad}^{t_2};M_{eq,v}} \end{pmatrix} \begin{pmatrix} J_{eq,u} \\ J_{eq,v} \\ M_{eq,u} \\ M_{eq,v} \end{pmatrix} \tag{6}$$

The field samples on S_{obs} can be expressed by means of 3 orthogonal components: $\boldsymbol{E}_{\mathbf{rad}} = E_{rad,t1}\boldsymbol{t}_1 + E_{rad,t2}\boldsymbol{t}_2 + E_{rad,n}\boldsymbol{n}$. The majority of antenna measurement systems only acquire the tangential field components, as the normal component decay much faster than the tangential ones. Thus, only the tangential field components ($E_{rad,t1}$, $E_{rad,t2}$) are used to recover the equivalent currents. These tangential components can be referred to the corresponding canonical coordinate systems. For example, in the case of cylindrical acquisition system, $\boldsymbol{t}_1 = \boldsymbol{\phi}$, $\boldsymbol{t}_2 = \boldsymbol{z}$. In spherical system: $\boldsymbol{t}_1 = \boldsymbol{\theta}$, $\boldsymbol{t}_2 = \boldsymbol{\varphi}$.

The resulting linear system of equations has 2xN equations and 4xM unknowns. Sometimes, it is possible to use just one type of equivalent currents if the reconstruction domain S' is a flat surface (Petre & Sarkar, 1992; Las-Heras et al., 2006; Sarkar & Taaghol, 1999). In this case, the observation domain is restricted to one of the two halves of the space divided by the virtual infinite plane containing S'.

2.4. Numerical solvers

Different techniques can be used to solve the linear system of equations (6). In general, the number of equations (N_e) and unknowns (M_u) may be different; $N_e \neq M_u$, i.e. the system of equations may not be square. In addition, depending on the Z-matrix rank and the number of unknowns, the system can be overdetermined (*Rank*(Z-matrix)>M_u) or undetermined (*Rank*(Z-matrix)< M_u). The numerical solvers considered in this section seek for the least mean squares solution, that is, the equivalent currents distribution that minimizes the error (ε) between the acquired field samples and the field radiated by the reconstructed equivalent currents evaluated on the field sampling points. The issue of the solution accuracy is also discussed in detail in (Araque & Vecchi, 2010), analyzing the system conditioning and discussing several regularization techniques.

Table 1 summarizes the solvers that have been tested. Regarding computational cost considerations, $N \approx M \approx N_e \approx M_u$ is assumed[1].

1 Notation: N denotes the number of field samples, and N_e the number of equations. M defines the number of equivalent currents facets, and M_u the number of unknowns.

Singular Value Decomposition (SVD) calculates the singular values of the Z-matrix (Álvarez et al., 2009b; Araque & Vecchi, 2010), providing a valuable analysis about its conditioning: if the problem is ill-conditioned (high condition number) (Sarkar et al., 1981b), that means that the solution of the problem (equivalent currents) is quite sensitive to small changes in the inputs (i.e. the field samples). Lower singular values can be neglected (Truncated-SVD) (Álvarez et al., 2009b; Araque & Vecchi, 2010), improving the Z-matrix conditioning, and then stabilizing the solution. The main drawback of this technique is the moderate computational cost ($O(N^3)$) as the SVD requires the calculation of the pseudoinverse of the Z-matrix.

Method	Type	Time O(N)	Memory O(N)
Singular Value Decomposition (SVD)	Matrix inversion	$O(N^3)$	$O(N^2)$
Conjugate Gradient (GC)	Iterative	$O(N^2)$	$O(N^2)$
Memory Saving Technique (MST) (1)	Iterative (1)	$O(N^2)$	$O(N)$
Fast Multipole Method (FMM) (1)	Iterative (1)	$O(N^{1.5})$	$O(N^{1.5})$
Adaptive Cross Approximation algorithm (ACA) (1)	Iterative (1)	$O(N^{4/3} \log N)$	$O(N^{4/3} \log N)$

(1) These techniques are implemented over the CG method.

Table 1. Numerical techniques and computational cost.

A more affordable implementation in terms of calculation time relies on iterative solvers. Among different techniques, the Conjugate Gradient (CG) (Sarkar & Arvas, 1981a) has been found to provide a good trade-off between calculation time $O(N^2)$ and accuracy, and in particular, the implemented CG algorithm (Wang & Hwang, 1995) has a monotonically decrease of the error (ε). Iterative solvers require the definition of stopping rules or criteria. In this case, the first rule is the definition of a maximum number of iterations, K. The second rule is the variation of the error ($\Delta\varepsilon$) between two consecutive iterations: $\Delta\varepsilon(k) = \varepsilon(k-1)-\varepsilon(k)$. If this variation is smaller than a fixed threshold, the CG finishes. Typical threshold values can be $\Delta\varepsilon = 0.001 - 0.01$ (Álvarez et al., 2009b; López et al., 2012). In order to provide a reference, convergence is usually reached in no more than $K = 20$ iterations with $\varepsilon < 0.05$ for the given $\Delta\varepsilon$ threshold values in practical antenna measurements.

Electrically large antenna diagnostics may require a significant amount of calculation time. To reduce it, different computational acceleration schemes have been applied over the CG formulation. The first one was the Fast Multipole Method (FMM) (Eibert & Schmidt, 2009; Engheta et al., 1992; Coifman et al., 1993): in brief, the idea is to approximate field and source samples which belong to a certain region in terms of plane and spherical waves. In general, FMM is applied to problems in which all the regions belong to the same physical domain, so interactions between adjacent regions have to be analyzed without the FMM approach. In the case of the SRM, the source (equivalent currents) and field samples belong to different physical domains (Fig. 2) (Eibert & Schmidt, 2009; Álvarez et al., 2008b), avoiding the presence of adjacent regions.

Another tested acceleration technique was the Adaptive Cross Approximation (ACA) algorithm. While the FMM takes advantage of the physical properties of the problem (i.e. the fields and currents approach in terms of plane and spherical waves), the ACA explodes the algebraic properties of the Z-matrix (Álvarez et al., 2009a). Assuming that the Z-matrix is rank deficient, then, it can be approximated as Z-matrix = U x V. Considering that the Z-matrix has N_e rows and M_u columns, then U has N_e rows and R columns, and V, R rows and M_u columns, where R is defined as the effective rank of the Z-matrix (Zhao et al., 2005). ACA is an efficient acceleration technique both in time and memory provided that $(N_e+M_u)\ R \ll N_e x\ M_u$ (Zhao et al., 2005; Álvarez et al., 2009a).

Finally, the inner CG implementation allows a quite efficient memory saving technique at the expense of increasing the calculation time. This method relies on the fact that the Z-matrix elements can be calculated using mathematical expressions (4), (5) if the indexes m,n are known. In consequence, instead of calculating and storing the entire Z-matrix, the Z-matrix elements are calculated every time they are needed, reducing the memory consumption from $O(N^2)$ to $O(N)$. As an example, consider a problem with similar number of field samples and facets, $N_e \sim M_u \sim 100{,}000$. The Z-matrix storage would require 150 GB of memory. The memory saving technique requires less than 100 MB of memory (Álvarez et al., 2007; López et al., 2012).

Graphics Processing Units (GPU) is the ultimate technology for hardware acceleration in those problems demanding intensive calculations. The above mentioned memory saving technique fulfils these requirements for GPU implementation, providing a speed-up greater than 70 times with respect to previous CPU implementation (Álvarez et al., 2009b; Álvarez et al., 2007). As an example, an antenna diagnostics problem with ~20,000 field samples and ~20,000 equivalent currents facets is solved in just 30 s (López et al., 2012).

2.5. Extension to inverse scattering problems

2.5.1. Inverse scattering methods for geometry reconstruction

The SRM extension to inverse scattering problems can be classified as an inverse source problem technique (Habashy et al., 1990; Caorsi et al., 1990; Qin & Ciric, 1993; Van den Berg, Kleimann, 1997; Álvarez et al., 2010b), that pursuits the geometry and constitutive parameters estimation by retrieving a set of equivalent sources. While the geometry reconstruction problem yields a linear system of equations, constitutive parameters retrieval involves a non-linear system of equations, increasing the inverse problem complexity.

Focusing on the geometry reconstruction case, other techniques have been developed. One is the Linear Sampling Method (Colton et al., 2003), which determines if a point on the region of interest belongs to the object-under-test or not. Its interest lies on its simplicity, as just a linear system of equations has to be solved.

Other methods are based on contour parameterization: starting from an initial guess, the parameterized geometry is modified until one that satisfies an imposed boundary condition (e.g. zero tangential field in case of perfect electric conductors (PEC) or tangential field continuity on dielectric surfaces) is found. The resulting system of equations relating the contour coordinates and the scattered field is non-linear, and it is necessary to use non-linear optimization methods to solve it (Lin & Kiang, 1996; Farmahini, 2009; Martínez et al., 2011; Hajihashemi & El-Shenawee, 2011). As the entire contour is optimized at the same time, these inversion methods usually require a remarkable amount of calculation time. Additionally, as the cost function to be optimized is non-linear, the inversion process can easily get trapped in local minima.

2.5.2. The Volumetric SRM

As indicated before, the analysis of the extreme near-field provides information about the physical currents flowing on the metallic structure of the antenna. For example, consider the case depicted in Fig. 5: the equivalent currents recovered on the radome enclosing the Yagi-Uda antenna elements provide information about their position inside the radome.

Taking into account this fact, it has been considered the possibility of extending the SRM for inverse scattering problems (Volumetric SRM) (Álvarez et al., 2010b); in particular, for metallic objects geometry reconstruction. The idea is to retrieve the electric currents induced by an incident field. As in the case of the Yagi-Uda antenna in Fig. 5, it is expected that the highest intensity of the reconstructed equivalent currents will be related to the placement of the physical induced currents.

The direct and inverse scattering problem using equivalent currents are depicted in Fig. 6. The main difference with respect to the SRM for antenna diagnostics is the use of a volumetric equivalent currents distribution ($\mathbf{J}_{\mathbf{eq,V}}$): now, the reconstruction domain is a volume V′ enclosing the metallic scatterers, so the field radiated by the reconstructed equivalent currents $\mathbf{J}_{\mathbf{eq,V}}$ is the same as the field scattered by the original geometry (Volumetric Equivalence Principle (Livesay & Chen, 1974)). The integral equation relating the scattered field ($\mathbf{E}_{\mathbf{scatt}}$) with the volumetic equivalent currents ($\mathbf{J}_{\mathbf{eq,V}}$) is (7) (Álvarez et al., 2010a; Álvarez et al., 2010b).

$$\vec{E}_{scatt}(\vec{r}) = -\frac{j\eta_0}{4\pi k_0}\int_{V'}\left\{k_0^2\frac{e^{-jk_0R(\vec{r};\vec{r}')}}{R(\vec{r};\vec{r}')}\vec{J}_{eq,V}(\vec{r}')+\nabla\left(\nabla\cdot\left(\frac{e^{-jk_0R(\vec{r};\vec{r}')}}{R(\vec{r};\vec{r}')}\vec{J}_{eq,V}(\vec{r}')\right)\right)\right\}dV' \qquad (7)$$

One of the main drawbacks of the Volumetric SRM is the problem dimensionality, which is increased from $O(N^2)$ to $O(N^3)$. For electrically large scatterers, the domain V′ may be too large to be analyzed in a reasonable time even with a powerful workstation. To overcome this drawback, an inverse Fast Multipole Method (iFMM) has been developed in (Álvarez et al., 2011) (2D) and (Álvarez et al., 2012a) (3D) allowing a significant speed-up (up to 500 times) without compromising the accuracy of the retrieved geometry. Even more, the use of a Dyadic formulation (7) provides an advantage with respect to scalar formulations (López-Sánchez & Fortuny-Guash, 2000) as polarization information can be efficiently used to improve the quality of the reconstruction.

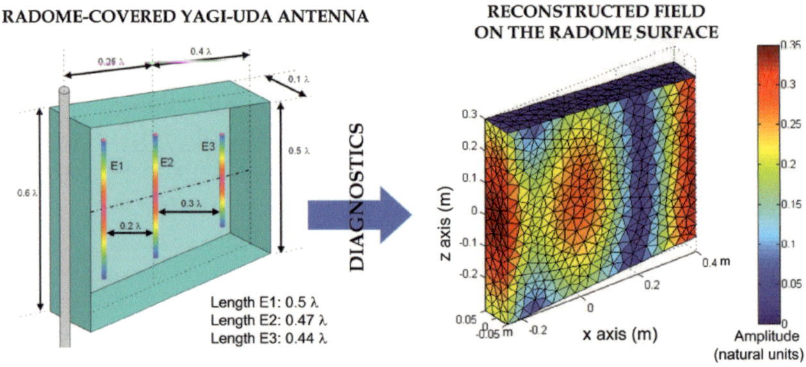

Figure 5. Antenna diagnostics of a 3-element Yagi-Uda antenna covered by a radome.

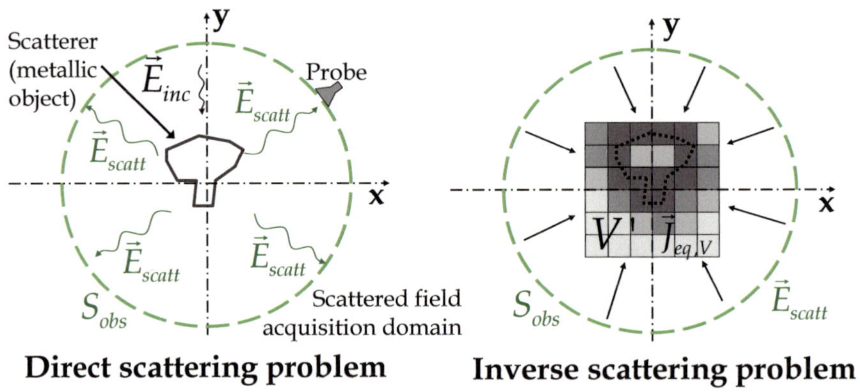

Figure 6. Direct and inverse scattering problems using equivalent currents.

3. Application examples

3.1. Antenna array diagnostics

The first application example is devoted to clearly show the antenna diagnostics capabilities of the SRM. For this purpose, a 3x3 antenna array of $\lambda/2$ dipoles has been designed and simulated. Nominal excitations are plotted in Fig. 7. In order to simulate a failure in two antenna array elements, dipoles 2 and 4 are distorted, reducing their feeding from 0 dB to -6 dB and -30 dB respectively. A Method-of-Moments code is used to analyze the antenna and to evaluate the radiation pattern, which is depicted in Fig. 7: it is clear that the distortion introduced in a two elements array antenna degrades the radiation pattern, but no information about the malfunctioning elements can be extracted from the pattern.

The SRM is applied to retrieve an equivalent currents distribution on the antenna array aperture plane. The reconstructed equivalent currents will be related to the extreme near-field and hence, to the nominal excitations of the antenna, so any failure in the array elements can be identified.

The working frequency is $f = 300$ MHz. The antenna size is defined in terms of the minimum sphere enclosing the array (R_{min}), which is R_{min} = ~1.5 m, so according to (Hansen, 1988; Yaghian, 1986) the minimum number of radiation pattern samples for this antenna is N=441: $\theta = [-90° +90°]$, with $\Delta\theta = 9°$, and $\varphi = [0° 180°]$, with $\Delta\varphi = 9°$. The equivalent currents are reconstructed in a planar domain S′ = (3 x 3 m)², discretized in M = 441 facets (ΔS′ = (0.15 x 0.15 m)²) in order to have, for this particular case, the same number of field samples and equivalent currents.

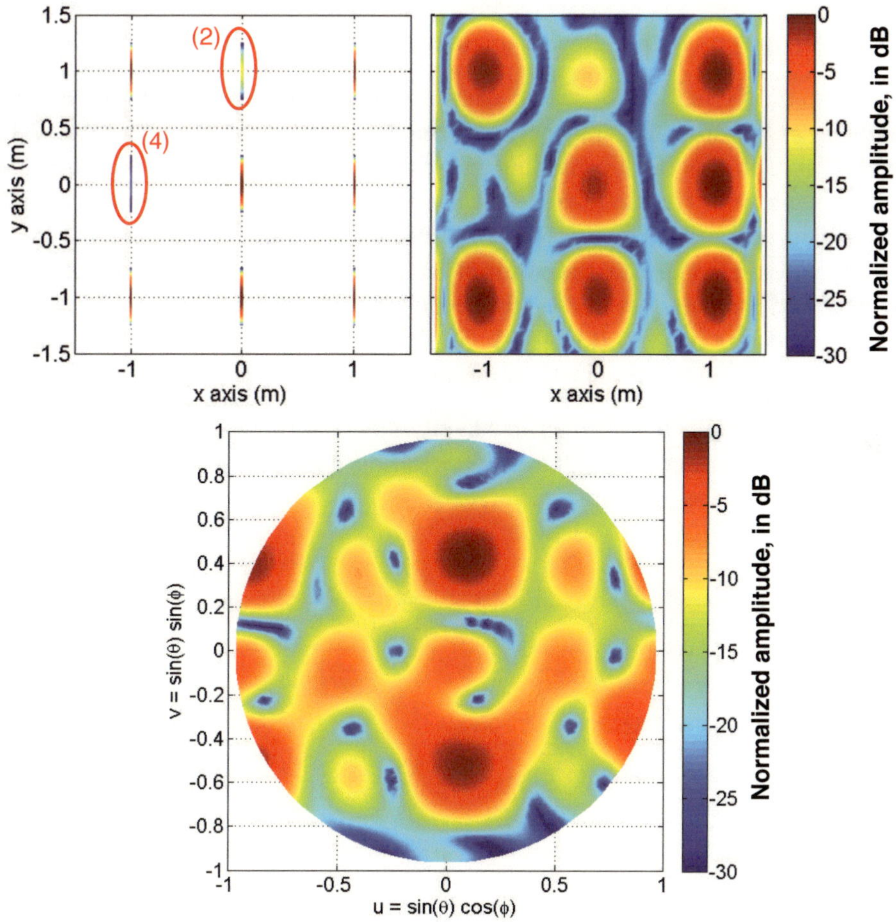

Figure 7. Array of dipoles. Top left: nominal excitations, in which dipoles 2 and 4 are distorted. Top right: reconstructed $M_{eq,x}$. Bottom: array of dipoles radiation pattern.

The application of Image Theory for this problem allows neglecting equivalent electric currents in the formulation, using just magnetic currents. The integral equation relating the spherical field tangential components E_θ, E_φ, with the magnetic currents components $M_{eq,u}$, $M_{eq,v}$ ($M_{eq,x}$, $M_{eq,y}$, for the particular case) can be expressed as a linear system of equations to be solved (8):

$$\begin{pmatrix} E_{rad}^{\theta} \\ E_{rad}^{\varphi} \end{pmatrix} = \begin{pmatrix} Z_{E_{rad}^{\theta};M_{eq,x}} & Z_{E_{rad}^{\theta};M_{eq,y}} \\ Z_{E_{rad}^{\varphi};M_{eq,x}} & Z_{E_{rad}^{\varphi};M_{eq,y}} \end{pmatrix} \begin{pmatrix} M_{eq,x} \\ M_{eq,y} \end{pmatrix} \tag{8}$$

The system has N_e=441 x 2 equations, and M_u=441 x 2 unknowns. As indicated before, one of the main drawbacks of inverse problems is the intrinsic ill-conditioning. The condition number of the Z-matrix in (8) has been calculated, and it is 3.9 x 10^{25}, or in other words, this is the relation between the highest and lowest singular values, which are plotted in Fig. 8.

The linear system of equations is solved using the Truncated SVD (T-SVD) (Álvarez et al., 2009b; Araque & Vecchi, 2010) and the Conjugate Gradient (CG) (Álvarez et al., 2009b; Álvarez et al., 2007), both of them providing the same results. In the case of the T-SVD, different thresholds are selected: the idea is that, by filtering the lowest singular values, the Z-matrix condition number decreases, then improving the matrix conditioning. With respect to the CG, the metric to evaluate the solution accuracy is based on the error between the simulated field and the field radiated by the reconstructed equivalent currents. The stopping criterion is set to $\Delta\varepsilon = \varepsilon\,(k\text{-}1) - \varepsilon\,(k) < 0.001$. For this problem, convergence was reached after 19 iterations ($\varepsilon = 0.019$), requiring just one minute of calculation in a conventional laptop.

To conclude this example, the reconstructed equivalent magnetic currents ($M_{eq,x}$) are plotted in Fig. 7. It is observed that the 9 elements (including the 2 distorted dipoles) are correctly identified, predicting also the amplitude on each.

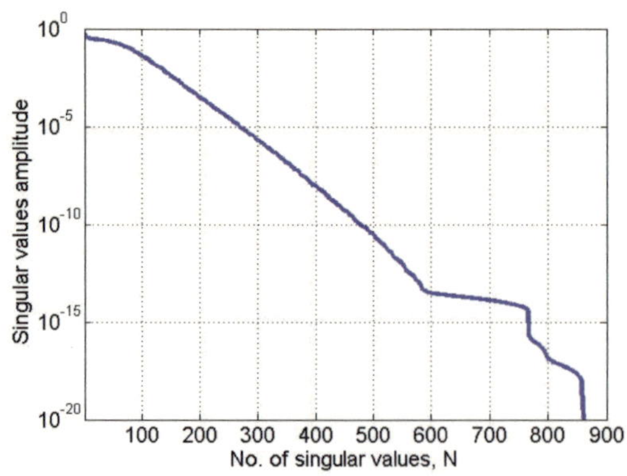

Figure 8. Singular values of the Z-matrix appearing in (8).

3.2. Phaseless measurements

The following application example is focused on antenna diagnostics and near-field to far-field transformation using phaseless information. The AUT is a logarithmic-periodic antenna working in the 400 MHz-3 GHz frequency band (Álvarez et al., 2010c). This antenna has been measured at the spherical range in anechoic chamber of the University of Oviedo, at two different distances: $R_1 = 3.20$ m, and $R_2 = 5.25$ m (see Fig. 9), and at the working frequency of $f = 1{,}030$ MHz. The field has been sampled on a full sphere: $\theta = [-180° \; 180°)$, $\Delta\theta = 4°$, and $\varphi = [0° \; 180°]$, $\Delta\varphi = 4°$, resulting in N=4,140 field samples on each acquisition sphere.

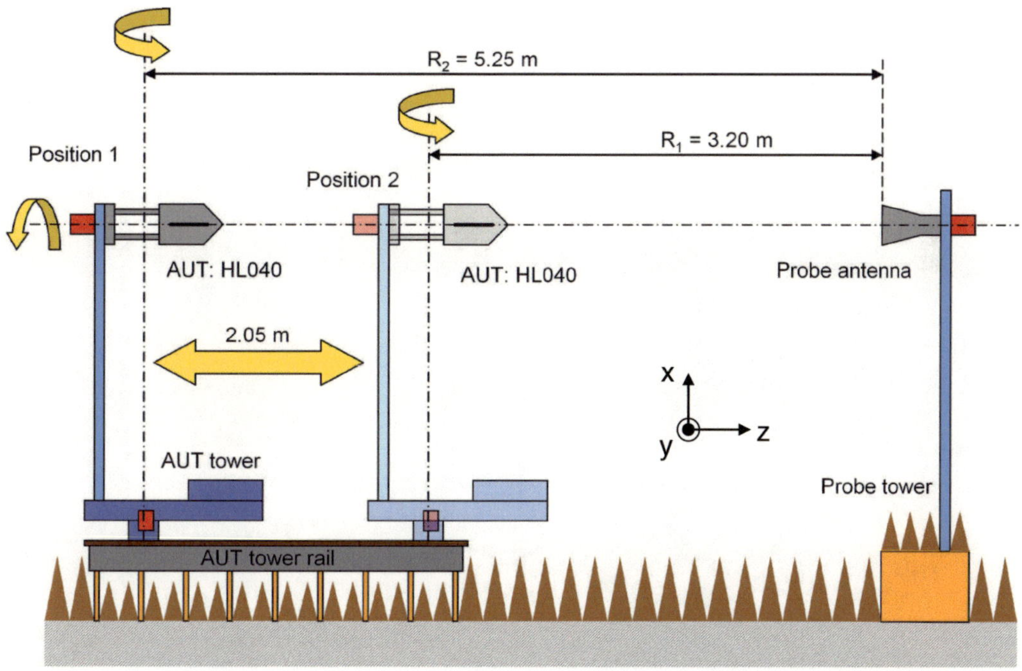

Figure 9. Antenna measurement at spherical range in anechoic chamber at two different distances.

The equivalent currents reconstruction domain is chosen to be a surface S′ fitting the log-periodic antenna radome (see Fig. 10, (a),(b)). This surface is discretized into M=3,298 triangular facets. In consequence, taking into account the number of field samples on each sphere and the number of facets, the number of equations is N_e=4,140 x 2 field components x 2 measurement distances = 16,560, and the number of unknowns, M_u = 3,298 x 2 type of currents x 2 current components = 13,192.

Two different kinds of problems are now solved. First, the system of equations relating the field samples (amplitude and phase) with the equivalent currents (9) is solved. The retrieved currents are plotted in Fig. 10 (d),(f): it is observed that for the selected working frequency, the resonant elements are placed in the middle of the log-periodic antenna (Álvarez et al., 2010c; Álvarez et al., 2009b). As the elements are x-polarized (Fig. 10 (b)), only the $J_{eq,x}$ component is plotted. Next, from the reconstructed equivalent currents, the field at any point of the space can be calculated, and hence, far field radiation pattern of the log-periodic antenna which is plotted in Fig. 11 (dark and gray solid lines).

$$\begin{pmatrix} E^{\theta}_{rad,R_1} \\ E^{\varphi}_{rad,R_1} \\ E^{\theta}_{rad,R_2} \\ E^{\varphi}_{rad,R_2} \end{pmatrix} = \begin{pmatrix} Z_{E^{\theta}_{rad,R_1};J_{eq,u}} & Z_{E^{\theta}_{rad,R_1};J_{eq,v}} & Z_{E^{\theta}_{rad,R_1};M_{eq,u}} & Z_{E^{\theta}_{rad,R_1};M_{eq,v}} \\ Z_{E^{\varphi}_{rad,R_1};J_{eq,u}} & Z_{E^{\varphi}_{rad,R_1};J_{eq,v}} & Z_{E^{\varphi}_{rad,R_1};M_{eq,u}} & Z_{E^{\varphi}_{rad,R_1};M_{eq,v}} \\ Z_{E^{\theta}_{rad,R_2};J_{eq,u}} & Z_{E^{\theta}_{rad,R_2};J_{eq,v}} & Z_{E^{\theta}_{rad,R_2};M_{eq,u}} & Z_{E^{\theta}_{rad,R_2};M_{eq,v}} \\ Z_{E^{\varphi}_{rad,R_2};J_{eq,u}} & Z_{E^{\varphi}_{rad,R_2};J_{eq,v}} & Z_{E^{\varphi}_{rad,R_2};M_{eq,u}} & Z_{E^{\varphi}_{rad,R_2};M_{eq,v}} \end{pmatrix} \begin{pmatrix} J_{eq,u} \\ J_{eq,v} \\ M_{eq,u} \\ M_{eq,v} \end{pmatrix} \tag{9}$$

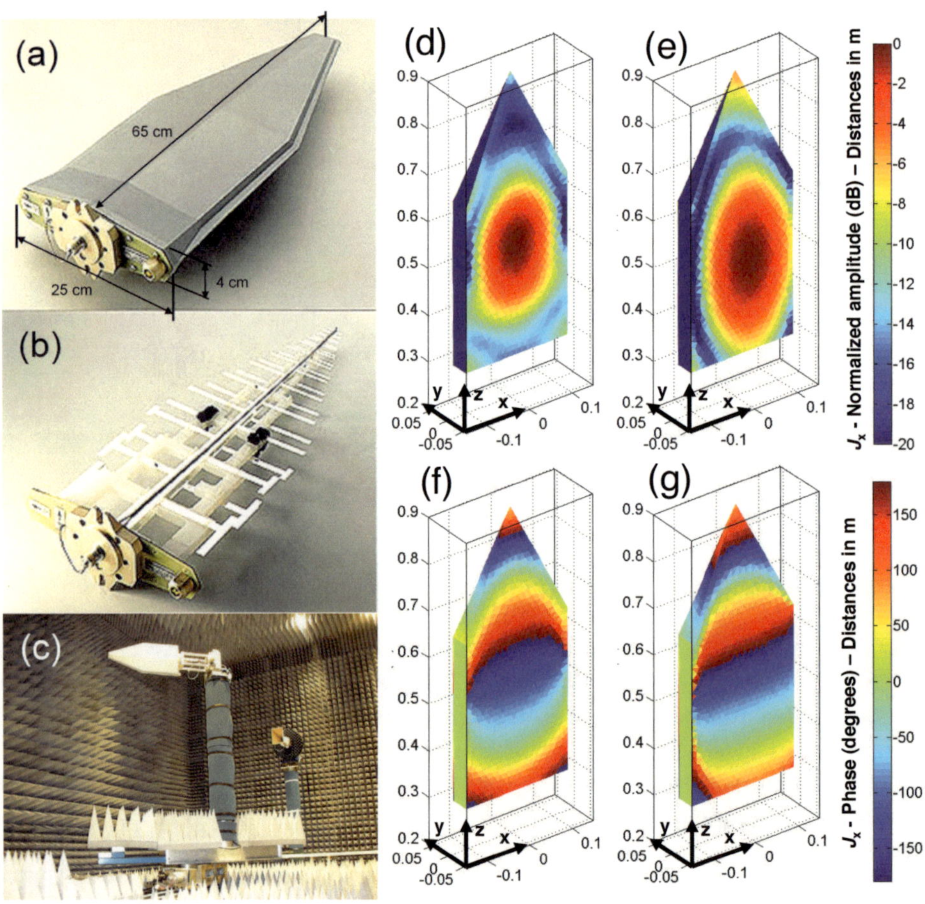

Figure 10. Photos: (a) log-periodic antenna array with radome. (b) Without radome. (c) Measurement at spherical range in anechoic chamber. Reconstructed equivalent currents, $J_{eq,x}$ component. (d),(f) using amplitude and phase information. (e),(g) using amplitude-only information.

The second problem to be solved neglects the phase information of the field samples. A non-linear system of equations, relating the amplitude of the field samples with the amplitude of the field radiated by the equivalent currents (10), has to be solved.

$$\begin{pmatrix} \left|E^{\theta}_{rad,R_1}\right| \\ \left|E^{\varphi}_{rad,R_1}\right| \\ \left|E^{\theta}_{rad,R_2}\right| \\ \left|E^{\varphi}_{rad,R_2}\right| \end{pmatrix} = \left|\begin{pmatrix} Z_{E^{\theta}_{rad,R_1};J_{eq,u}} & Z_{E^{\theta}_{rad,R_1};J_{eq,v}} & Z_{E^{\theta}_{rad,R_1};M_{eq,u}} & Z_{E^{\theta}_{rad,R_1};M_{eq,v}} \\ Z_{E^{\varphi}_{rad,R_1};J_{eq,u}} & Z_{E^{\varphi}_{rad,R_1};J_{eq,v}} & Z_{E^{\varphi}_{rad,R_1};M_{eq,u}} & Z_{E^{\varphi}_{rad,R_1};M_{eq,v}} \\ Z_{E^{\theta}_{rad,R_2};J_{eq,u}} & Z_{E^{\theta}_{rad,R_2};J_{eq,v}} & Z_{E^{\theta}_{rad,R_2};M_{eq,u}} & Z_{E^{\theta}_{rad,R_2};M_{eq,v}} \\ Z_{E^{\varphi}_{rad,R_2};J_{eq,u}} & Z_{E^{\varphi}_{rad,R_2};J_{eq,v}} & Z_{E^{\varphi}_{rad,R_2};M_{eq,u}} & Z_{E^{\varphi}_{rad,R_2};M_{eq,v}} \end{pmatrix} \begin{pmatrix} J_{eq,u} \\ J_{eq,v} \\ M_{eq,u} \\ M_{eq,v} \end{pmatrix}\right| \tag{10}$$

The problem of antenna diagnostics using phaseless field samples has been addressed by different approaches. For example (Yaccarino & Rahmat-Samii, 1999; Razavi & Rahmat-Samii, 2007) introduces a plane-to-plane iterative backpropagation method for phase retrieval, that can be used for both near-field to far-field transformation and antenna diagnostics applications. Phaseless SRM is described in (Las-Heras & Sarkar, 2002): in this case, an equivalent magnetic currents distribution is calculated by minimizing a cost function which relates the amplitude of the measured field and the contribution due to the equivalent currents, similarly to Eq. (10). However, these techniques are restricted to flat reconstruction surfaces S′, whereas the problem to be solved comprises a 3D reconstruction domains S′. Thus, in order to solve Eq. (10), the two-stage algorithm described in (Álvarez et al., 2010c; Álvarez et al., 2012b) is applied.

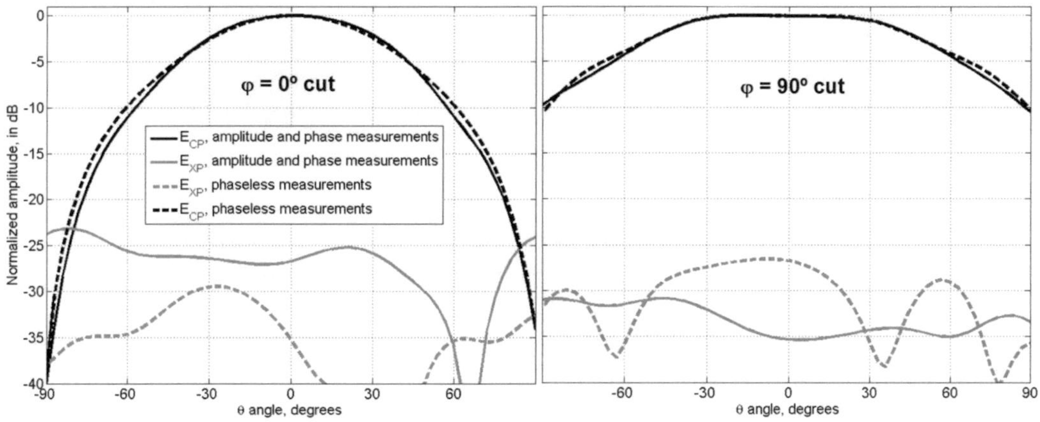

Figure 11. Log-periodic antenna far field pattern. Cuts φ = 0° and φ = 90°. Comparison of the results when considering amplitude and phase field samples, and phaseless field samples.

The reconstructed equivalent currents from the phaseless field samples are plotted in Fig. 10 (e),(g), observing a good agreement with the ones retrieved from amplitude and phase information. It is important to remark that not only the amplitude, but also the phase of the equivalent currents (Fig. 10, (f),(g)) are in good agreement, proving the phaseless SRM for phase retrieval.

The latest step is the calculation of the log-periodic antenna radiation pattern from the reconstructed equivalent currents. The comparison between the radiation pattern from amplitude and phase information (Fig. 11, solid lines) and using phaseless information (Fig. 11, dashed lines) shows an acceptable agreement in the copolar component (black line). With regard to the crosspolar component, the amplitude levels are in the same order of magnitude (20-25 dB below the copolar component).

3.3. Geometry reconstruction applications

The latest example illustrates the SRM application for geometry reconstruction of metallic scatterers. A 2D problem is considered (Álvarez et al., 2010b) in which the object-under-test (depicted in Fig. 12, left plot) is supposed to have translation symmetry along one axis. The object is illuminated by 12 TM-polarized plane waves coming from 0° to 330° in 30°-steps. Four working frequencies are considered: 300, 450, 700 and 1,000 MHz. The scattered field is observed on a circumference of R = 10 m, sampled every 6°, which results in 60 field samples. Finally, noise is added according to a Signal-to-Noise (SNR) radio of 30 dB.

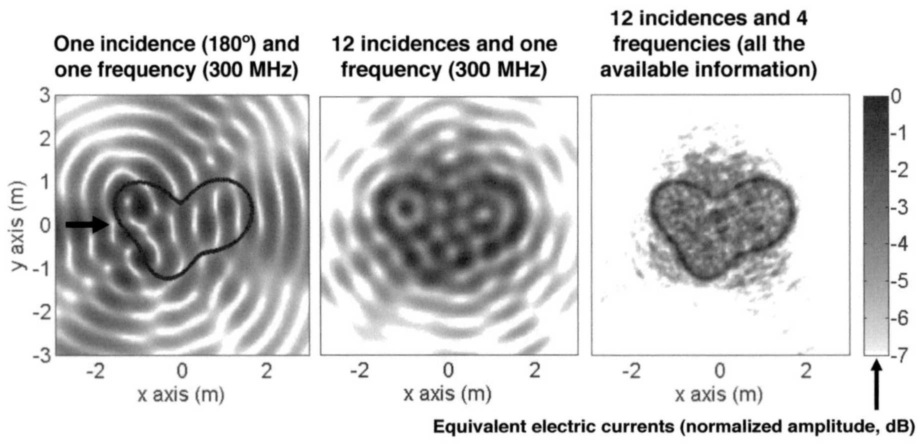

Figure 12. Geometry reconstruction application example. The solid line on the left plot indicates the true profile of the object-under-test.

The reconstruction domain V' is a square region of 6 x 6 m discretized into 121 x 121 square subdomains. Thus, for each incident wave and frequency, the system of equations to be solved has N_e=60 equations, and M_u=14,641 unknowns. The Conjugate Gradient algorithm is used, requiring about K=9-11 iterations to reach convergence ($\varepsilon < 0.01$, $\Delta\varepsilon < 0.001$).

First, the inverse problem for one frequency (300 MHz) and one incidence (180°) is solved. The reconstructed equivalent electric currents are plotted in Fig. 12 (left). The highest intensity of the reconstructed equivalent currents is located on the boundary of the object-under-test and, in particular, the part of the contour illuminated by the plane wave.

Combination of multiple incidences yields to a complete retrieval of the profile of the object-under-test. For example, the center plot of Fig. 12 represents the combination of the equivalent currents reconstructed for every incident plane wave at 300 MHz. The object-under-test profile can be now guessed. Higher resolution is provided by combining multifrequency information: the right plot of Fig. 12 clearly shows a well-defined profile.

4. Conclusion

The Sources Reconstruction Method for antenna diagnostics, near-field to far-field transformation, and geometry reconstruction applications has been described. The interest on this non-invasive technique is supported by several features which, in summary, are:

i) Establishment of an accurate equivalent electromagnetic model of the AUT that can be used to evaluate the electromagnetic field at any point of the space.

ii) The capability of handling arbitrary field acquisition and extreme near-field reconstruction domains, overcoming other diagnostics techniques limitation to canonical domains.

iii) Reliability and accuracy as the method uses full-wave integral equation formulation.

iv) The possibility of implementation on low cost GPU cards, enabling fast calculation even in conventional computers as those ones that can be found in antenna measurement facilities. Low memory footprint is also kept thanks to a developed memory saving technique (López et al., 2012).

v) Extension to phaseless measurements, which is of special interest for sub-mm and THz systems in which phase measurements are especially complex and expensive.

Acknowledgements

This work has been supported by the European Union under COST Action IC1102 (VISTA); by the "Ministerio de Ciencia e Innovación" of Spain/FEDER under projects CONSOLIDER-INGENIO CSD2008-00068 (TERASENSE), TEC2011-24492/TEC (iSCAT), IPT-2011-0951-390000 (TECNIGRAF), and under PhD grant BES-2009-024060; and by the "Gobierno del Principado de Asturias" PCTI-FEDER under project PC10-06 (FLEXANT).

Author details

Yuri Álvarez[1*], Fernando Las-Heras[1] and Cebrián García[1]

*Address all correspondence to: yurilope@gmail.com

1 Área de Teoría de la Señal y Comunicaciones – Universidad de Oviedo, Spain

References

[1] Álvarez, Y., Las-Heras, F., & Pino, M. R. (2007). Reconstruction of Equivalent Currents Distribution Over Arbitrary Three-Dimensional Surfaces Based on Integral Equation Algorithms. *IEEE Transactions on Antennas and Propagation*, 54, 3460-3468.

[2] Álvarez, Y., Las-Heras, F., & Pino, M. R. (2008). On the Comparison between the Spherical Wave Expansion and the Sources Reconstruction Method. *IEEE Transactions on Antennas and Propagation*, 56(10), 3337-3341.

[3] Alvarez, Y., Las-Heras, F., Pino, M. R., & López, J. (2008). Acceleration of the Sources Reconstruction Method via the Fast Multipole Method. *2008 IEEE International Symposium on Antennas and Propagation (APS'08)*, San Diego, CA (USA) 5-12 July 2008.

[4] Alvarez, Y., Cappellin, C., Las-Heras, F., & Breinbjerg, O. (2008). On the comparison of the Spherical Wave Expansion-to-Plane Wave Expansion and the Sources Reconstruction Method for Antenna Diagnostics. *Progress In Electromagnetics Research*, PIER 87, 245-262.

[5] Alvarez, Y., Las-Heras, F., & Pino, M. R. (2009). Application of the Adaptive Cross Approximation Algorithm to the Sources Reconstruction Method. *3rd European Conference on Antennas and Propagation (EUCAP'09)*, Berlin (Germany), 23-27 March 2009, 1-5.

[6] Álvarez, Y., Las-Heras, F., Pino, M. R., & Sarkar, T. K. (2009). An Improved Super-Resolution Source Reconstruction Method. *IEEE Transactions on Instrumentation and Measurement*, 58(11), 3855-3866.

[7] Álvarez, Y., Casas, B. A., García, C., & Las-Heras, F. (2010). The Volumetric Sources Reconstruction Method and its Applications to Geometry Reconstruction. *4th European Conference on Antennas and Propagation (EUCAP'10)*, Barcelona (Spain), 12-16 April 2010, 1-5.

[8] Álvarez, Y., Casas, B. A., García, C., & Las-Heras, F. (2010). Geometry Reconstruction of Metallic Bodies Using the Sources Reconstruction Method. *IEEE Antennas and Wireless Propagation Letters*, 9, 1197-1200.

[9] Álvarez, Y., Las-Heras, F., & Pino, M. R. (2010). The Sources Reconstruction Method for Amplitude-Only Field Measurements. *IEEE Transactions on Antennas and Propagation*, 58(8), 2776-2881.

[10] Álvarez, Y., Martínez, J. A., Las-Heras, F., & Rappaport, C. (2011). An Inverse Fast Multipole Method for Imaging Applications. *IEEE Antennas and Wireless Propagation Letters*, 10, 1259-1262.

[11] Álvarez, Y., Martínez, J. A., Las-Heras, F., & Rappaport, C. (2012). An inverse Fast Multipole Method for geometry reconstruction using scattered field information. *IEEE Transactions on Antennas and Propagation*, 60(7), 1-9, In press.

[12] Álvarez, Y., Las-Heras, F., & Pino, M. R. (2012). Antenna Diagnostics Using Phaseless NF Information. *AUTOMATIKA Journal for Control, Measurement, Electronics, Computing and Communications*, 53(1), 49-55.

[13] Araque, J., & Vecchi, G. (2009). Removal of unwanted structural interactions from antenna measurements. *Antennas and Propagation Society International Symposium, 2009*, APSURSI'09. Charleston, South Caroline, USA. June 2009.

[14] Araque, J., & Vecchi, G. (2010). Field And Source Equivalence In Source Reconstruction On 3D Surfaces. *Progress In Electromagnetics Research*, PIER 103, 67-100.

[15] Balanis, C. A. (1997). *Antenna Theory. Analysis and Design*, John Wiley & Sons.

[16] Cano, F. J., Sierra-Castañer, M., Burgos, S., & Besada, J. L. (2010). Applications of sources reconstruction techniques: Theory and practical results. *Proceedings of the Fourth European Conference on Antennas and Propagation (EuCAP 2010)*, Barcelona, Spain. 12-16 April 2010, 1-5.

[17] Caorsi, S., Gragnani, G. L., & Pastorino, M. (1990). Two-dimensional microwave imaging by a numerical inverse scattering solution. *IEEE Transactions on Microwave Theory and Techniques*, 38(8), 981-989.

[18] Cappellin, C., Frandsen, A., & Breinbjerg, O. (2008). Application of the SWE-toPWE antenna diagnostics technique to an offset reflector antenna. *IEEE Antennas and Propagation Magazine*, 50(5), 204-213.

[19] Cappellin, C., & Meincke, P. (2011). *Diatool* by TICRA. Brochure: http://www.ticra.com/products/software/diatool-antenna-diagnostics.

[20] Coifman, R., Rokhlin, V., & Wandzuraz, S. (1993). The Fast Multipole Method for the Wave Equation: A Pedestrian Prescription. *IEEE Antennas and Propagation Magazine*, 35(3), 7-11.

[21] Colton, D., Haddar, H., & Piana, M. (2003). The Linear Sampling Method in Inverse Electromagnetic Scattering Theory. *Inverse Problems*, 19, 105-137.

[22] Eibert, T. F., & Schmidt, C. (2009). Multilevel fast multipole accelerated inverse equivalent current method employing Rao-Wilton-Glisson discretization of electric and magnetic surface currents. *IEEE Transactions on Antennas and Propagation*, 57(4), 1178-1185.

[23] Engheta, N., Murphy, W. D., Rokhlin, V., & Vassiliou, M. S. (1992). The Fast Multipole Method (FMM) for Electromagnetic Scattering Problems. *IEEE Transactions on Antennas and Propagation*, 40(6), 634-641.

[24] Farmahini, M. (2009). Fast and Accurate Cascaded Particle Swarm Optimization Method for solving 2-D Inverse Scattering Problems. *Journal of Applied Computational Electromagnetic Society, ACES*, 24(5), 1-4.

[25] Foged, L. J., Scialacqua, L., Saccardi, F., Sabbadini, M., Araque, J. L., & Vecchi, G. (2011). Practical Application of the Equivalent Source Method as an Antenna Diagnostics Tool. *2011 Antenna Measurement Techniques Conference (AMTA 2011)*, Denver, Colorado (USA). 17-20 October, 1-5.

[26] Foged, L. J., Scialacqua, L., & Saccardi, . (2011). *Microwave Vision Group*, Brochure:, http://www.microwavevision.com/sites/www.microwavevision.com/files/files/plaquette_insight_BD.pdf.

[27] García, C., Álvarez, Y., Casas, B. A., & Las-Heras, F. (2011). Characterization of Antenna Interaction with Scatterers by means of Equivalent Currents. *Progress In Electromagnetics Research*, PIER 116, 185-202.

[28] Gatti, M. S., & Rahmat-Samii, Y. (1988). FFT applications to plane-polar near-field antenna measurements. *IEEE Transactions on Antennas and Propagation*, 36, 781-791.

[29] Habashy, T. M., Chow, E. Y., & Dudley, D. G. (1990). Profile inversion using the renormalized source-type integral equation approach. *IEEE Transactions on Antennas and Propagation*, 38(5), 668-682.

[30] Hajihashemi, M. R., & El-Shenawee, M. (2011). Inverse Scattering of Three-Dimensional PEC objects using the level-set method. *Progress In Electromagnetics Research PIER*, 116, 23-47.

[31] Hansen, J. E. (1988). *Spherical Near-field Antenna Measurements*, IEE Electromagnetic Waves Series 26, London: Peter Peregrinus Ltd., 10.1049/PBEW026E.

[32] Harrington, R. F. (2001). *Time-Harmonic Electromagnetic Fields*, John Wiley & Sons, 10.1109/9780470546710.

[33] Hellicar, A. D., Hanham, S. M., Hislop, G., & Du, J. (2009). Terahertz Imaging with Antenna Coupled Detectors. *Proceedings on 3rd European Conference on Antennas and Propagation (EuCAP'09)*, Berlin, 23-27 March 2009, 1-5.

[34] Hernando, M., Fernández, A., Arias, M., Rodríguez, M., Álvarez, Y., & Las-Heras, F. (2008). EMI Radiated Noise Measurements Using the Sources Reconstruction Technique. *IEEE Transactions on Industrial Electronics*, 55(9), 3258-3265.

[35] Hislop, G., Li, L., & Hellicar, A. (2009). Phase Retrieval for Millimeter- and Submillimeter-Wave Imaging. *IEEE Transactions on Antennas and Propagation*, 57(1), 286-290.

[36] Kaplan, L., Hanfling, J. D., & Borgiotti, G. V. (1979). The backward transform of the near field for reconstruction of aperture field. *IEEE Antennas and Propagation Society Symposium Digest*, 764-767.

[37] Las-Heras, F. (2001). Using Equivalent Currents to Analyze Antennas in Complex Environments. *Microwave and Optical Technology Letters*, 31(1), 62-65.

[38] Las-Heras, F., Pino, M. R., Loredo, S., Alvarez, Y., & Sarkar, T. K. (2006). Evaluating near field radiation patterns of commercial antennas. *IEEE Transactions on Antennas and Propagation*, 54, 2198-2207.

[39] Las-Heras, F., Galocha, B., & Álvarez, Y. (2009). On The Sources Reconstruction Method Application For Array And Aperture Antennas Diagnostics. *Microwave and Optical Technology Letters*, 51(7), 1664-1668.

[40] Las-Heras, F., & Sarkar, T. K. (2002). A direct optimization approach for source reconstruction and NF-FF transformation using amplitude-only data. *IEEE Transactions on Antennas and Propagation*, 50(4), 500-510.

[41] Laviada, J., Álvarez, Y., & Las-Heras, F. (2010). Efficient determination of the near-field in the vicinity of an antenna for the determination of its safety perimeter. *Progress In Electromagnetics Research*, PIER 103, 371-391.

[42] Lin, C. Y., & Kiang, Y. W. (1996). Inverse scattering for conductors by the equivalent source method. *IEEE Transactions on Antennas and Propagation*, 44(3), 310-316.

[43] Livesay, D. E., & Chen, K.-M. (1974). Electromagnetic fields induced inside arbitrarily shaped biological bodies. *IEEE Transactions on Microwave Theory and Techniques*, 22(12), 1273-1280.

[44] López, J. A., López, M., Álvarez, Y., García, C., Martínez, D., & Las-Heras, F. (2012). Fast Antenna Characterization Using The Sources Reconstruction Method on Graphics Processors. *Progress In Electromagnetics Research*, PIER 126, 185-201.

[45] Lopez-Sanchez, J. M., & Fortuny-Guash, J. (2000). 3-D Radar Imaging Using Range Migration Techniques. *IEEE Transactions on Antennas and Propagation*, 48(5), 728-737.

[46] Martinez, J. A., Gonzalez, B., Rappaport, C., Meana, J. G., & Pino, A. G. (2011). Reconstructing Distortions on Reflector Antennas With the Iterative-Field-Matrix Method Using Near-Field Observation Data. *IEEE Trans. on Antennas and Propagation*, 59(6), 2434-2437.

[47] Mikki, S., & Kishk, A. (2007). Theory and Applications of Infinitesimal Dipole Models for Computational Electromagnetics. *IEEE Transactions on Antennas and Propagations*, 55(5), 1325-1337.

[48] Persson, K., & Gustafsson, M. (2005). Reconstruction of equivalent currents using a near-field data transformation- with radome applications. *Progress In Electromagnetics Research*, PIER 54, 179-198.

[49] Petre, P., & Sarkar, T. K. (1992). Planar near-field to far-field transformation using an equivalent magnetic current approach. *IEEE Transactions on Antennas and Propagation*, 40(11), 1348-1356.

[50] Qin, Y. M., & Ciric, I. R. (1993). Inverse scattering solution with current modeling and Tikhonov regularization. *Proceedings on IEEE International Antennas and Propagation Symposium*, June 1993, 1, 492-495.

[51] Rahmat-Samii, Y. (1984). Surface diagnosis of large reflector antennas using microwave holography metrology. *Radio Science*, 19, 1205-1217.

[52] Rao, S. M., Wilton, D. R., & Glisson, A. W. (1982). Electromagnetic Scattering by Surfaces of Arbitrary Shape. *IEEE Transactions on Antennas and Propagation*, AP-30(3), 409-418, May 1982.

[53] Razavi, S. F., & Rahmat-Samii, Y. (2007). A new look at phaseless planar near-field measurements: limitations, simulations, measurements, and a hybrid solutions. *IEEE Antennas and Propagation Magazine*, 49(2), 170-178.

[54] Sarkar, T. K., & Arvas, E. (1981). On the class of finite step iterative methods (Conjugate Directions) for the solution of an operator equation arising in electromagnetics. *IEEE Transactions on Antennas and Propagation*, 29, 370-373.

[55] Sarkar, T. K., Weiner, D., & Jain, K. V. (1981). Some Mathematical Considerations in Dealing with the Inverse Problem. *IEEE Transactions on Antennas and Propagation*, AP-29(2), March 1981.

[56] Sarkar, T. K., & Taaghol, A. (1999). Near-field to near/far-field transformation for arbitrary near-field geometry utilizing an equivalent current and MoM. *IEEE Transactions on Antennas and Propagation*, 47(3), 566-573.

[57] Sijher, T. S., & Kishk, A. (2005). Antenna modeling by infinitesimal dipoles using genetic algorithms. *Progress In Electromagnetics Research*, PIER 52, 225-254.

[58] Stratton, J. (1941). Electromagnetic theory. McGraw-Hill, New York.

[59] Van den Berg, P. M., & Kleinman, R. E. (1997). A contrast source inversion method. *Inverse Problems*, 13, 1607-1620.

[60] Wang, H.-C., & Hwang, K. (1995). Multicoloring of grid-structured PDE solvers for parallel execution on shared-memory multiprocessors. *IEEE Transactions on Parallel and Distributed Systems*, 6(11), 1195-1025.

[61] Yaccarino, R. G., & Rahmat-Samii, Y. (1999). Phaseless bi-polar planar nearfield measurements and diagnostics of array antennas. *IEEE Transactions on Antennas and Propagation*, 47(3), 574-583.

[62] Yaccarino, R. G., Rahmat-Samii, Y., & Williams, L. I. (1994). The bi-polar planar near-field measurement technique, part II: near-field to far-field transformation and holographic imaging methods. *IEEE Transactions on Antennas and Propagation*, 42, 196-204.

[63] Yaghjian, A. D. (1986). An overview of near-field antenna measurements. *IEEE Transactions on Antennas and Propagation*, 34(1), 30-45.

[64] Zhao, K., Vouvakis, M. N., & Lee, J.-F. (2005). The Adaptive Cross Approximation Algorithm for Accelerated Method of Moments Computations of EMC Problems. *IEEE Transactions on Electromagnetic Compatibility*, 47(4), 763-773.

Section 3

Microwave Applications

Chapter 7

Multi-Hole Waveguide Directional Couplers

Mahmoud Moghavvemi,
Hossein Ameri Mahabadi and Farhang Alijani

Additional information is available at the end of the chapter

http://dx.doi.org/10.5772/51355

1. Introduction

The directional couplers are inherently assumed as four-port devices, which consisted of two transmission lines that are electromagnetically coupled to each other. The first port is named as input, and the second one as output or transmitted, the third one as sampling or coupled and the fourth one as isolated or terminated. By using a special design the input power is divided between output and coupled port in a certain ratio named coupling factor. The required value for coupling factor P_1/P_3 defines the range of applications for directional couplers. Based on the application, coupling factor could be any value like 3, 6, 10, 20 dB and even more. The performance of the directional coupler is usually evaluated by its directivity between port 3 and 4.The directivity is a calculated parameter from isolation and coupling factor and shows how the two components of wave cancel each other at port 4. Though we prefer to have high value for directivity as much as possible, but in real situation this could be happened only around center frequency of designing band. The waveguide directional couplers have a good directivity compared to microstrip or stripline couplers and in spite of their bulky size, give us a low loss, high power handling, good characteristics and low cost due to use of just a simple waveguide.

Nowadays the numerical methods are widely used for simulation and optimization of high frequency structures. Some of them such as HFSS and FEKO, are well commercialized and used widely by researchers and engineers.. But for designing procedure and for starting point we need an initialization value to input into simulator and then optimize the parameters by its internal routines.

In this chapter we focus on the waveguide directional couplers and we try to give a good reference as well as finalized designing formulas in closed form and tables to be used individually or as initial values for numerical software. The full generalized field theory and its

INTECH
open science | open minds

© 2012 Moghavvemi et al.; licensee InTech. This is an open access article distributed under the terms of the Creative Commons Attribution License (http://creativecommons.org/licenses/by/3.0), which permits unrestricted use, distribution, and reproduction in any medium, provided the original work is properly cited.

equations for designing based on multi-hole coupling structure will be introduced, though more detailed basic information could be found in given references.

Moreover, by solving the equations, the recursive procedure is employed in a computer program to adjust the required directivity, coupling and frequency or waveguide type to specify the number of coupling holes, individual diameters and locations of holes in waveguide's coordinate. Besides of those parameters, the length of coupler, matched load and other sizes of structure for fabricating, will be defined too. By using the different methods like binomial, Chebyshev, and super imposed to calculate the coupling of each hole, the wide bandwidth response is achieved. At the end, a number of books and papers are given as good references for further study.

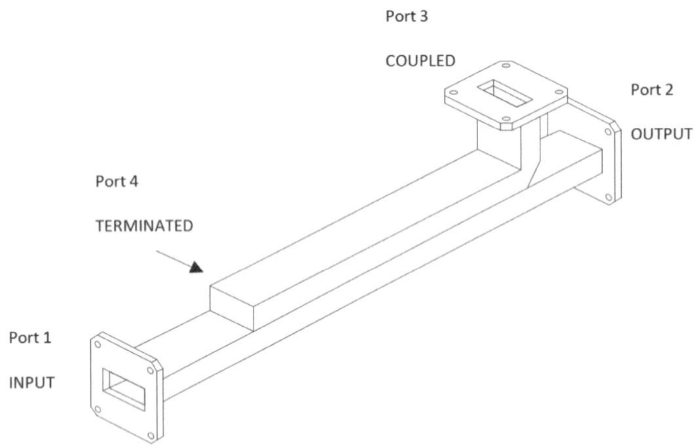

Figure 1. An Ordinary Broad-Wall Waveguide Directional Coupler and its ports

1.1. Definitions

As mentioned, couplers are considered as 4-port passive devices in which, a part of input wave reaches to output port 2 and the remained would be coupled to the coupled port 3. Port 4 usually internally is matched to damp the residual internally reflected waves from port 2 and port 3. Ideally there is no wave reach port 4. Port 4 usually is terminated by a full band load as shown in Figure1 and Figure2.

In waveguide couplers, the coupling method is done by putting a waveguide on top of another one and by making some aperture holes in their common wall a determined portion of wave would be leaked into the other waveguide. Though the waveguides axis and coupling apertures can be chosen arbitrary [2, 3], but for adequate specific usage and for easy derivation of design equations, we consider that two waveguide's are lay exactly on each other. Here the broad-wall coupling configuration is more interested and concentrated. Though, for side-wall the derivation of the design equations are so simple but the bandwidth is limited in spite of higher power handling.

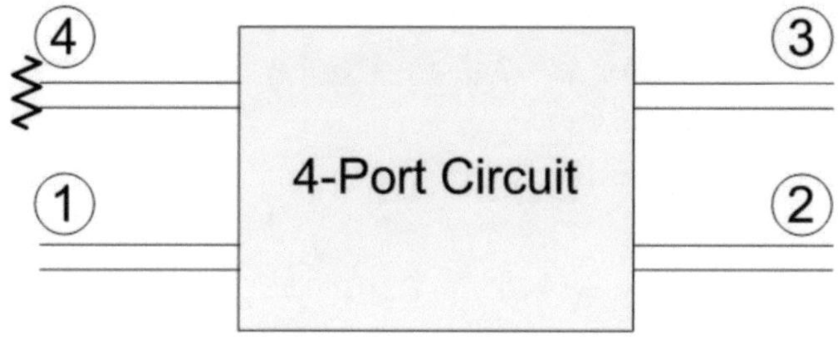

Figure 2. An equivalent 4-Port configuration for a directional coupler

Since 1945, extensive studies have been conducted on the issue and many researchers have tried to optimize the designing equations to make the result more accurate. For instance, different slot shapes had been introduced to increase the bandwidth and modifying the specifications. [12, 25, 27]

In order to start the calculations, there are three major parameters that we need to define for each coupler:

- Coupling factor "C" in dB, that represents the power received by port 3 as:

$$C = 10\log\frac{P_i}{P_f} \quad \text{or} \quad C = -10\log\frac{P_f}{P_i} \tag{1}$$

  The coupling factor shows the ratio of power at port 3 to input power at port 1. Typically we prefer to have one of the 3, 6, 10, 20 or 30 dB as standard values but for specific application it also can be defined freely.

- Directivity "D", is the ratio of output power at port 3 to received power at port 4. Since we prefer to eliminate the power at port 4, therefore the high values for "D" is more interested. The "D" in terms of dB is defined as:

$$D = 10\log\frac{P_f}{P_b} \quad \text{or} \quad D = -10\log\frac{P_b}{P_f} \tag{2}$$

- Bandwidth "BW", which depends on directivity. By increasing the number of coupling apertures, the order of coupler increases (similar to the order of filter) hence the directivity is increased. Meanwhile, higher bandwidth is also achieved. Therefore, by choosing the required minimum directivity, D_{min}, the available bandwidth is calculated.

For a 10 dB coupling or having a 0.1 of input power to port 3, we would have:

$$10\log\frac{P_f}{P_b} = 10 \;\rightarrow\; P_f = \frac{1}{10}P_i \tag{3}$$

In the same way, for a 3dB coupling, half of input power will receive to port 3:

$$10\log\frac{P_i}{P_f}=3 \quad \rightarrow \quad P_f=\frac{1}{2}P_i \tag{4}$$

And if we consider $D = 40$ dB for directivity:

$$10\log\frac{P_f}{P_b}=40 \quad \rightarrow \quad P_f=10000P_b \tag{5}$$

It is the adequate value for designing a good directional coupler.

A number of references, which have studied the couplers and have given the relationship between number of aperture holes "n" and directivity "D" are listed in references [12, 25, 23]. In addition to number of aperture holes"n", in the designing procedure for directional couplers, certain parameters should be well defined as:

- Distances between the holes
- Distances between holes to side-wall (holes center offset from waveguide axis)
- The holes dimensions (diameter of holes for circular holes).

It has been shown that to have an optimum coupling around a certain frequency, the criteria (6) should be kept in which "x" is the distance between the holes centers to the side-wall and "a" is the broad wall size of waveguide: [24]

$$\frac{x}{a}\leq 0.25 \tag{6}$$

Furthermore, by precise study, the best design value for ratio of (6) is given as [23]

$$\frac{x}{a}=0.203 \tag{7}$$

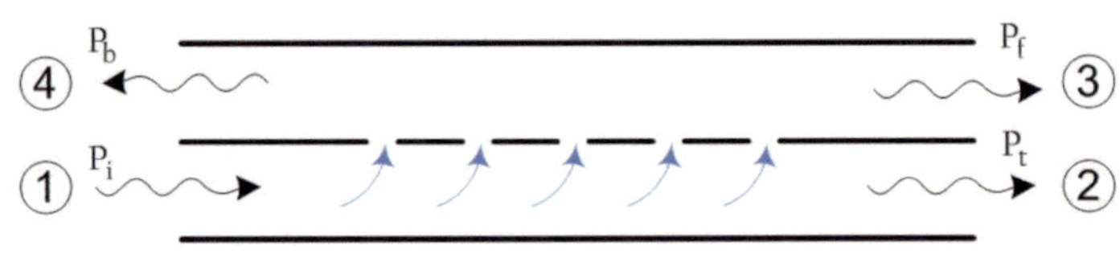

Figure 3. Cross section of the directional coupler and coupling holes

The distances between holes should be about $\lambda_g/4$ however a question remains, what is the proper value of λ_g when the bandwidth is limited to the λ_{g_1} to λ_{g_2} interval? To answer the question, there are three definitions used for λ_g:

1. λ_g is the average of wavelength of lower band λ_{g1} and upper band λ_{g2} so:

$$\lambda_g = \frac{\lambda_{g1} + \lambda_{g2}}{2} \quad \rightarrow \frac{\lambda_g}{4} = \frac{1}{8}\left(\lambda_{g1} + \lambda_{g2}\right) \tag{8}$$

2. λ_g can be considered as geometric mean between λ_{g1} and λ_{g2}:

$$\lambda_g = \sqrt{\lambda_{g1} \cdot \lambda_{g2}} \rightarrow \frac{\lambda_g}{4} = \frac{1}{4}\sqrt{\lambda_{g1} \cdot \lambda_{g2}} \tag{9}$$

3. λ_g can be considered as mean value between λ_{g1} and λ_{g2}:

$$\frac{2}{\lambda_g} = \frac{1}{\lambda_{g1}} + \frac{1}{\lambda_{g2}} \quad \rightarrow \lambda_g = \frac{2\lambda_{g1}\lambda_{g2}}{\lambda_{g1} + \lambda_{g2}} \quad \rightarrow \quad \frac{\lambda_g}{4} = \frac{\lambda_{g1}\lambda_{g2}}{2\left(\lambda_{g1} + \lambda_{g2}\right)} \tag{10}$$

The best choice for defining the centers of two holes is the 3rd definition since it has been practically approved too [23]. So the wavelength would be derived from (10). Therefore, in order to define the dimensions of each hole (or diameter in case of circular hole type), the each hole's coupling should be calculated first, and the hole's diameter would be derived consequently.

1.2. Fields Equations

In order to calculate the coupling of each hole and by using the required D_{min} that we need-here, the number of holes "n" will be derived in two different ways:

i- The coupling coefficient mapped to coefficients of nth order of Chebyshev polynomial.

ii- The coupling coefficient mapped to coefficients of nth order of Binomial polynomial.

By assuming the same order for polynomials "*i*" and "*ii*" and by noticing that the directivity slope in case of "*i*" is higher, we expect to have higher bandwidth in comparizon to "*ii*" and limited ripple in pass-band. In case of "*ii*" though there is no ripple in pass-band but the slope of directivity is lower than "*i*" with same order of polynomial, therefore the band-width is lower than "*i*". For years many of manufacturers chose the "*i*" method and considering the number of holes n = 20. Here, the "*i*" method is chosen, however the number of holes "n" would be defined from D_{min} and it will be not fixed anymore.

In fabricating the couplers, any arbitrary shape for holes can be used but the circular; elliptic and rounded-edge rectangle has been widely studied, simulated and used in research reports. [30] Here, the circular holes have been adopted. The circular holes can be aligned in one, two or three parallel rows, but in our case, 2-rows are used. In order to calculate the coupling coefficients and related field equations the "Bethe's small-hole coupling theory" is used as the main computational method. Further, by using a correcting function, the theory is expanded to use big-size holes as well [12, 26, 27]. In that way, Levi's work would be followed to find the effect of wall-thickness "*t*" and also the relationship between variations of directivity "*D*" and coupling error "ΔC". [27]. Levi showed that if "*D*" increases, "ΔC" will decrease.

In special case, if we require high directivity "D", like "D = 50 dB" for small bandwidth like 8.9 to 9 GHz, only 2 holes are needed to synthesis the coupler.

For calculating the distance of holes' centers to side-wall "x", the equation $\sin\frac{\pi x}{a}=\frac{\lambda_0}{\sqrt{6}a}$ is used in which "a" is broad-side of waveguide and λ_0 is the wavelength in the middle of the band. [2]

The coupled wave equations for incident wave A_1 and reflected B_1 by assuming the same amplitude for waves are as follows:

$$A_1=j\frac{2\pi}{ab\lambda_g}\left[M_x H_x^{(1)} H_x^{(2)}+M_z H_z^{(1)} H_z^{(2)}-PE_y^{(1)} E_y^{(2)}\right] \quad (11)$$

$$B_1=j\frac{2\pi}{ab\lambda_g}\left[-M_x H_x^{(1)} H_x^{(2)}+M_z H_z^{(1)} H_z^{(2)}-PE_y^{(1)} E_y^{(2)}\right] \quad (12)$$

In which, "a" and "b" are the waveguide dimensions, M_x and M_z are the magnetic polarization components in "x" and "z" axis and P is electrical polarization. The $H_x^{(1)}$ is the amplitude's wave component in the first waveguide and $H_x^{(2)}$ is for second and so on. If two waveguides are identical then $H_x^{(1)}=H_x^{(2)}$ and $H_z^{(1)}=H_z^{(2)}$. Then the fields' components are expressed as:

$$H_x=-Sin\frac{\pi x}{a}e^{-j\gamma z} \quad (13)$$

$$H_z=j\frac{\lambda_g}{2a}Cos\frac{\pi x}{a}e^{-j\gamma z} \quad (14)$$

$$E_y=\frac{\lambda_g}{\lambda}Sin\frac{\pi x}{a}e^{-j\gamma z} \quad (15)$$

The field equations are given separately for "Narrow wall" and "Broad wall" cases. Here, we briefly introduce them and give the relations for our interested one (i.e., Broad wall):

1.2.1. Narrow wall

By referring to Figure 5, since $x=0$, the equations could be simplified as:

$$A_1=A_2=-\frac{j\lambda_g M_z}{2a^3 b} \quad (16)$$

In which, M_z is independent from frequency. In other words, "Narrow wall" coupling has significant difference comparing to "Broad wall" coupling.

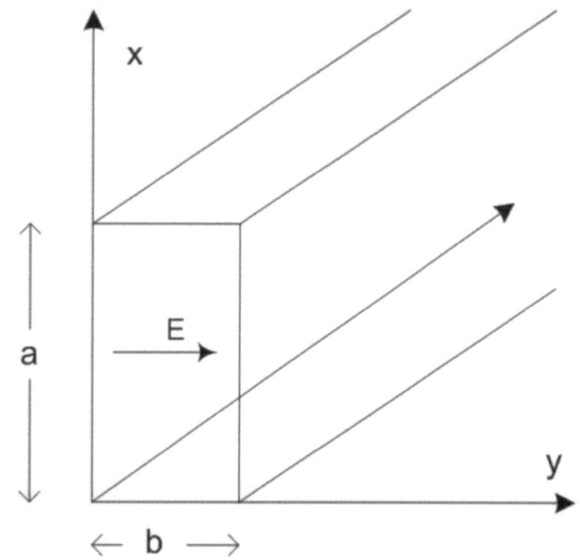

Figure 4. The geometry of field equations and waveguide

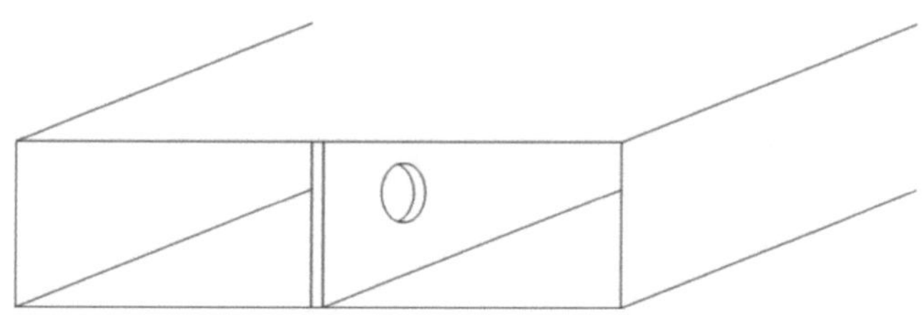

Figure 5. Narrow wall coupling

1.2.2. Broad wall

This case is shown in Fig.6 and equations are expressed as following:

$$A_1 = \frac{j2\pi}{ab\lambda_g}\left[\left\{M_x - \left(\frac{\lambda_g}{\lambda}\right)^2 . P\right\}\sin^2\frac{\pi x}{a} + M_z\left(\frac{\lambda_g}{2a}\right)^2\cos^2\frac{\pi x}{a}\right] \tag{17}$$

$$B_1 = \frac{j2\pi}{ab\lambda_g}\left[\left\{M_x + \left(\frac{\lambda_g}{\lambda}\right)^2 . P\right\}\sin^2\frac{\pi x}{a} - M_z\left(\frac{\lambda_g}{2a}\right)^2\cos^2\frac{\pi x}{a}\right] \tag{18}$$

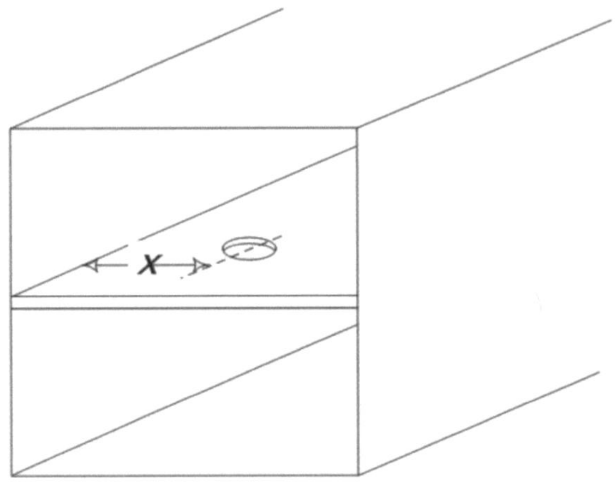

Figure 6. Broad-wall coupling

If the holes are circular with a diameter "d" the parameters "P, M_x and M_z " are calculated as:

$$M_x = M_z = \frac{d^3}{6} \tag{19}$$

$$P = \frac{d^3}{12} \tag{20}$$

By putting (19 & 20) in (17 &18) we will have simplified form:

$$A_1 = \frac{j\pi d^3}{12a^2b}\left[\frac{2a}{\lambda_g} + \frac{\lambda_g}{2a} - \frac{a\lambda_g}{\lambda^2}\right] \tag{21}$$

$$B_1 = \frac{j\pi d^3}{12a^2b}\left[\frac{2a}{\lambda_g} - \frac{\lambda_g}{2a} + \frac{a\lambda_g}{\lambda^2}\right] \tag{22}$$

And if we put $\frac{1}{\lambda^2} = \frac{1}{\lambda_g^2} - \frac{1}{(2a)^2}$ we will have:

$$A_1 = \frac{j\pi d^3}{24a^2b}\left[\frac{2a}{\lambda_g} + \frac{\lambda_g}{2a}\right] \tag{23}$$

$$B_1 = \frac{j\pi d^3}{24a^2b}\left[\frac{6a}{\lambda_g} - \frac{\lambda_g}{2a}\right] \tag{24}$$

By putting $\lambda_g = 2a$ then minimum value for A_1 is calculated as:

http://dx.doi.org/10.5772/51355

$$A_{10} = \frac{j\pi d^3}{12a^2 b} \tag{25}$$

And coupling variation to frequency changing is:

$$\Delta C = 20\log\left[\frac{\frac{2a}{\lambda_g} + \frac{\lambda_g}{2a}}{2}\right] \tag{26}$$

On the other hand, variation of directivity to frequency is given by:

$$D = 20\log\frac{A_1}{B_1} = 20\log\left[\frac{\frac{2a}{\lambda_g} + \frac{\lambda_g}{2a}}{\frac{6a}{\lambda_g} - \frac{\lambda_g}{2a}}\right] \tag{27}$$

The solved equations for R100 waveguide is given in Fig.7

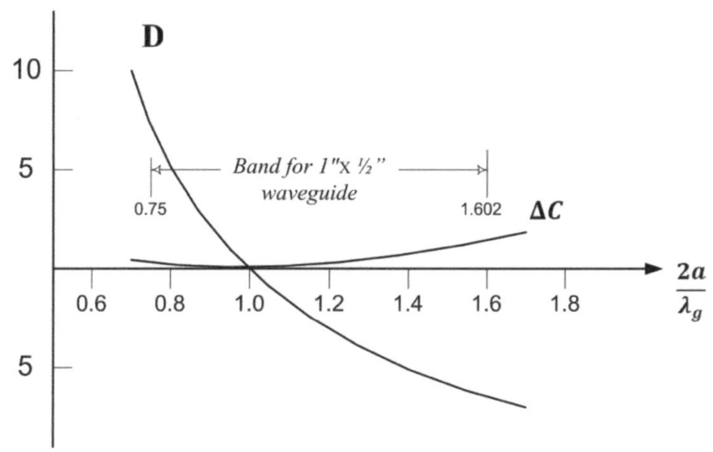

Figure 7. Theoretical coupling variation and directivity for transverse arrangements of 1, 2 or 3 holes in common broad-wall of R100 waveguide.[14]

When a number of single holes are aligned in a row, the array's directivity will be added to basic value of directivity.

As it has been shown earlier, the Broad-wall couplers have better characteristics in wide frequency spans comparing to Narrow-wall couplers while the power handling characteristic of Narrow-wall couplers are better.

1.3. Coupling by big holes

The equations given in section 1.3 were valid for small holes. By considering the wall-thickness "*t*" and big hole's by surface size of "*A*", the equations should be corrected. This has already done and the results will be used now. [15]

1.3.1. Corrections for wall thickness "t"

When the coupling holes have thickness "t", it equals to a short-length circular waveguide, which is working under its cut-off frequency (when frequency is lower than resonance frequency). Therefore the amplitude of the coupled wave attenuates by following term:

$$exp\left\{\frac{-2\pi tA}{\lambda_c}\left[1-\left(\frac{\lambda_c}{\lambda}\right)^2\right]^{1/2}\right\} \tag{28}$$

In which, the λ_c is the cut-off wavelength of hole and λ is the wavelength in operating frequency, "*t* " is the wall-thickness and "*A*" is the correction factor which includes the interactions of adjacent fields and is defined practically. [15] For elliptic holes (Narrow slot) the "A" factor is in 3rd order but for circular holes it is close to first order. If the fields have multi components, correction factor (28) should be applied on all of the Bethe's equations. It means that in equations (17) and (18) the magnetic field correction factor for coupling have a parameter λ_c that is due to excitation by TE_{11} mode so for this mode $\lambda_c \to \lambda_H = 1.705d$ and correction factor for electric field coupling uses a λ_c, which relates to TM_{01} mode and equals to $\lambda_c \to \lambda_E = 1.305d$.

1.3.2. Corrections for big size holes

A good study for such big size holes have been done by an equivalent circuit. [15] In this way, the hole's effect will be simulated by a two port network in parallel or in series between two waveguides which have mutual coupling. This network is a lossless and should be defined by Foster's reactance theorem. Here, the impedance will be defined by locations of its zeroes and poles in addition to the multiplications coefficients.

In simple expression of Bethe's small sized coupling hole theory, this impedance has a simple form. For example the reactance of a small hole in a thin diaphragm at the cross section of a rectangular waveguide, working in TE_{10} is:

$$X = \frac{4\pi M Z_0}{ab\lambda_g} \tag{29}$$

In which, the "a" and "b" are the waveguide dimensions that have been shown in Fig.4

The term λ_g is the guided wavelength, "M" magnetic polarization, Z_0 characteristic impedance of the waveguide. The Z_0 has a direct relation to term $\frac{\lambda_g}{\lambda}$, which shows that lumped reactance "X" has a direct relation to frequency ($X \propto f$) too.

Therefore the small hole coupling theory assumes that the "X" would be a constant reactance but it is not true. Because, there are a few numbers of unwanted resonances occurred in the frequency band. For this reason the (29) would be a good definition when the operating frequency is somehow lower than the first resonance. For considering the resonance effect in equation (29), the corrected "M" would be expressed by introducing a new term that considers the effect of cut-off wavelength $\frac{M}{1 - \left(\frac{f^2}{f_c^2}\right)}$ and the result is as follows:

$$\frac{X}{Z_0} = \frac{4\pi M}{ab\lambda_g\left(1 - \left(\frac{f^2}{f_c^2}\right)\right)} \tag{30}$$

From measurements, it has been shown that the above correction factor gives a good approximation.

The attenuation definition (28) can be combined to (30) to give us a general correction factor for big size holes:

$$\frac{exp\left\{\frac{-2\pi tA}{\lambda_c}\left[1 - \left(\frac{f^2}{f_c^2}\right)\right]^{1/2}\right\}}{1 - \left(\frac{f^2}{f_c^2}\right)} \tag{31}$$

1.4. Multi holes coupling

A longitudinal coupling consists of a series of holes by center distance of $\frac{\lambda_g}{4}$ that has a great coupling in forward and weak coupling in backward direction.

The slight coupling for a single hole has been studied and the directivity introduced by:

$$Directivity\ (dB) = 20\log\frac{A_1}{B_1} \tag{32}$$

In which: [15]

$$\frac{A_1}{B_1} = \frac{\int_{-L/2}^{L/2}\phi_{(x)}dx}{\int_{-L/2}^{L/2}\phi_{(x)}\exp(-2j\beta x)dx} \tag{33}$$

In the Fig.8 a series of "n" holes in one row is shown. The coupling voltage of the series is named $a_1, a_2, \ldots, a_n$.

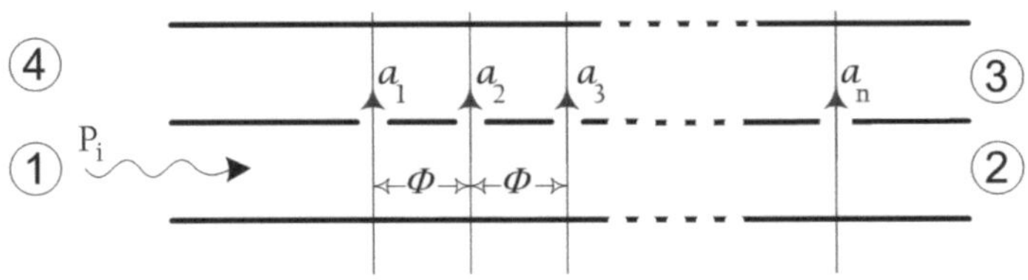

Figure 8. The cross section of n-Hole array and coupling coefficients

All the hole's center distances and electrical length are the same and are considered in the middle of the band. If the input wave to port 1 has constant amplitude and matched to other 3 ports, the reflected wave can be expressed by:

$$B_1 = a_1 + a_2 Exp(-2j\phi) + a_3 Exp(-4j\phi) + \ldots + a_n Exp(-2(n-1)j\phi) \tag{34}$$

The interesting and useful case is when the coefficients of the series being symmetrical from center. Therefore:

$$a_1 = a_n, \qquad a_2 = a_{n-1}, \qquad a_K = a_{n-K+1} \tag{35}$$

So by putting the values (35) into (34):

$$B_1 = \begin{cases} [2a_1 \cos(n-1)\phi + 2a_2 \cos(n-2)\phi + \ldots + 2a_{n/2}\cos\phi]e^{j(n-1)\phi} & n \text{ even} \\ \left[2a_1 \cos(n-1)\phi + 2a_2 \cos(n-2)\phi + \ldots + \frac{a_{n+1}}{2}\right]e^{j(n-1)\phi} & n \text{ odd} \end{cases} \tag{36}$$

The direct coupled wave at port 3 will be:

$$A_1 = \sum_{r=1}^{n} a_r \, e^{j(n-1)\phi} \tag{37}$$

And the directivity "D" is calculated by normalizing B_1 to A_1 in (36) by dividing the sum of each coupling voltages. In special case if there are "n" identical holes, therefore:

$$\frac{B_1}{A_1} = \frac{\sin n\phi}{n \sin \phi} \tag{38}$$

2. Design methods based on arrays

2.1. Chebyshev Array

If the minimum voltage over the full bandwidth to reach a good directivity "D" is needed, the Chebyshev polynomial can be used for distribution function of each hole's voltage. Such coefficients are derived by putting the B_1 in (36) by considering the following equal ripple's directivity function as following:

$$B_1 = a_m T_{n-1}\left(\frac{\cos \phi}{\cos \phi_0}\right) \tag{39}$$

In which the a_m is the maximum of B_1 over the coupling bandwidth that given by following:

$$\phi_0 \le \phi \le \pi - \phi_0 \tag{40}$$

The a_m is calculated by putting the $\phi = 0$ in (39):

$$a_m = \frac{\sum_{r=1}^{n} a_r}{T_{n-1}\left(\frac{1}{\cos \phi_0}\right)} = \frac{|A_1|}{T_{n-1}\left(\frac{1}{\cos \phi_0}\right)} \tag{41}$$

In (36) if we put $\phi = 0$:

$$B_1 = \sum_{r=1}^{n} a_r \tag{42}$$

Therefore the minimum directivity over the bandwidth would be:

$$D_{min} = 20\log\left[T_{n-1}\left(\frac{1}{\cos \phi_0}\right)\right] \tag{43}$$

Comparing this method to method of Binomial polynomial is very informative that has been done by Levi. In this case we should have: [16]

$$B_1 = a_m' \cdot \left(\frac{\cos \phi}{\cos \phi_0}\right)^{n-1} \tag{44}$$

In which:

$$a_m' = \frac{\sum_{r=1}^{n} a_r}{\left(\frac{1}{\cos \phi_0}\right)^{n-1}} = \frac{|A_1|}{\left(\frac{1}{\cos \phi_0}\right)^{n-1}} \tag{45}$$

The minimum directivity at the edge of the band for this case is:

$$D_{min}=20\log\left[\left(\frac{1}{\cos\phi_0}\right)^{n-1}\right] \tag{46}$$

Obviously the (46) always is significantly lower than the value for Chebyshev case (43).

The coupling equation for Chebyshev case is derived by putting the identical coefficients of $\cos\phi$ in (36). Young gave such coefficients for $3\leq n\leq 8$ [3]. But here, the generalized case is obtained by a computer program for $1\leq n\leq 25$.

For coupling $C=0$ these coefficients are changed into Pascal's triangle that for $C=0$ the infinite directivity over a zero bandwidth obtained.

The hole's size is derived by coupling of each hole in dB. That relation for r th hole is as follows:

$$C_r=20\log\left(\frac{\sum_{r=1}^{n} a_r}{a_r}\right) \tag{47}$$

Since $\sum_{r=1}^{n} C_r=1$, all the theoretically given hole couplings, transferred all power by assuming the 0dB in the formula. Therefore in order to design a "C dB" coupler the "C" is added to C_r in (47). The entire hole sizes by this way and by given theory for small size holes (or if we need by using the correction coefficient curves given by references) can be computed. [17, 18]

In addition to both mentioned series for calculating the coefficients (Chebyshev and Binomial), there is another method that actually derived from them. It is named "Super Imposed Arrays".

2.2. Super Imposed Array

When the strong coupling is needed, i.e. 3dB or 6dB, it is not possible to use the one row of holes (single array), since diameter of holes will be increased. Therefore it is more convenient to have approximately same diameter for all to get good coupling quality. For this case the super imposed array is used. As first step, we need the coefficient series in which the holes get bigger. It would be happened when $n>4$. For starting we can use Chebyshev or binomial coefficient series in one line. Then the same series should be written in second line but in shifted position. It means, first coefficient of line 2 in under the 4th coefficient of line 1 and so on. By adding the two lines we would have a new series that its coefficients (or holes' diameters) alternately are the same. For example by a 6-element binomial series, we can make a 9-element super imposed series:

As it has shown, the elements in new series are alternately identical. This can be done by any other number of elements or polynomials. If we wanted to add more number of holesTable 1, the same way is chosen:

http://dx.doi.org/10.5772/51355

1	5	10	10	5	1			
			1	5	10	10	5	1
1	5	10	11	10	11	10	5	1

Table 1. A 9-element super imposed array based on 6-element binomial array

1	5	10	11	10	11	10	5	1			
						1	5	10	10	5	1
1	5	10	11	10	11	11	10	11	10	5	1

Table 2. Added more number of elements to table 1.

For special characteristics of Chebyshev series, hereinafter, it will be used as basic polynomial for further design. In super imposed method the number of elements can be any value and we would have longer coupler. If we chose other two methods the the holes diameter get bigger and bigger and it may exceeds the broad-wall size of waveguide. By considering the $\sum_{r=1}^{n} C_r = 1$it is cleared that each $C_r$must be less than 1 and in super imposed this would be happened.

For strong couplings and the holes with the same shape the following relation is proposed by Cohn: [14]

$$A_m = \sin\left(\frac{m\pi}{4m_{3dB}}\right) \tag{48}$$

In (48) the m_{3dB} is the number of series for obtaining the 3 dB coupling. For example, if a 6-element series is needed for 3dB coupling, therefore the coupling for one series will be:

$$A_1 = \sin\left(\frac{\pi}{24}\right) = 0.1305 \tag{49}$$

That equals to 17.68 dB. The directivity for Chebyshev-based super imposed array is greater than the single array. The reason can be explained by this fact that returned waves are added in phase and amplitude, so the maximum amplitude for returned wave will not exceed from single array in any case and it will damped soon.

2.3. Transverse groups of holes

It maybe concluded that by using two or three rows of holes in broad wall, the stronger coupling would be obtained. It is true if there are two rows and the distance of center of holes to side walls being equal to $x = 0.25a$ but for three rows the result is not good. For two rows the coupling and directivity are derived from (23) and (27). See Fig. 9

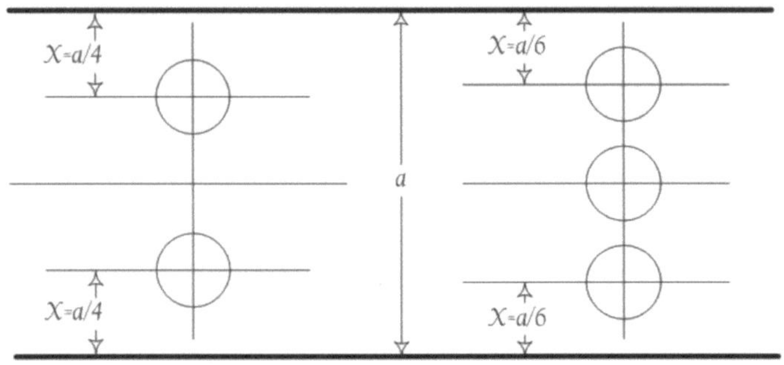

Figure 9. Groups of two or three rows holes.

The test results of multi-hole couplers are as following:

A 6dB coupler by $1" \times 1/2"$ waveguide in 8.2~12.4GHz band is designed by 21-elements super imposed array based on 6-element Chebyshev in two rows (totally 42 holes). The distance from side-wall for circular-holes is $x = 0.25a = a/4$ and distance between hole centers is $\lambda_g/4$. The obtained directivity "D" is more than 40dB and coupling deviation ΔC is about $\pm 0.5dB$. [19]

Shelton has tested multi-rows couplers and has given the coupling curve in terms of holes diameters for X and Ku bands waveguides. His efforts by using 3 rows were not successful. [20] For $1" \times 1/2"$ waveguide, Cohn used rounded rectangular holes in two rows. His research approved that 2-rows is better than 3-rows and for shortening the length it is not possible to use 3-rows holes.

It was also declared that in case of 3-rows, the resonance in upper band will happen. The reason is, where the electric vector is in parallel to broad wall, the even mode is excited and coupler acts as side wall coupler. Such case is not happened for 2-rows couplers. By reducing the height of "b" it is possible to put the resonance frequency of even modes out of operating frequency band. Only a slight reduction in "b" is needed since the resonance is occurred when the coupling region has length of $\lambda_g/2$ and it is near cut-off for even modes. Indeed, as an example, the "b" should be reduced from 0.4" to 1/16". This reduction in "b" increases the coupling of each hole. See (17). So a few number of holes needed to make 3dB coupling. As an example, a 2nd order Chebyshev transformer has a theoretical VSWR of 1.01 over its bandwidth. For better matching the waveguide height should be reduced at the two ends. Each series of holes at each side of coupler should be located inside of transformer in the way that it does not change the length. The final length would be 3.9 inches consisting of 2-rows of 10 holes that gives a coupling of $3 \pm 0.5dB$ and Isolation of more than 30dB for X-band 8.2~12.4GHz.

3. Practical designing

3.1. A real sample

After reviewing the basics of directional coupler, we start to design a coupler practically. First of all it is better to introduce the abbreviations that we use. They are listed in following table:

C	= Coupling in dB
λ_{g_1}	= Guided wavelength at the lower end of the required bandwidth (mm)
λ_{g_2}	= Guided wavelength at the upper end of the required bandwidth (mm)
$\lambda_{g_{mid}}$	= Mean guided wavelength
N	= number of coupling elements in basic array
D_{min}	= minimum directivity (dB)
	$=180 \Big/ 1+\frac{\lambda_{g_1}}{\lambda_{g_2}}$ (deg)
λ_g	= Guided wavelength (at the center frequency of the wave-guide bandwidth) (mm)
X	= Axis across broad dimensional of waveguide
A	= Broad dimension of a waveguide wall (mm)
B	= Narrow dimension of waveguide wall (mm)
d	= Diameter of hole in millimeter (mm)
A'	$=1-\left(\frac{1.71d}{\lambda_0}\right)^2$ Term giving correction of resonance phenomena
λ_0	= free space wavelength (mm)
T	= wall thickness (mm)
A''	$=32\left(\frac{t}{d}\right)\left[1-\left(\frac{1.71d}{\lambda_0}\right)\right]^{1/2}$ term giving correction to the attenuation effect on a finite wall thickness
X_0	= 1 / cos

Table 3. The terms and abbreviations that used in design procedure.

As it is mentioned before in (10):

$$\frac{\lambda_{g_{mid}}}{4}=\frac{\lambda_{g1}\lambda_{g2}}{2(\lambda_{g1}+\lambda_{g2})} \tag{50}$$

The number of holes can be defined by minimum directivity D_{min} as: [23]

$$n = 1 + \cosh^{-1}\left[\frac{10^{\frac{D_{min}}{20}}}{\cosh^{-1}\left(\frac{1}{\cos\phi}\right)}\right] \tag{51}$$

The starting coefficient in Chebyshev argument is calculated as:

$$X_0 = \frac{1}{\cos\phi} = \frac{1}{\cos\left(\frac{180}{1+\frac{\lambda_{g_1}}{\lambda_{g_2}}}\right)} \tag{52}$$

For X-band that we have $\lambda_{g_1} = 6.089\ Cm$ and $\lambda_{g_2} = 2.489\ Cm$the $X_0 = 1.853$ is obtained. Next step is to find the Chebyshev polynomial coefficients by computer program that gives:

{40.507, 172.277, 355.449, 445.373} (Notice that, only a half of the coefficients are enough due to symmetric specification of Chebyshev polynomial).

Then the coefficients are normalized to least element that gives following table:

A	B	C	D
1.0	4.253	8.775	10.995

Table 4. Normalizing the Chebyshev coefficients.

Therefore the whole structure of the holes will be as follows:

A	B	C	D	C	B	A

Table 5. Sequence of holes and its related Chebyshev coefficients for coupler synthesis.

Now we add them all together:

$$2(A + B + C) + D = 39.051 \tag{53}$$

The coupling for each hole will be defined in dB as follows:

Coupling for Holes $A = 20\log\frac{39.051}{1}\ dB = 31.832\ dB$

Coupling for Holes $B = 20\log\frac{39.051}{4.253}\ dB = 19.259\ dB$

Coupling for Holes $C = 20\log\frac{39.051}{8.775}\ dB = 13.968\ dB$

Coupling for Holes $D = 20\log\frac{39.051}{10.051}\ dB = 11.776\ dB$

Now, consider that we want to design a 10dB coupler, so we add a 10dB to each coefficient:

C_A=41.832 dB, Holes C_B=29.259 dB, Holes C_C=23.968 dB, Holes C_D=21.776 dB

Finally the achieved numbers should be inserted into Bethe's formula for small size holes: [12]

$$C = 20\log\left\{\frac{12a^2b}{\pi d^3}\left[1-\left(\frac{1.71d}{\lambda_0}\right)^2\right]\right\} + 20\log\left\{32\left(\frac{t}{d}\right)\left[1-\left(\frac{1.71d}{\lambda_0}\right)\right]^{1/2}\right\} \tag{54}$$

Now we solve the above equation (for each hole) by iteration method and the diameter of each hole would be determined. By considering the distance of circle centers to side wall as $x = 0.203a$ (7) following values for diameters would be obtained:

A=0.234 inch, B=0.343 inch, C=0.397 inch, D=0.421 inch

Note that the solved example is for single array. If we wanted to have the double rows we should put the (*C*+6) dB instead of *C* dB (that we considered 10 dB in above example).

Notice: an approximation way to define the number of holes "n" is using the D_{min} in equal to maximum coupling between holes plus 3 ~ 5 dB. For instance in the solved example, the maximum coupling is belonged to "*A*" that was C_A=41.832 dB. So:

$$D_{min} = C_A + 5 = 47\ dB \tag{55}$$

And the number of holes would be:

$$n = 1 + \cosh^{-1}\left[\frac{10^{\frac{47}{20}}}{\cosh^{-1} X_0}\right] \approx 6.9 \rightarrow n = 7 \tag{56}$$

Therefore if we wanted to have a good directivity, a directivity higher than 47dB then we should have 7 holes in the coupler.

In practice, for eliminating the effect of wall thickness "*t*", it is possible to remove one broad wall of a waveguide and mill- the next wall to have half thickness between to waveguides. [23]

The real designed 20 dB coupler by R70 waveguide and 14 holes in two rows (each row has 7 holes) is fabricated and tested. The results are given in Fig.10

In Fig. 11 and 12, another directional coupler for *C* = 10 dB is sketched. The diameter of circles for its five categories of holes, are:

Holes number 4 =4.16 mm

Holes number 3 =6.45 mm

Holes number 2 =8.00 mm

Holes number 1 =8.66 mm

Holes number 0 =8.94 mm

For further information see Fig. 11 and 12.

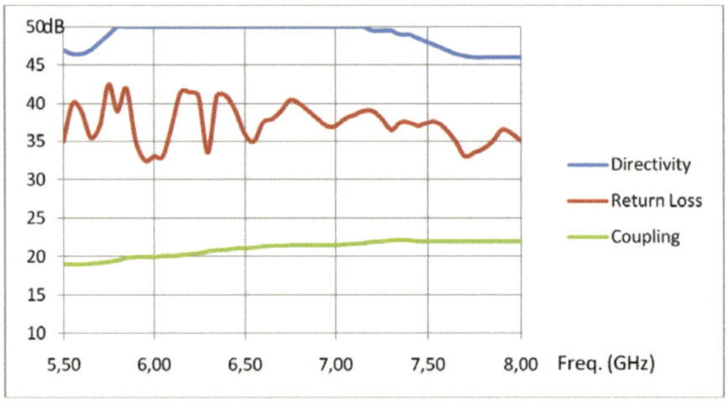

Figure 10. The test result for a multi-hole waveguide R70, directional coupler C=20dB

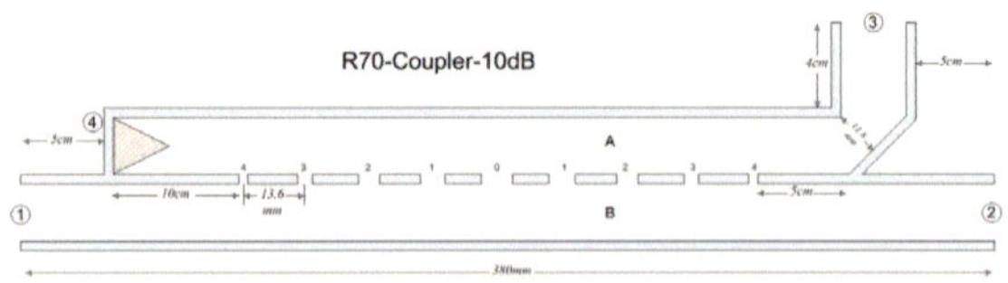

Figure 11. The R70 directional coupler, C = 10 dB by two rows of 9-elements

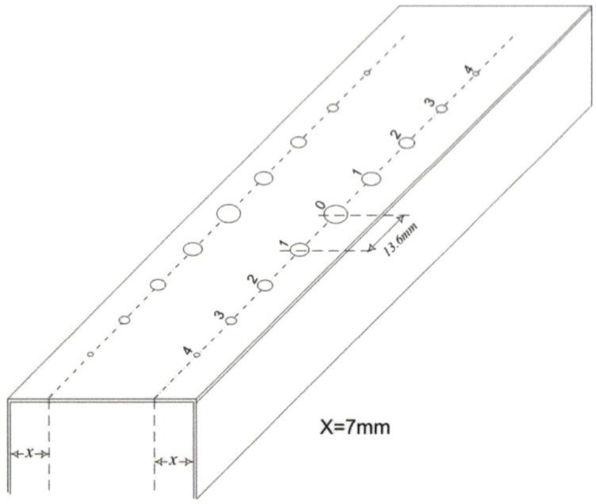

Figure 12. The circle centers and distance to side walls are the same.

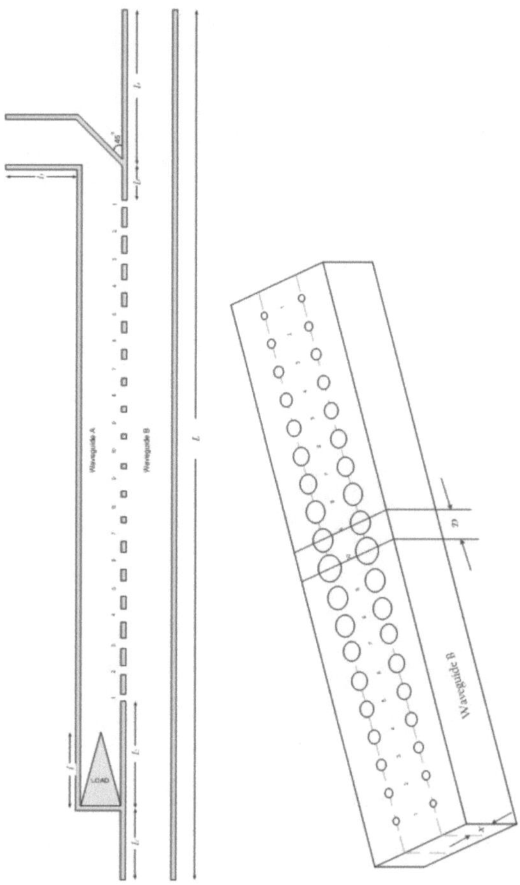

Figure 13. Cross section for a 38-holes (double rows of 19-elements) directional coupler. The lossy load maching using ferro-based materials (courtesy H. Mottaghi: hossain_mtg@yahoo.com)

The port-4 in the couplers is matched by a conical or pyramidal load. To obtain the lowest reflection from this port, the tapered structure is used to make a slight change in characteristic impedance of the waveguide along its length. Since the power loss at this port, naturally is not significant, the medium or low power absorber or load is enough.

Author details

Mahmoud Moghavvemi, Hossein Ameri Mahabadi and Farhang Alijani

Department of Electrical Engineering, University of Malaya (UM), Malaysia

References

[1] Gandhi, O. P. (1981). *Microwave Engineering and Applications*, Pergamon Press.

[2] Collin, R. E. (1966). *Foundations for Microwave Engineering*, Mc. Graw Hill.

[3] Young, Leo. (1966). *Advances in Microwave, volume I*, Academic Press.

[4] Miller, S. E. (1954). Coupled Wave Theory and Waveguide Applications. *Bell system Tech. J.*, 33, 661.

[5] Riblet, H. J. (1954). The short- slot hybrid Junction. *Proc, IRE*, 40, 180.

[6] Tomiyasu, K., & Cohn, S. B. (1953). The Transvar Directional Coupler. *Proc. IRE*, 41, 922.

[7] Cook, J. S. (1955). Tapered velocity couplers. *Bell system Tech. J.*, 34, 807-822.

[8] Fox, A. G. (1955). Wave coupling by warped normal modes. *Bell System Tech. J.*, 34, 823.

[9] Louisell, W. H. (1955). Analysis of a single tapered mode. *Bell System Tech. J.*, 34, 853.

[10] Miller, S. E., & Mumford, W. W. (1952). Multi-element directional couplers. *Proc. IRE*, 40, 1070.

[11] Bethe, H. A. (1943). Theory of side windows wave guides. *Mass. Inst. Technology Radiation Lab Report* [43].

[12] Bethe, H. A. (1944). Theory of diffraction by small holes. Phy. Rev. , 66, 163.

[13] Surdin, M. (1946). Directive couplers in wave guides. J. Inst. Elec. Engrs (London) , 93(IIIA), 725.

[14] Cohn, S. B., Knootz, R. H. S., & Wehn, S. L. (1962). Microwave hybrid coupler study program. *Rantec- Corporation Report* [61], 361-363.

[15] Cohn, S. B. (1952). Microwave coupling by large aperture. *Proc IRE*, 40, 697.

[16] Levy, R. H. (1959). Guide to the practical application of Chebyshev functions to the design of microwave components. *Proc. Inst. Elec. Engrs. (London)* C, 106, 193.

[17] Harrison, J. R. (1945). Design considerations for directional couplers. *MIT Radiation Lab. Rept.* [724].

[18] Saad, T. (1964). The Microwave Engineers handbook. *109 Boston, Massachusetts*.

[19] Hensperger, E. S. (1959). The design of multi-hole coupling arrays. *Microwave J.*, 2, 38.

[20] Shelton, W. (1961). Compact multi-hole waveguide directional couplers. *Microwave J.*, 4, 89.

[21] Geranpaye. (1985). Waveguide standards. *ITRC, Tehran, Iran*.

[22] Alison, W. B. W. A. (1987). A handbook for the mechanical tolerance of waveguide components. *Artech house.*

[23] Hancock, K. E. (1967). The design and manufacture of waveguide Tchebyshev directional-coupler. Electron. Eng , 39, 292.

[24] Liang, C. H., & Cheng, D. K. (1982). On performance limitations of aperture coupling between rectangular waveguide. *MTT-30* [5].

[25] Levy, R. (1968). Analysis and synthesis of waveguide multi-aperture directional couplers. *MTT-16* [12].

[26] Mc Donald, N. A. (1972). Electric and Magnetic coupling through small apertures in shield wall of any thickness. *MTT-20* [10].

[27] Levy, R. (1980). Improved single and multi- aperture waveguide coupling theory including explanation of mutual interactions. *MTT-28* [4].

[28] Riblet, H. J. (1947). A Mathematical theory of differential coupler. *Proc. IRE*, 35, 1307.

[29] Crompton, J. W. (1957). A contribution to the design of multi-element directional couplers. *Proc. IEE*, 104C(6TD), 398-402.

[30] Matthaei, G. L. (1964). *Microwave Filters impedance matching networks and coupling structures*, Mc Graw Hill.

[31] Wheeler, H. A. (1964). Coupling holes between resonant cavities or waveguides evaluation terms of volume ratios. *MTT*, 231.

[32] Sangster, A. J. (1973). Slot Coupling between uniform rectangular waveguides. *MTT-27* [7], 705.

[33] Pandhuri, Pande. V. H., & Das, B. W. (1979). comments on coupling of waveguides through large aperture. *MTT-27* [7], 702.

[34] Pandhuri, Pande. V. H., & Das, B. W. (1978). coupling of waveguide through large aperture. *MTT-26* [3], 209.

[35] Schmiedel, H., & Arndt, F. (1986). Field Theory Design of Rectangular Waveguide Multiple-Slot Narrow-Wall couplers. *MTT-34* [7], 791.

[36] Oliner, A. H. (1960). Equivalent circuits for small symmetrical longitudinal aperture and obstacles, . *IRE MTT*, 72.

[37] Hakkak, M., & Barakat, F. (1984). Waveguide bend characteristics. *ITRC publicationsTehran, Iran.*

[38] Mottaghi, H. (2010). Measuring the Load absorption parameters in waveguide. *IT&mmwaves SDN BHD, Kuala Lumpur.*

Chapter 8

The Diffraction of Electromagnetic Waves on the Periodic Heterogeneities and Its Use for Realization of Practical Technical and Electronic Devices of Millimeter and Submillimeter Wavelength Range

Gennadiy Vorobyov, Larissa Vietzorreck,
Ivan Barsuk and Aleksandr Rybalko

Additional information is available at the end of the chapter

http://dx.doi.org/10.5772/50694

1. Introduction

Among the open structures, which are used in millimeter and submillimeter (MSM) wave engineering, diffraction gratings (DG) made in different modifications (periodic metal and metal-dielectric structures (MDS)) are of primary importance along with open cavities and open waveguides. Such systems are basic in the design of electromagnetic oscillation sources and electronic components of different instrumentation of such wavelength range. If there is a diffraction of electromagnetic fields by DG, "two-act" wave transformation usually takes place. When homogeneous plane wave falls on the plane one-dimensionally periodic grating, scattered field can be considered as a spectrum of homo- and heterogeneous plane waves. In this case body (incident) plane wave is transformed into body (scattered) homogeneous plane and heterogeneous (surface) waves and, thus, "two-act" transformation occurs. This type of the boundary-value problems has been thoroughly studied in the work [1] and partly realized in the experiment [2]. In addition, processes of surface wave transformation of distributed sources into body waves by periodic heterogeneities are of special interest. Such phenomenon can be watched when an electron beam (EB) moves uniformly near the metal DG or periodic MDS. In this case self-surface field of the EB is scattered by DG and at least one of its harmonics is transformed into body wave of the diffraction radiation or Cherenkov radiation. It should be noted, that transformation of the surface wave of EB by DG into the diffraction radiation is also an example of the "two-act" diffraction process. In addition, phenomena, connected with the transformation of DG of the surface waves of a dielectric waveguide (DW), play a great role in

INTECH
open science | open minds

© 2012 Vorobyov et al.; licensee InTech. This is an open access article distributed under the terms of the Creative Commons Attribution License (http://creativecommons.org/licenses/by/3.0), which permits unrestricted use, distribution, and reproduction in any medium, provided the original work is properly cited.

MSM engineering. In this case surface waves of a DW are transformed by the DG into surface waves of the DW or into body waves separated from them.

Multilinked quasi-optical systems in different modifications are very perspective in designing fundamentally new electronic devices and electronic components of instrumentation of MSM wavelength range including infrared waves. One should mentioned, that they are partly investigated by numerically-analytic methods in the approximation of constant current [1] and using the experimental modelling. It was determined that systems with MDS included in the open cavities structure can have qualitatively new properties [3] that makes it possible to propose high-frequency filters, frequency stabilizers, energy output devices from the volume of open cavities and new types of semiconductor oscillators on its basis. A possibility of the development of a Smith-Parcell amplifier was considered in paper [4].

Huge number of the previously proposed systems isn't accompanied with its complex research. This fact, therefore, makes the process of practical realization of new modifications of electronic device schemes and electronic components of MSM wavelength range on its basis very slow. So far existed numerical methods of optimization of the three-dimensional superhigh frequency structures [5, 6] enable to analyze effectively electrodynamic characteristics only of some multilinked quasi-optical system elements such as reflecting MDS. Meanwhile, a comprehensive solving problem of optimization of such structures requires the huge expenditure of computer time and computational power with usually ambiguous results in the end. That is why problems of development of the universal experimental facility, general technique of experimental modelling of the electromagnetic phenomena in multilinked quasi-optical systems and its realization for studying electromagnetic effects in complex quasi-optical systems using so far existed numerically-analytic methods, are very topical.

In the current work general electrodynamic characteristics of the multilinked quasi-optical system coupling elements have been determined on the basis of previous theoretical and experimental results in research of the simplest types of radiating system, formed by single-row DG, and using experimental modelling of transformation of the DW surface waves into body waves by two-row DG of different modifications. Schemes of practical microwave engineering devices have been suggested on basis of these characteristics.

2. An experimental setup and measuring technique

The use of periodic MDS formed by a strip metal grating on a dielectric layer is promising for both constructing extremely high-frequency electric vacuum devices (optical coupler and diffraction radiation generator [7, 8]) and integrated forms of various functional assemblies and devices, operating in MSM ranges, including the terahertz range [9]. In contrast to reflecting metal gratings, the MDS have a number of specific features related to a possibility of exciting spatial waves of Cherenkov, normal, and abnormal diffraction radiation, when an

EB moves along the MDS [1]. Therefore, it is necessary to have exhaustive information on electrodynamic characteristics of the DG and MDS, when they are used in various devices. Paper [1] presents a method for identifying properties of waves of the space charge of an EB traveling along a periodic structure, to the surface wave of the DW. This method has found a wide utility in simulating electrodynamic characteristics of various radiation modes of spatial waves on periodic MDS and on normal metal gratings. The setup for studying conversions of DW surface waves into body waves on normal metal gratings is described in [10]. However, its application for studying the MDS calls for modernization of the measuring part of the setup, taking into account special features of the analyzed object (possibility of the presence of body waves both in dielectric and outside).

Below, a universal setup that can be used for analyzing electrodynamic characteristics of both the MDS and traditional metal periodic structures is described. Results of test measurements are compared to the numerical analysis.

The complex experimental test bench for determining electrodynamic characteristics of periodic heterogeneities (MDS, strip and reflecting metal gratings), when they are excited by the DW surface wave, consists of two main modules (figure 1): (1) module for measuring waveguide characteristics (standing wave ratios (SWR), attenuation constants, etc.), and (2) module for measuring spatial characteristics of periodic structures (directional radiation patterns in the far-field zone and amplitude distributions of fields in the near-field zone). The module for measuring spatial characteristics includes the studied object, which is generally dielectric prism 1 with the strip DG 2 imposed on its side surface. Prism 1 is attached to the special adjusting unit intended to spatially orient it in the x-, y-, and z-directions with an error of ±0.1 mm. DW 3 joins in matching junctions 4, which through waveguides of specified sections, determined by the studied wavelength range, are mated with the waveguide characteristics recording unit and matched load 5. Depending on characteristics of the DW (phase velocity of the surface wave v_{ph}) and MDS (strip grating 2 with period l and dielectric permittivity ε of prism 1), there are two possible modes of transforming the DW surface waves into body waves, which are excited with indices $n = 0, \pm1, \pm2, \ldots$ [1]. The main ones are shown schematically in figure 1 as directional radiation patterns (γ are the angles of radiation of the MDS-DW system, and ϕ are the angles of corresponding radiation harmonics in the free space after transmission through the dielectric prism).

The module for measuring spatial characteristics of the object consists of two versatile horn antennas 6, rotation axes of which pass in the E-plane through the radiation aperture and are brought into coincidence with the vertical axis of grating 2, and the rotation axis in the H-plane coincides with the DW longitudinal axis 3. This is intended to record radiation angles in interval $\phi = 10°–170°$ with error $\Delta\phi = \pm0.25°$, when the horns are mounted on special precision movable units and installed in the far-field zone, which is determined by the known ratio $z = a_m / \lambda$, where a_m is the maximal size of the antenna aperture, and λ is the radiation wavelength. In the course of measuring radiation patterns, the signal received by horn 6 arrived at input Y of the two-coordinate GP, and input X of the GP was connected to the rate-of-turn sensor of the receiving horn. Thus, when the movable antenna travels, the directional radiation pattern is fixed on the GP, and, after conversion into digital form, it is used for the computer processing of the data received.

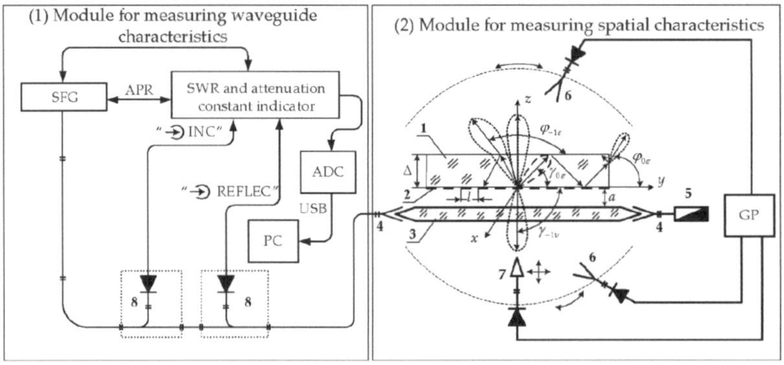

Figure 1. Block diagram of the experimental setup: (SFG) sweep-frequency generator, (APR) automatic power regulator, (ADC) analog-to-digital converter, (PC) personal computer, (GP) two-coordinate graph plotter. 1 — dielectric prism, 2 — strip DG, 3 — DW, 4 — matching junctions, 5 — matched load, 6 — horn antenna, 7 — probe, 8 — (DC) directional couplers, γ — radiation angles of the *MDS-DW system*, φ — angles of corresponding radiation harmonics in the free space after transmission through the dielectric prism, *x*, *y*, and *z* — spatial coordinates, Δ — thickness of the dielectric layer, *a* — distance between the DW and strip grating 2 (aiming parameter).

The amplitude field distributions along the axes of the radiating system are studied in the near-field zone ($z \approx \lambda$) by probe 7, made as a dielectric wedge ($\varepsilon = 2.05$), conjugated with the standard waveguide measuring section via the matching junction. Upon detecting, the signal from probe 7 arrives at the GP and is processed by the computer. Typical sizes of the probe of ~ $(0.1–0.2)\lambda$ ensured minimal distortions of the fields during measurements. The surface field indication system was installed on the transportation carriage, and, in this case, samples in the x-, y-, and z-directions were obtained with an error of ~ 0.1 mm.

On the whole, all indication systems of spatial characteristics and studied object were placed on a single solid laboratory platform, illuminated by special precision adjusting units. This allowed us to orient the regular part of the DW with respect to the MDS plane with an error of ~ 0.1 mm and ensure the corresponding monitoring of the coordinates of measuring elements of the setup and studied object.

The module for measuring waveguide characteristics (figure 1) is based on a standard panoramic SWR and attenuation constant measurer, consisting of the SFG, SWR and attenuation constant indicator, and DC 8 with detector sections connected to corresponding connectors of the SWR indicator. Depending on the method of bringing the couplers into the measuring line, plots of either transmission gain or SWR were determined in the specified frequency range. The obtained results were processed by the designed ADC and arrived via the USB bus at the PC for further processing. The constant power level at the input to the studied object was kept by the APR, being the part of the panoramic measurer, and the minimal reflections at the DW output were ensured by inserting matched load 5 into the measuring section (when absolute power levels were measured in the transmission line, standard watt-meters inserted directly into the measuring section instead of matched load 5 were used).

Procedure of performing measurements on the setup (figure 1):

(1) The matching of DW 3 with the module for measuring waveguide characteristics, which consists in reaching 1.1- to 1.2 SWR values for the specified frequency band by optimizing parameters of matching junctions 4. (2) Determination of the relative velocity of DW surface wave $\beta_w = v_{ph}/c$ (c is the velocity of light) and stray radiation level from matching elements 4. For this purpose, the calibration reflecting DG, characteristics of which are intended for radiation in the normal direction (angle γ_{-1v} in figure 1) in accordance with the procedure [10], and indication system of fields in the far-field zone (horn antennas 6) are used. (3) Determination according to the procedure [1] of the optimal value of the aiming parameter a from the minimal distortion of directional lobes at the central frequency. (4) Determination of MDS and DW characteristics for the specified excitation modes of body waves in accordance with relationships of [1]. (5) Replacement of the calibration reflecting grating by the studied MDS. (6) Measurements of spatial MDS characteristics with the simultaneous automatic control of its waveguide characteristics.

The described experimental setup (figure 1) is implemented for a 53- to 80-GHz band, and this fact determined the selection, as a panoramic SWR and attenuation constant measurer, of a corresponding device and waveguide sections with a 3.6×1.8-mm^2 cross section. The fluoroplastic DW with a 5.2×2.6-mm^2 cross section permitted us to obtain relative phase velocities of the surface wave in interval β_w = 0.78–0.81 and to excite on the fluoroplastic MDS two main spatial radiation harmonics, namely, $n = 0$ (l = 1.17 mm) and $n = -1$ (l=3.07 mm). The performed test measurements of SWR, directional radiation patterns, and amplitude field distributions gave satisfactory results, as compared with the numerical analysis by the finite difference method [5]. Thus, in particular, directional radiation patterns of the studied MDS are shown in figure 2.

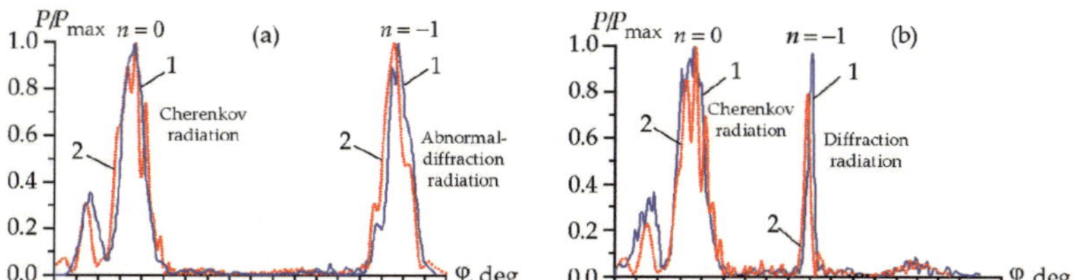

Figure 2. Typical directional radiation patterns of spatial waves on the MDS: (a) Cherenkov and abnormal-diffraction modes; (b) diffraction-Cherenkov mode. 1 — experiment, 2 — numerical simulation.

MDS parameters implemented in the experiment for three main operating modes of free-space wave excitation at the central frequency are given in the Table 1.

No	Operating mode of exciting spatial waves	*l* (mm)	*d* (mm)	*k*	*u*	β
1	Cherenkov $n = 0$ — in dielectric	1,17	0,39	0,30	+0,5	0,788
2	diffraction-Cherenkov $n = 0, -1, -2$ — in dielectric $n = -1$ — in free space	3,07	1,535	0,79	0	0,788
3	abnormal diffraction $n = -1$ — in dielectric	1,36	1,084	0,35	-0,8	0,598

Table 1. Parameters of the investigated MDS

The experimentally obtained data has been compared to the results of the numerical experiment based on solving Maxwell equations in a form of partial derivative using finite-difference approach and taking into account constitutive relations.

It follows from the given plots that the experiment satisfactorily correlates with the numerical analysis, and this, in turn, confirms serviceability of the described setup for studying electrodynamic characteristics of periodic structures belonging to a new class, i.e., planar MDS, which can find application in producing devices of MSM and terahertz wavelength ranges.

3. Practical devices of microwave technology and electronics

3.1. A quasi-optical directional coupler

The general principle of designing DC is to use two energy transmission lines coupled to each other [11, 12], along one of which the main power flow is transmitted; the auxiliary line is intended for interference and separation of forward and backward waves.

To date, depending on the imposed requirements, a great number of DC modifications are used in MSM wavelength measuring circuits.

The systems based on DW or dielectric planar waveguides [13] with distributed coupling, local coupling, and reemission into the secondary channel are the most close to the proposed DC.

In this work, the design of the DC based on two diffraction-coupled transmission lines with distributed radiation sources is studied, the sources being formed by periodic structures and DW placed along them. Figure 3 shows a general DC drawing and two possible reflector configurations, namely, plane-parallel and plane-cylindrical.

The basic DC section is formed from periodic structure 1, along the longitudinal axis of which DW 2 is placed at distance a. The second section is made similarly; it also consists of periodic structure 3 and DW 4. The periodic structures are applied on surfaces of flat and cylindrical mirrors with aperture *D*, the axes of which are placed in parallel at distance *H* and form a quasi-optical transmission line producing a beam of spatial waves. One of the radiators of the system is installed so that it can move smoothly along axis z, changing distance H and correcting the phase characteristics of waves.

The principle of operation of the diffraction-coupled DC is based on exciting in-phase and antiphase spatial waves. As a result of the propagation of these waves along the longitudinal DC axis, the radiators exchange energies and some power is directed into the secondary channel. Results of theoretical and experimental studies of transformations of DW waves into spatial waves and of spatial waves into surface waves on periodic structures are described in detail in [1].

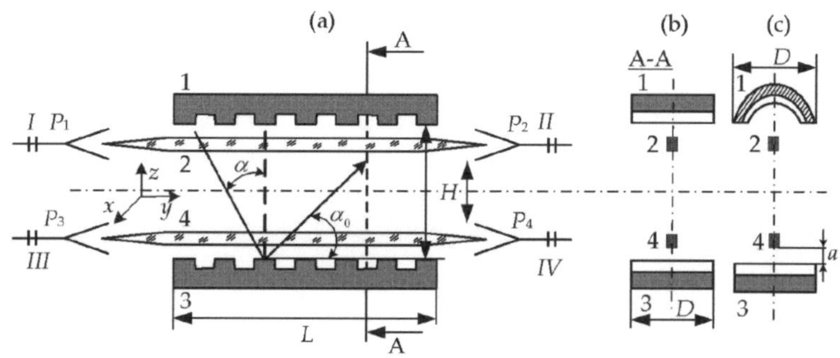

Figure 3. Schematic drawing of the design of the DC based on diffraction-coupled transmission lines: (a) cross section of components along the longitudinal axis; (b, c) cross section of components in the A–A plane at the (a) plane-parallel and (b) plane-cylindrical configurations of reflectors. 1, 3 — periodic structures, 2, 4 — DW, I–IV — arms of the DC, P_1— power at the DC input, P_3 — power of the backward derived wave, P_2 , P_4 — transmitted powers.

Let us dwell on the special features of the wave processes in the proposed quasi-optical DC (figure 3). When the microwave signal is applied to input I, the delayed wave propagates in DW 2 and is scattered at periodic structure 1. In this case, a diffraction field arises, which is a superposition of plane waves. Some waves go into the DC volume at angles α as spatial waves, and the remaining ones are localized near the grating as slow harmonics propagating to the output of waveguide 2. The radiation angle of the spatial waves is determined by the relation [1]:

$$\alpha = \arccos(1/\beta_w + n/k) \tag{1}$$

where $\beta_w = v_w/c$ is the relative wave velocity in the DW; v_w is the phase velocity of the wave; n = -1, -2,... is the number of the spatial radiation harmonic; and $k = l/\lambda$ is the wave number (lis the period of the grating, and λ is the radiation wavelength). It is possible to obtain prevalence of one or another type of waves by selecting the radiator characteristics, namely, the period of the structure, the wave velocity in the waveguide, and impact parameter α.

The second excitation stage of the system in figure 3 is the incidence of the spatial wave formed at angle α on structure 3. As a result of diffraction, the complete field over the periodic structure consists of the incident and spatial harmonics of the scattered field. If $l<\lambda/$

(1+sinα), the zero spatial harmonic is only reflected from structure 3 at angle α_0=90°-α, while the remaining field is the superposition of surface spatial harmonics propagating along the grating with the phase velocities

$$v_f = kc / (n + k\sin\alpha) \quad (2)$$

When $v_f \approx v_w$, the delayed wave is also excited in DW 4 and is transformed into a spatial wave on periodic structure 3. When surface waves are successively transformed into spatial ones and spatial waves are transformed into surface ones, in-phase and antiphase waves are formed along the axis of reflectors of the DC and, as a result of their interference, part of the incident and reflected power is branched off to the secondary section.

The geometric sizes of the considered DC are selected from the inequalities [6] that specify the fulfillment of laws of ray optics in the double-mirror quasi-optical systemml:

$$\frac{d^2}{H\lambda} \ll \left(\frac{H}{d}\right)^2; l / \lambda \geq 10 \quad (3)$$

where H and L are the distance between the mirrors and their length, respectively, and d=D/2 is the aperture radius of the radiator mirrors (distance from the longitudinal axis of the system to the mirror periphery).

The fulfillment of the first inequality allows one to represent the field of the studied system as paraxial wave beams (figure 3), which are in many ways similar to a plane wave [14]. The second inequality minimizes the resonance phenomenon display along the longitudinal axis y. The structure period l is selected from (1) for specified values of λ, n, β_w, and α. The values of impact parameter $a < \lambda$ are corrected experimentally by studying spatial characteristics of radiating systems, based on the criterion of the minimum parasitic DC effect on the field of the periodic structure.

When cylindrical mirrors with quadratic correction are used, the optimal values of their curvature radii, R, are in interval 2/R = 4÷6. This is attributed to the fact that the focusing action of the mirrors decreases at $2R/\lambda > 6$ and, for $2R/\lambda < 4$, the mirror aperture intercepts only part of the radiation of the periodic system–DW system.

We consider an example of selecting parameters of the DC and its components in frequency band f = 30÷37 GHz. As a source of the surface wave, a polystyrene waveguide with a cross section of 7.2×3.4 mm^2 was used. It provided for relative velocity $\beta_w \approx 0.9$ at the wavelength λ = 9 mm. The main lobe of the radiation pattern (n = -1) of the spatial wave was formed at angle $\alpha \approx 70°$, and, from relation (1), it corresponded to l = 11.5mm. The waveguide length L = 150 mm and radii of radiating apertures d = 30 mm ensured the fulfillment of conditions (3), and value R = 25 mm ensured the optimal focusing of the quasi-optical waveguide beam along the DC axis, when one of the mirrors was cylindrical (figure 3 (c)).

The experimental studies of the DC prototype were performed on the described above setup by measuring directional diagrams of the radiating systems and their near-zone fields and also by measuring waveguide characteristics of both separate DC components and the system as a whole. The mechanical part of the setup allowed one to move DC radiators in three planes with an accuracy of ±0.1 mm and, by varying over wide ranges values a and *H*, to adjust the coupler according to optimal output parameter values.

As an example, the characteristics of the above design of the DC for two distances between the radiating mirrors are shown in figure 4.

It follows from these characteristics that, for the specified frequency interval, at distances between radiators H = 65 mm, the attenuation constant (P_4/P_2) values are in a 3- to 12-dB interval and the directivity factors (P_3/P_4) are in a 15- to 25-dB interval. It is possible to improve the DC output characteristics by correcting the phases of propagating waves, while H is varied; this is demonstrated by curves 2 in figure 4 ($P_4/P_2 \approx 3$–7 dB, and $P_3/P_4 \approx 30$ dB).

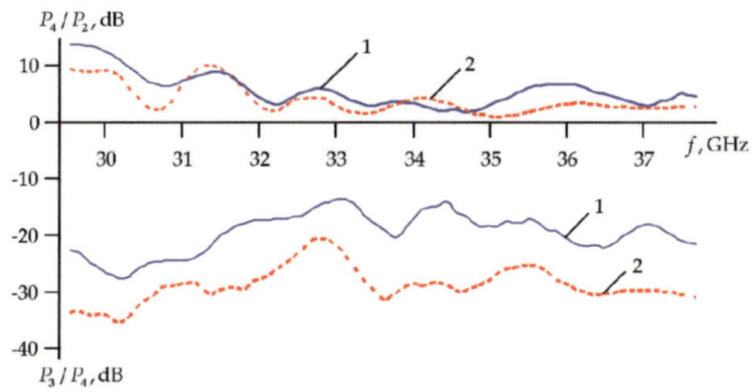

Figure 4. Characteristics of the coupler at (1) *H* = 65 and (2) 80 mm.

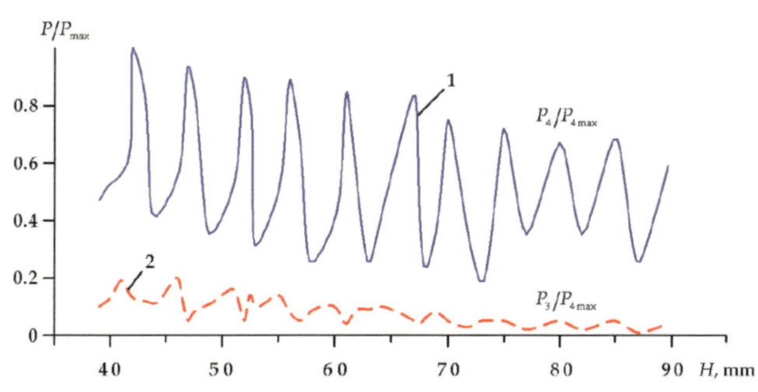

Figure 5. Relative power levels in the secondary channel as function of the distance between the mirrors. P_{4max} is the maximal power arriving at the secondary channel.

It is determined that the quasi-optical wave behavior of the studied system remains unchanged for $H = (5–10)\ \lambda$, for which inequality (3) is true.

Figure 5, in particular, illustrates the relative dependences of power levels $P_4/P_{4max} = f(H)$ and $P_3/P_{4max} = f(H)$, which are detected in arms IV and III, respectively (figure 3). It can be seen from the plots that the optimal (from the viewpoint of decoupling the branched and reflected signals) values are H = 60–90 mm. At $H > 10\lambda$, the diffraction loss increases and the system becomes critical to the mirror alignment.

By comparing the characteristics of the described DC with couplers on coupled DW [15], the following conclusions can be drawn. The DC characteristics in the working frequency band Δf = 5 GHz are comparable with those of the coupler on the effect of the directional signal reradiation from the section of a sharp DW bend [15], the DC design being significantly simpler. In addition, this DC has two variants for correcting the transient attenuation (by changing impact parameter a and distance between radiators H), thus decreasing process requirements for manufacturing this system in the MSM wavelength range.

3.2. Quasioptical power dividers

In this subsection the results of studying the power divider (PD) design based on a two-row semi-transparent periodic structure with a distributed radiation source (main section) in the form of the DW, which is placed along its longitudinal axis are presented. In this system, the surface wave of the DW is transformed on periodic grids into spatial harmonics of body waves [16]. Figure 6 shows two PD modifications with longitudinal (Δd) and angular (ϕ) shifts of grid bars.

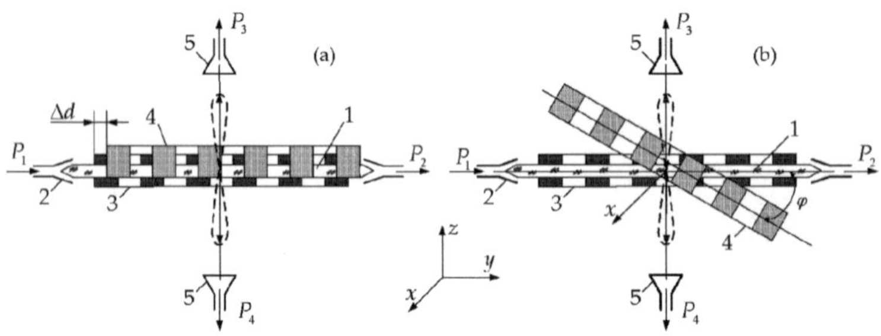

Figure 6. Embodiment scheme of the PD on two-row periodic structures: (a) with longitudinal Δd; (b) with angular φ shifts of grid bars. 1 — main section (DW), 2 — matching junctions, 3, 4 — grids, 5 — radiation receivers.

The PD contains DW 1, which is embedded into the main microwave section through matching junctions 2, grids made of bars 3 and 4, and radiation receivers 5. Grid 3 is fixed in position with respect to the y axis, and grid 4 is placed on the positioner that ensures either its linear movement along the y axis with accuracy $\Delta d = \pm 0.01$ mm or angular movement in the plane of the x, y axes with accuracy $\Delta\phi = \pm 1°$. The radiation receivers operating in MSM wavelength ranges can be made, e.g., as horn or lens antennas and are intended for detecting the power diverted from the main section.

The principle of operation of these PD is based on transforming the DW surface wave into the spatial (body) wave, which is excited on the grids. The radiation power level of this wave can be regulated by changing phase relations of slow harmonics of waves excited on the grids of the two-row structure when its bars are shifted or when the rotation angle with respect to the DW axis is changed.

We consider general features of wave processes in the described quasi-optical PD (see figure6). When the microwave signal with power P_1 is applied to the input of DW 1, the delayed wave propagates over its surface and scatters on grids 3 and 4. In this case, the diffraction field is formed, which is the superposition of plane waves. The part of these waves goes away into the free space (powers P_3 and P_4 extracted from the main power section) at angles γ_n in the form of spatial waves, and the remaining waves are localized near the grids in the form of slow harmonics propagating toward DW output 1 (power P_2 . The angle and conditions of emission of the spatial waves in the DW-grid system are determined from the relation [10]:

$$|\cos\gamma_n| = \frac{1}{k}\left(\alpha - \frac{2\pi n}{l}\right) \le 1 \tag{4}$$

where γ_n is the emission angle of the harmonic with index $n < 0$, 1 is the period of the grid, $k = 2\pi/\lambda$, and $\alpha = k - (2\pi n)/l$ is the propagation constant.

From equation (4), it follows that the waves with $n \ge 0$ and $|\cos\gamma_n| > 1$ are a spectrum of heterogeneous plane waves existing near the grid and propagating along the longitudinal axis of PD system with phase velocities $v_{ph} < c$, where the v_{ph} is the phase velocity of the wave, and c is the speed of light. For all remaining waves with indexes $n < 0$, the condition of their emission into the ambient space is met, which characterizes them as spatial (body) waves. By selecting appropriate parameters of the electrodynamic system and the surface wave propagating along the DW, it is possible to attain the predominance of one or another type of waves.

The geometrical sizes of the considered PD are selected from equation (4) when the fundamental ($n = -1$) spatial wave is emitted at angle $\gamma_{-1} = 90°$ to the plane of grids 3 and 4 and the latter are mirrored with respect to the DW axis ($\Delta d = 0$ and $\Delta\phi = 0°$). In this case, it is necessary to take into account the influence of bar thickness $2h$ on the radiation intensity [17]. The maximum bar thickness is determined as:

$$2h = N\frac{\lambda}{2} + d\frac{4}{\pi}\ln\sin\frac{\pi\theta}{2}, \ (N = 1,2,...) \tag{5}$$

where d is the bar width, and θ is the relative distance between the grid bars.

In the grid with the optimal profile, somewhat smaller than the integer number of half-waves should be fitted in bar thickness $2h$. The correction for half-wave thickness $\Delta h = d\frac{4}{\pi}\ln\left(\sin\frac{\pi\theta}{2}\right)$ is equal for all harmonics and determined only by the geometrical parameters of the grid regardless of the wavelength and angle of emission. This correction is at maximum ($\Delta h_{\max} = -0.23l$), when the slot width $\theta \cong 0.3$.

In addition, the intensity of the spatial waves in the considered systems can be regulated by the duty factor of the semi-transparent grids made of bars $u = \cos(\pi d/l)$ and sighting parameter a, which is the distance from the DW to the surface of the grids and determines the degree of relationship between the surface waves and heterogeneous waves of the grids.

The interval of the optimal value region of parameter Δd for the first PD modification (see figure 6(a)) was determined experimentally due to the absence of the theory of these systems. For the second PD modification (see figure 6(b)), the ranges of angles ϕ within which grid 4 is shifted with respect to grid 3 can be estimated from the limiting (critical) values of the angle of rotation of the single grid [10]:

$$\phi_{cr} = \arccos\left[\frac{1}{2}\left(\frac{k\beta_w}{n} - \frac{n\beta_w}{k} - \frac{k}{n\beta_w}\right)\right] \tag{6}$$

Hence, the surface wave mode occurs in the open structure of the DW and grid 4 at $\phi = \phi_{cr}$, and the further increase in ϕ results in the Bragg diffraction, for which $\phi_{Br} = \arccos\frac{N_B\beta_w}{2k}$, $N_B = 1,2,3,...$ is the order of diffraction. In this mode, the heterogeneous surface wave of the DW is transformed into the surface wave of the grid. The formed new surface wave propagates at an angle of (180°–ϕ) with the direction of the original heterogeneous wave. The degree of transformation of one wave into the other is determined by the longitudinal dimensions of the grid and the coupling between the DW and grid field.

As an example, let us select the parameters of the PD and their main elements in the four-millimeter wavelength range (f = 60–80 GHz). As a source of the surface wave, a fluoro-plastic waveguide with a 5.2×2.6 mm^2 cross section is used, which ensures the relative wave velocity $\beta_w = 0.788$ at wavelength $\lambda = 4$ mm. The main lobes of the radiation pattern (n=–1) of the spatial waves were formed at angle $\gamma = 90°$. According to equations (4) and (5), this corresponded to values $l = 3$ mm, $u = 0$, and $2h = 1$ mm. The grid aperture $D = 60$ mm and their length along the DW axis $L = 54$ mm ensured the meeting of the main quasioptic conditions [15].

The experimental studies of the PD prototypes (figures 6 (a), (b)) were performed in accordance with the described above procedure.

As an example, the amplitude (relative values P_3 / P_{3max}) and waveguide (transmission coefficients $K_{tr} = P_2 / P_1$) characteristics of the above PD are shown in figure 7. They indicate that it is possible to extract 80% of the power from the main section ($P_2 = 0.2P_1$) and divide it equally when the grids are placed symmetrically with respect to the DW ($P_3 = P_4 = 0.4P_1$) for the specified sighting parameter $a = 0.7$ mm.

The analysis of the plots in figure 7 shows that the power in the arm P_3 can be smoothly regulated either by shifting the bars or changing the rotation angle of grid 4 with respect to axis of grid 3. In this case, for the PD modification in figure 6 (a), it is expedient to select the $\Delta d/\lambda$ values in an interval of 0.2–0.4, in which the dependence of the radiated power is close to the linear one and has no resonance effects.

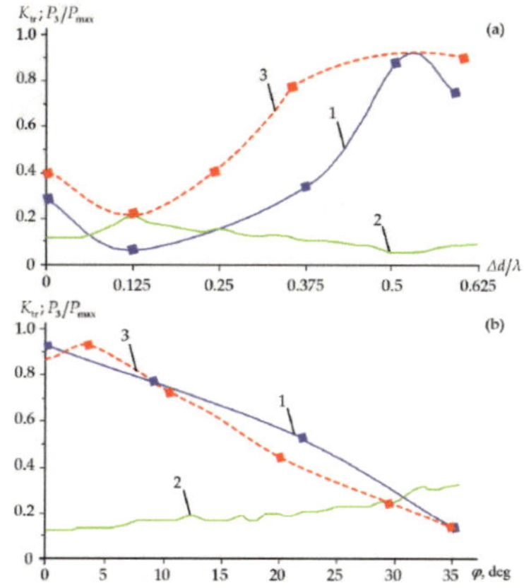

Figure 7. Amplitude and waveguide characteristics of the PD with (a) longitudinal and (b) angular shifts of grid bars with respect to the axis of the waveguide. 1 — relative power values P_3, 2 — K_{tr} values, 3 — calculated curve.

From the viewpoint of regulating and calibrating the emitted power P_3, the second PD modification is preferable (figure 6 (b)). Its amplitude characteristic is virtually linear for the selected angle interval ϕ = 0–35°. For both PD modifications, the standing-wave ratio values were 1.1–1.3 in frequency range f = 60–80 GHz. The characteristics shown in figure 7 were numerically simulated by the finite difference method [5], and the calculated results are illustrated in figure 7 by dotted lines 3. The comparison of the experimental results and numerical analysis indicates their correlation.

By analyzing the characteristics of the described PD, it is possible to make the following conclusions. In contrast to the waveguide analogs and PD on DW [15], the described PD possess wider functional capabilities, allowing one to divide the power coming from the main section into equal parts between two channels and smoothly regulate it in one of the output arms of the divider. In this case, the powers arriving at the output arms of the divider can be regulated by changing the sighting distance of the DW with respect to the grid surfaces. The presented PD can be also manufactured in the planar form if periodic MDS are used [11].

3.3. Possible variants of implementation of MSM radiation sources

Nowadays MSM microwave devices (backward-wave tubes, taveling-wave tubes, klystrons) with high level of output power are widely used in transmitting equipment of communication systems, radars and radio countermeasures devices. Thus, there is the rapid development in production of traveling-wave tubes with slow-wave structures as a chain of coupled resonators in different geometrical modifications [8], creation amplifiers based on the multielectron-beam tubes [18] etc. Axially symmetric EB as the main working element of such devices virtually defines their basic working parameters. Therefore, special attention is given to the improvement and optimization of electron-optical systems of MSM microwave devices.

As a rule, an optimization of electron-optical system parameters bases on the information about EB characteristics. Currently, because of the rapid development in computational resources and approaches, different techniques of electromagnetic simulation based on various numerical algorithms are mostly used to get this information. Such approach allows avoiding carrying out time-consuming and expensive experiment. Usually modelling techniques consist of two stages: computation of electromagnetic fields of the investigated system and subsequent trajectory analysis of charged particles in these fields.

In the current subsection the example of the model of the three-electrode electron gun which is used in real microwave devices such as travelling-wave tube has been presented, the optimization problems of its operating modes and axially symmetric EB characteristics have been outlined.

Axially-symmetric EB, as a rule, is formed by three-electrode electron gun with converging optics and introduced into slowing-down system where it is focused by periodic magnetic field. The cathode is usually produced in the form of a core made of tungsten-rhenium blend with activated surface. The typical electrode configuration of axially symmetric electron-optical system is demonstrated in figure 8.

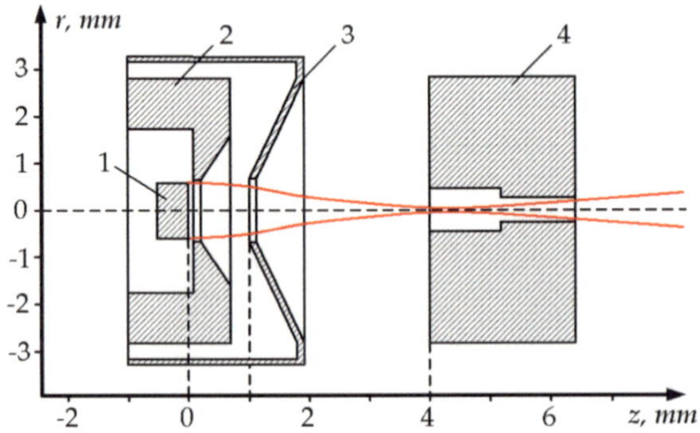

Figure 8. Scheme of electron-optical system of axially symmetric electron gun: 1 — hot cathode (U_k), 2 — focusing electrode (U_f), 3 — first anode (U_{a1}), 4 — second anode (U_{a2}).

The guns of such type allow forming EB with diameter of about 0.3mm in the crossover, the beam current of 1–25 mA with accelerating voltage 2000–6000 V. For numerical simulations the finite integration technique (in literature is known as FIT [19]) was chosen as the optimal numerical algorithm for analyzing the above mentioned systems.

The combination of electrode potentials, described in [20], was taken as the initial parameters with a beam perveance P = 0.045 mkA/V$^{3/2}$, the number of emitted particles N = 2965, average kinetic energy of electrons E_k = 5 eV with energy spread dE_k = ±0.33 E_k, maximal angular deflection α = ±10° from the axis of the system, that corresponds to the real guns of such type. Meanwhile, the emitting area has been defined as a perfectly plane surface with the uniform current density distribution.

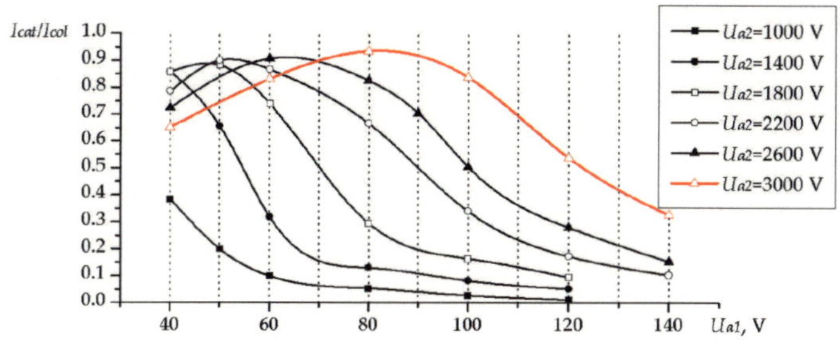

Figure 9. Coefficient of the EB passage with the different potential values U_{a1} and U_{a2}.

When simulation of the electron-optical system is carried out, the main parameter of optimality is the coefficient of EB passage. That is why in the first step of the optimization of the electron gun operating modes the research of a beam passage coefficient $K = I_{cat}/I_{col}$ was conducted, where I_{cat} and I_{col} are cathode and collector currents respectively. The simulation has shown that too high potential of focusing electrode ($U_f \approx$ -40V) impairs the passage of EB and causes unsatisfactory crossover position between two anodes of electron-optical system but not after the second anode. In addition, as the focusing electrode is placed in the immediate vicinity to the cathode surface the partial EB blocking effect is created and the kinetic energy of the emitted electrons is not high enough to overcome thus formed potential barrier. This blocking effect decreases considerably an EB passage ratio and disturbs beam laminarity. Changing the potential of the focusing electrode might partly solve mentioned problems. When U_f is decreased to zero potential the beam crossover is shifted to the region of the second anode transit channel. In this case the diameter of the crossover decreases and the passage of EB with U_f = -10÷0V might reach 93-95%. Thus, to avoid any impact of the focusing electrode on the cathode edge U_f = 0V was taken as the most appropriate for the experimental research, filament current of the tungsten- rhenium (W-Re) cathode for the whole set of measurements was constant — 0.7 A.

Experimental measuring results have shown that the coefficient of the EB passage increases with increasing U_{a2} nd riches the value of K = 0.934 with U_{a2} = 3000V, that corresponds to the current value I = 9.1 mA. However, further increasing of the U_{a1} and U_{a2} potentials stimulates electron emission and causes rapid growth of the cathode current (I_{cat} > 15 mA). This, in turn, leads to the strong heating of the gun elements, bombarded by electrons, which considerably disturb the consistency of the device operation and causes different measurement errors.

Figure 9 shows that the coefficient K reaches its maximum value at the accelerating potentials of U_{a1} = 80 V and U_{a2} = 3000 V. Thus, such electron-optical system operating mode can be considered as the optimal for the passage of the EB.

Besides the information about the electron deposition on the gun electrodes, which allows carrying out a preliminary analysis of its operating modes, the EB quality parameters (such as type of a particle distribution in the cross-section, laminarity, spread of the velocity transverse components etc.) are also very important. In the current work root-mean-square emittance (statistical emittance) was used to describe such characteristics:

$$\varepsilon = \left(\langle x^2 \rangle \cdot \langle x'^2 \rangle - \langle x \cdot x' \rangle^2 \right)^{1/2} = \left(\langle x - \langle x \rangle^2 \rangle \cdot \langle (x' - \langle x' \rangle)^2 \rangle - \langle (x - \langle x \rangle) \cdot (x' - \langle x' \rangle) \rangle^2 \right)^{1/2} \tag{7}$$

where

$$x' = \frac{v_x}{v_\perp}, \ \langle x \rangle = \frac{1}{N} \sum_{n=1}^{N} x_i \tag{8}$$

It should be noted, that formula of this type in contrast to other definitions makes it possible to express an EB emittance in a simple numerical form. In addition, we do not find it reasonable to consider a normalized beam emittance in this case because of short drift duct and low accelerating potentials of the system (EB is not relativistic) [21].

The calculated emittance with the maximal passage of the EB at a distance of 10 mm from the cathode made up $\varepsilon_x = 1.71$ mrad·mm and $\varepsilon_y = 1.85$ mrad mm in x- and y-directions respectively, the beam diameter — $D = 0.8$ mm.

However, as the numerical experiments to study transversal electron dynamics (figure10 (a)) has shown, value of the emittance can be decreased by variations of the potential U_{a1}.

The analysis of two-dimensional diagrams of the transversal emittance has demonstrated that the increase of the U_{a1} to 110 V considerably disturbs the beam laminarity (figure10(b)) while its decrease improves the EB quality (type of the diagram corresponds to the perfectly divergent beam) with slight growth of its diameter ($D \approx 0.8$ mm). On the basis of the analysis of experimental and numerical simulation results the optimal operating mode with the potentials $U_{a1} = 60$ V и $U_{a2} = 3000$ V has been chosen. It allows getting the passage of the EB at the level of $K \approx 0.85$ with transversal emittance values $\varepsilon_x = 1.67$ mrad·mm and ε_y=1.76 mrad·mm.

Development of high-resolution sensitive elements for terahertz frequency bandwidth is an actual problem due to a number of existing international projects of radio astronomy as well as the projects with the objective of studying the Earth atmosphere. The basic issues of arrangement of receivers in the given bandwidth are solved by application of solid-state heterodyne oscillation sources and the mixers based on the effect of electron heating in the superconductor (Hot Electron Bolometer) because such mixers have no competition analogs in this bandwidth. By now these mixers have been successfully realized at the frequencies of the order of 3 GHz. However, several research projects are related to development of the mixers optimized for higher frequencies; e.g. within the framework of the SOFIA project it is developed a heterodyne receiver for 4.8 THz. Principal opportunity for creation of the mixer for the given bandwidth is shown in the paper [22] where the method of electron and photo lithography is applied for its realization. As an example, figure 11 provides photos of the central part of the mixer obtained with the scanning electron microscope.

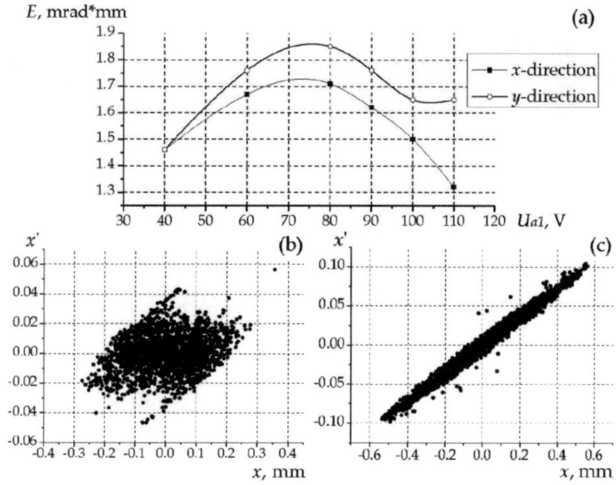

Figure 10. Dependence of EB emittance on the U_{a1} potential: (a) numeral values in x- and y-directions; (b), (c) two-dimensional diagrams for U_{a1} = 110 V and U_{a1} = 40 V, respectively.

The mixers were made of NbN films with the thickness of 2 to 3.5 nm upon a silicon substrate with a MgO buffer sublayer and possessed the temperature of superconducting transition of 9 to 11 K. The best value of the noise temperature of the receiver on the basis of the electron heated mixer amounted to 1300 K and 3100 K at the heterodyne frequencies of 2.5 and 3.8 THz correspondingly.

In the radio astronomy devices and tools the preference is given to solid state heterodyne sources within the terahertz band due to their small dimensions, low weight and power consumption requirements, despite low output power level, which is not exceeding 1 μW at the frequency of 2 THz that to a significant extent complicates the problem of development of mixers for a low level of output power. Thus, there still remains actual the problem of realization of low-voltage electrovacuum oscillation sources within the terahertz wavelength band possessing higher values of power levels as compared to those of solid state oscillators.

By the present time this problem can be solved by means of using in the vacuum electronics of planar periodic MDS [1, 11]. Block diagrams of such devices that could be realized on the basis of the above described technologies and new types of dielectrics with larger values of dielectric permeability and low loss values at high frequencies [23], are provided in figure12.

Block diagram of the oscillator in figure 12(a) suggests modulation of the EB upon a backward wave of the periodic structure (position 2) with subsequent excitation of the Cherenkov oscillation harmonic in MDS (position 3). Within the oscillator in figure 12(b) it is applied the mode of abnormal diffraction oscillation, which is realized at substantially less values of accelerating voltages of the electron flux compared to those of the Cherenkov oscillation harmonic.

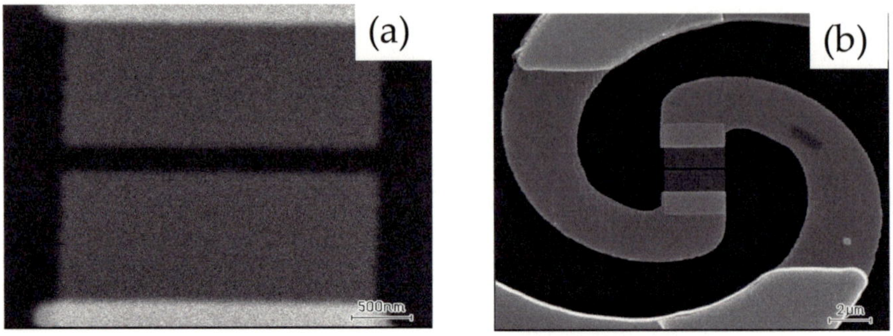

Figure 11. Image of the helix (a) and of the central part of the helix with the NbN bridge (b) obtained with the scanning electron microscope [22].

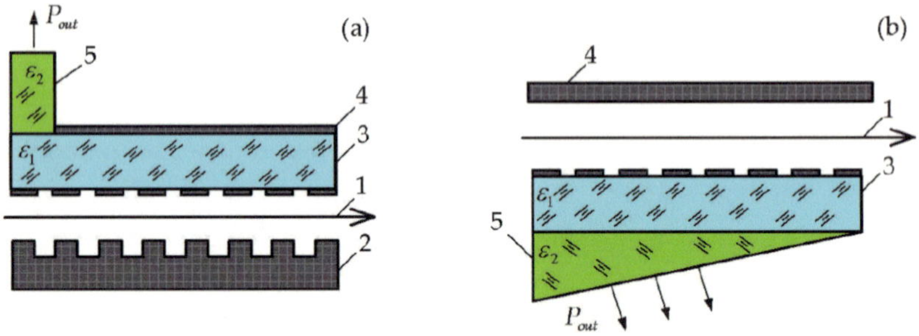

Figure 12. Block diagrams of terahertz frequency band electrovacuum oscillation sources.1 — EB, 2 — periodic structure, 3 — MDS with diffraction strip-like grating, 4 — screen (mirror), 5 — energy output device.

At practical realization of the above block diagrams of sources of oscillation it is necessary to solve a set of problems related to the technology of manufacturing of main units of the device – the electrodynamic system, the low-voltage source of electrons and the focusing magnetic system. The above considered technological processes – like, for instance, nano die forming – eliminate all problems related to manufacturing of both reflecting and ribbon (applied upon the dielectric layer) DG with the micron period.

Oscillator schemes presented in figure 10 are the ideological continuation of the so far developed low-voltage backward-wave tube with multi-row slow-wave structures [8, 24]. A schematic view of a low-voltage orotron based on parallel MDS and three EB are shown in figure 10 (a). The calculated MDS parameters have been determined for EB radiation at the angle of 90°. The diagram of a backward-wave tube based on an anomalous diffraction radiation for MDS is shown in figure13 (b). The arrows show the direction of wave radiation on the MDS.

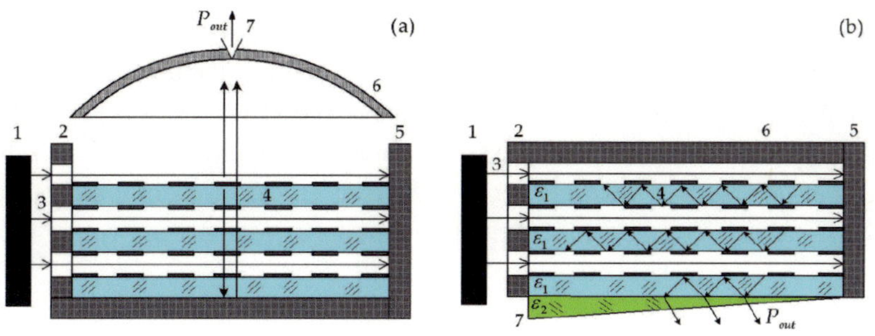

Figure 13. Schematic view of low voltage vacuum devices employing multi-linked MDS:(a) orotron, (b) backward wave oscillator. 1 — cathode, 2 — anode, 3 — EB, 4 — planar periodic MDS, 5 — collector, 6 — mirror, 7 — output port.

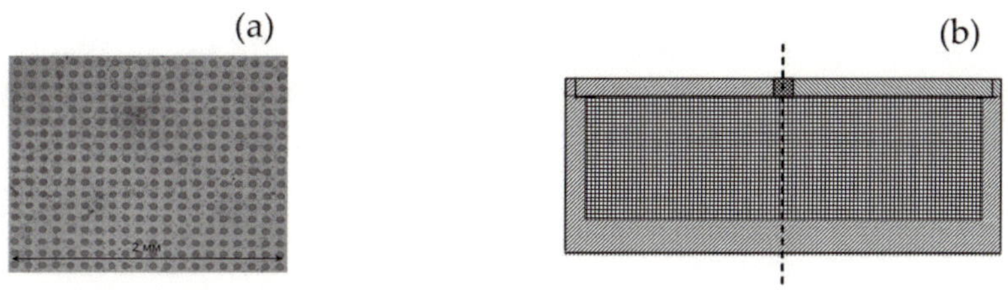

Figure 14. Microphotographic images of the field cathode surface [27] (a) and block diagram of the "slot" L-cathode device [28] (b).

The analysis of parameters of the discussed electrodynamic systems [11] and devices demonstrated in figure 13 shows that the DG/MDS or multi-row systems on basis of MDS at acceleration voltages $U \approx 1000$ V and $\varepsilon_r = 100$ (ceramics based on titanium oxide) can be realized for a wave length range from 1mm to 0.1 mm and periods lfrom 64 μm to 4 μm. Currently the fabrication technology for such structures is mature and has been used for fabrication of multiple parallel slow-wave structures in backward-wave tube. The fabrication technologies include electro-erosion machining, cold forming, photolithography, electron and x-ray lithography, vacuum and plasma deposition. The fabrication of slow-wave structures of vacuum electron tubes requiring nanoscale precision is performed by ion beam lithography [25] in combination with nanoforming [26].

The research conducted in [1] indicates that in order to realize the described devices in submillimeter and infrared wave ranges, an EB should be as thin as 0.04 mm. Nowadays the practical realization of such EB is possible using array type or slot type L-cathode, which allow getting uniformly distributed and stable electron emission with high current density at comparatively low field intensity.

The matrices of field emission cathodes possess the preset geometrical dimensions (diameter, step), they allow elimination of the screening effect and obtaining of homogeneous upon the surface and stable in time electron emission with the average current value of 40 μA from a separate cathode at relatively low electric field intensity values. The cathode represented in figure 14 (b) includes the cylinder container filled with a stock of the substance that decreases the output operation of the operating surface, which is represented by a continuous row of micro-elements forming up a "slot" L-cathode. Experimental results obtained while pilot testing of those cathodes at IRE NAS of Ukraine proved the opportunity for obtaining of such electron fluxes with high-value current density at not high values of accelerating voltages.

4. Conclusions

In the current work the results of development of the experimental setup and procedure for measuring electrodynamic characteristics of planar periodic MDS, which can be used for manufacturing practical equipment operating in MSM and terahertz wavelength ranges are presented. Serviceability of the setup is checked by comparing the waveguide and spatial characteristics obtained experimentally and by numerical methods in a 4-mm wavelength range. A circuit and principle of operation of DC based on diffraction-coupled transmission lines, the emitting apertures of which are formed from periodic structures and DW has been proposed on the basis of the conducted research.

The experimental studies of the coupler prototype in a 30- to 37-GHz band have shown that the coupling on spatial waves allows one to obtain transient attenuation values in a 3- to 20-dB interval at a ~ 30 dB directivity, this being roughly in conformity with similar characteristics of a coupler on DW. The main advantage of the described DC is that it is possible to correct its characteristics over wide ranges by changing the distance between the emitting apertures.

Circuits and principle of operation of quasi-optical PD based on two-row periodic structures, formed by grids of metal bars and a DW placed along their longitudinal axis, have been proposed and described as the second example of realization of described above characteristics. The experimental studies of PD prototypes in a frequency range of 60–80 GHz have shown a possibility of regulating the emitted power level in the main divider arm by changing longitudinal and angular coordinates of the two-row periodic structure, which can be used for designing quasi-optical attenuators.

The example of the model of the three-electrode electron gun of travelling-wave tube has been described, the optimization problems of its operating modes and axially symmetric EB characteristics have been outlined.

In addition, studied properties of electrodynamic diffraction characteristics of the surface waves on periodic heterogeneities can be realized in implementation of radiation sources on the Smith-Parcell effect.

Work is supported by the governmental programme No0112U001379.

Author details

Gennadiy Vorobyov[1*], Larissa Vietzorreck[2], Ivan Barsuk[1] and Aleksandr Rybalko[1]

*Address all correspondence to: g.vorobyov@sumdu.edu.ua

1 Sumy State University, Sumy, Ukraine

2 Technische Universität München, München, Germany

References

[1] Shestopalov, V.P. (1991). *Diffraction radiation oscillators*, Kyiv, Naukova Dumka.

[2] Vorobjov, G., Shulga, Yu., & Zhurbenko, V. (2011). Quasi-optical Systems Based on Periodic Structures. *Zhurbenko V. (ed.) Electromagnetic Waves*, Rijeka, InTech, 257-282.

[3] Vorobjov, G. S., Petrovskii, M. V., & Krivets, A. S. (2006). Possible applications of quasioptical open resonant metal-dielectric structures in EHF electronics. *Radioelectronics and Communications Systems*, 49(7), 38-42.

[4] Vorobjov, G. S., Tsvyk, A. I., Krivets, A. S., Shmatko, A.A., & Petrovsky, M. V. (2003). The Smith-Pursell Effect Amplification of the Electromagnetic Waves in a Open Waveguide with a Matal-Dielectric Layer. *Telecommunications and Radio Engineering*.

[5] Bankov, S.E. (2004). *Analysis and optimization of microwave three-dimensional structures with HFSS.*, Moscow, Solon-Press.

[6] Sirenko, Yu. K., Strom, S., & Yashina, N. P. (2007). Modeling and Analysis of Transient Processes in Open Resonant Structures: New Methods and Techniques. New York, Springer.

[7] Bratman, V. L., Glyavin, M., Yu, , Kalynov, Yu. K., Litvak, A. G., Luchinin, A. G., Savilov, A. V., & Zapevalov, V. E. (2010). Terahertz Gyrotrons at IAP RAS: Status and New Designs. *Journal of Infrared, Millimeter and Terahertz Waves* [8], 8-934.

[8] Yakovenko, V.M. (2007). *Vacuum devices millimeter waves*, Sevastopol, Weber.

[9] Federici, J., & Moeller, L. (2010). Review of terahertz and subterahertz wireless communications. *Journal of Applied Physics* [107], 107-1063.

[10] Shestopalov, V.P. (1985). *Physical basis for millimeter- and submillimeter-wave equipmen. 1Open-type structures)*, Kyiv, Naukova Dumka.

[11] Vorobyov, G. S., Petrovsky, M. V., Zhurba, V. O., Ruban, A. I., Belous, O. I., & Fisun, A. I. (2007). Perspectives of application of new modifications of resonant quasi-optical structures in EHF equipment and electronics. *Telecommunications and Radio Engineering*, 66(20), 1839-1862.

[12] Xiao-Ping, Chen., & Ke, Wu. (2008). Substrate Integrated Waveguide Cross-Coupled Filter With Negative Coupling Structure. *Microwave Theory and Techniques, IEEE Transactions on Microwave Theory and Techniques*, 56(1), 142-149.

[13] Demydchik, V.I. Microwave electrodynamics Minsk Universitetskoe (1992). .

[14] Weinstein, L.A. (1966). *Open resonators and open waveguides*, Moscow, Sov. Radio.

[15] Valitov, R.A. (1969). *Submillimeter-wave equipment*, Moscow, Sov. Radio.

[16] Quan, X., Kang, X., Yinghua, R., & Jun, T. (2011). 3dB power splitter design based on coupled cavity waveguides. *International Journal for Light and Electron Optic*, 122(2), 156-158.

[17] Shestopalov, V.P. (1976). *Diffraction electronics*, Kharkiv, Kharkiv UniversityPublishers.

[18] Sinicin, N. I., Zakharchenko, Y. F., & Gulyaev, Y. V. (2009). A new class of high-power low voltage multipath TWT elated to chains multiple-Resonators with a transversely-extendedtype of interaction for airborne radar and communication systems the short-millimeter waves. *Jornal of radioelectronics*, 50(10), 46-51.

[19] Weiland, T.A. (1977). Discretization Method for the Solution of Maxwell's Equations for Six-Component Fields.Electron. *Commun. (AEU)*, 31(3), 116-120.

[20] Belousov, Ye. V., Vorobyov, G. S., Korg, V. G., Pushkarev, K. A., & Chaban, V. Y. (1992). Experimental investigation of the static parameters axisymmetrical electron beams of small diameter. *Recent developments in applied physics*, 2(1), 87-100.

[21] Braun, Ian. G. (1998). *The physics and technology of ion sources*, New York, A Wiley and Interscience Publication.

[22] Finkel, M. I., Maslennikov, S. N., & Goltsman, G. N. (2005). Superheterodyne receivers with superconducting THz mixer on electronic heating. Trans. Higher Education. *Radio Physics*.

[23] Nanasheva, Ye.A., Trubitsina, O. N., Kartenko, N. F., & Usov, O.A. (1999). Ceramic materials for microwave electronics. *Solid State Physics*, 41(5), 882-884.

[24] (2002). *Vacuum Microwave Electronics: Collection of reviews*, Nizhny Novgorod, Institute of Applied Physics.

[25] Watt, F., Bettiol, A. A., van Kan, J. A., Teo, E. J., & Breese, M. B. (2005). Ion beam lithography and nanofabrication. *International Journal of Nanoscience*, 4(3), 269-286.

[26] Ansari, K., van Kan, J. A., Bettiol, A. A., & Watt, F. (2006). Stamps for nanoimprint lithography fabricated by proton beam writing and nickel electroplating. J. Micromech. *Microeng.*, 16-1967.

[27] Solovey, D. V., Sakharuk, V.N., & Novitsky, A.M. (2009). *19th International Crimean Conference "Microwave Equipment and Telecommunication Technologies CriMiCo 2009 September*, 14-18, Sevastopol National University, Sevastopol, Ukraine.

[28] Belousov, Ye. V., Zavertannyi, V. V., & Nesterenko, A. V. (2006). Diode electron gun with a slot L-cathode. *Radio Physics and Electronics*, 11(2), 275-280.

Section 4

Scattering, Radiation and Propagation

Chapter 9

Radiative Transfer: Application to Indoor Propagation

Ada Vittoria Bosisio

Additional information is available at the end of the chapter

http://dx.doi.org/10.5772/50871

1. Introduction

In recent years, with the increasing interest in indoor wireless communications systems, the development of appropriate tools for modeling the propagation within an indoor environment is becoming of utmost importance. A versatile technique for studying propagation in such a complex scenario is ray tracing [1-2]. Through this approach, a number of paths, stemming from the transmitter, are traced along their way to the receiver, accounting for reflection over the obstacles within the scenario. Other mechanisms of interaction between the wave and the environment, such as diffraction, can be accommodated in ray tracing procedures by appropriate generalization of the basic theory [2].

While this method is purely deterministic, in actual environments with many randomly placed scatterers of size comparable to the wavelength (Mie scattering), statistical characterization of the multipath channel [3-6] may be the only viable approach in order to have an accurate model of the propagation [7-8]. Statistical modeling built on iteration of ray tracing results suffers of its inner computational intensity. As for the radio channel design what it is of interest are the fluctuations about a mean value of the received power, one can use a simpler and efficient method to take into account variations due to randomly placed obstacles in the propagation environment. Radiative transfer theory seems to be appropriate, as it deals with the wave propagation within a random medium characterized by randomly placed scatterers. Based on a phenomenological description of the transfer of energy, the basic equation (referred to as radiative transfer equation, RTE) simply states the conservation of energy in terms of the specific intensity $\boldsymbol{I}(r, \hat{s})$, i. e. the power per unit area and per unit solid angle propagating along $\hat{s}$, and which is a function of position r in the random medium [9]. The development of the theory is heuristic and does not entail any information about the phase of the wave. Chandrasekhar first studied the RTE within the context of astrophysics [10]. Later, the same formulation was employed successfully for many other applications, such as propagation modeling in the atmosphere or in forested environments,

INTECH
open science | open minds

© 2012 Bosisio; licensee InTech. This is an open access article distributed under the terms of the Creative Commons Attribution License (http://creativecommons.org/licenses/by/3.0), which permits unrestricted use, distribution, and reproduction in any medium, provided the original work is properly cited.

heat transfer through insulating materials, neutron scattering, and power absorption within biological tissues [11-14].

In this chapter, the author reports the RTE results for the evaluation of the power fluctuations in an indoor environment described as a homogeneous medium filled with scatterers arbitrarily placed. The radiative transfer outcome is compared with the ray tracing predictions to assess its limits of applicability.

For the sake of simplicity and without introducing inessential complications in our analysis, the investigation is limited to a propagation environment that can be modeled as a layered parallel plane medium. This modeling is a first approximation of an indoor environment including people, benches or, for instance, a row of chairs in an auditorium, whose location is not fixed. The random medium layer accounts for the average condition out of many possible spatial configurations, given or assumed the number density of scatterers of a succession of layers where each layer is modeled as random medium containing randomly placed scatterers. Moreover, we consider infinite-length circular cylinder as scatterers so as to simplify the solution of the RTE. It should be clear that while the ray tracing approach can be in principle used for any geometry and provide information about the phase of the wave, RT is in practice only applicable to simple geometries and can only yield information about the second order statistics of the wave. The scatterer's number density is chosen so as interference and interaction between scatterers could be neglected. RT reliability under this condition has been thoroughly investigated [15-17]; here, it is shown how it can be a useful and simple tool for indoor propagation analysis regarding the spatial correlation as defined in Section 5.

The chapter is organized as follows. A review of the radiative transfer is presented in Section 2, together with the definition of main quantities. Section 3 and 4 are devoted to the description of the RTE and of the numerical techniques for its solution. Iterative procedure and its limits of applicability are discussed. Section 5 reports the numerical results in two study cases in a 2D geometry with focus on the impact of the system parameters on the specific intensity and spatial correlation. The results of the comparison between the radiative transfer results and the predictions obtained through ray tracing are also reported. Section 6 is devoted to comments and conclusions.

2. Radiative Transfer Theory: physical background

Two basic theories have been developed in order to approach the study of wave propagation within a random medium characterized by randomly placed scatterers. The first is the analytical theory, where taking into account the scattering and absorption characteristic of the particles solves the Maxwell equations. This approach is mathematically and physically rigorous since in principle the effects of the mechanisms involved in multiple scattering, diffraction and interference can be appropriately modeled. However, in practice, various approximations have to be made in order to obtain feasible solutions (see [9] or [10] for an overview). Recent developments [18] adopt the random medium as paradigm to describe the propagation channel. Stochastic Green's functions are computed to obtain the channel transfer matrix T in MIMO applications. On the other hand, the transport theory is based on a phenomenological description of the transfer of energy. The basic equation simply states the conservation of energy expressed in terms of the *specific intensity* and it is equivalent to the Boltzmann equation used in kinetic theory of gases [9]. The development of the theory is heuristic and it does not entail any information about the phase of the wave.

In this chapter, the basic concepts and the quantities of interest when dealing with radiative transfer theory are reviewed, while the reader is referred to literature for a deeper insight. As the problem is defined in a two-dimensional domain, the classic theoretical formulation is adapted to this framework. Accordingly, the chosen reference system is in cylindrical coordinates, as shown in Figure 1. The analysis deals with monochromatic signals with frequency *f* and the phasor notation is used.

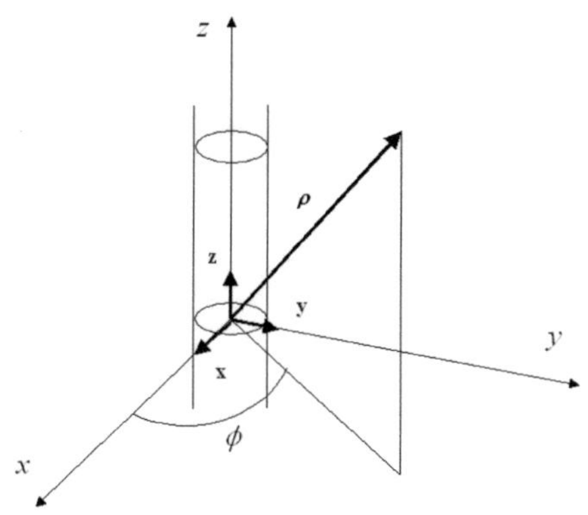

Figure 1. Coordinate system.

To study the propagation of a wave in presence of randomly distributed particles, the main results related to scattering and absorption of a single particle in vacuum are reviewed. Let us consider an elliptically polarized incident plane wave $E_i(\rho)$ in the direction t_i described by the azimuth angle ϕ_i:

$$\hat{t}_i = cos\phi_i \hat{x} + sin\phi_i \hat{y} \tag{1}$$

that is:

$$\boldsymbol{E}_i = \boldsymbol{E}_i^0 e^{jk_i.\rho} = \left[E_{v,i}^0 v + E_{h,i}^0 h \right] e^{jk_i \cdot \rho} \tag{2}$$

being ($E^0_{v,i}$, $E^0_{h,i}$) the field amplitudes for the vertical (*z*) and the horizontal (unit vector h lying on the *xy* plane) components, respectively. k is the wave number vector:

$$\mathbf{k}_i = k_0 t_i = \omega\sqrt{\mu_0 \varepsilon_0} t_i \quad (3)$$

and (μ_0, ε_0) represent the vacuum dielectric constant and its permeability. As the propagation takes place in the xy plane, the observed directions are denoted by a unit vector $\hat{t}$ or by the corresponding azimuth angle ϕ according to (1). The particle is characterized by a complex dielectric relative constant:

$$\varepsilon_r = \frac{\varepsilon}{\varepsilon_0} = \varepsilon_r^{'} + j\varepsilon_r^{''} \quad (4)$$

and it is assumed for simplicity to be homogeneous with:

$$\varepsilon_r^{''} = -\frac{\sigma}{\omega\varepsilon_0} \quad (5)$$

where σ is the conductivity [Sm^{-1}].

In far field, the scattered field behaves like a cylindrical wave:

$$\mathbf{E}_s = \frac{1}{\sqrt{r}} \mathbf{F}(\phi_i, \phi) \mathbf{E}_i^0 e^{j\left(k\rho - \frac{\pi}{4}\right)} \quad (6)$$

where $F(\phi_i, \phi)$ is the 2×2 scattering matrix accounting for the amplitude, the phase and the polarization of the scattered wave in direction ϕ when illuminated by a plane wave propagating in direction ϕ_i. For a 2D problem the polarizations are independent and the scattering matrix is diagonal:

$$\mathbf{F}(\phi_i, \phi) = \begin{bmatrix} F_{vv}(\phi_i, \phi) & 0 \\ 0 & F_{hh}(\phi_i, \phi) \end{bmatrix} \quad (7)$$

Therefore, for each polarization p ($p = v, h$) we can write:

$$\mathbf{E}_{p,s} = \frac{1}{\sqrt{r}} F_{pp}(\phi_i, \phi) E_{p,i}^0 e^{j\left(k\rho - \frac{\pi}{4}\right)} \quad (8)$$

In the following, where not stated otherwise the subscript p is dropped for simplicity of notation.

2.1. Scattering and absorption cross widths

The power dP_s scattered along direction ϕ within the differential width dl subtended by the differential angle $d\phi, dl = r\, d\phi$ is:

$$dP_s = S_s dl$$

where S_s [Wm^{-1}] is the amplitude of the Poynting vector of the scattered wave, $S_s = |E_S|^2 / 2\eta$ and $\eta = \sqrt{\mu_0 / \varepsilon_0}$ is the vacuum electromagnetic impedance. It follows from (8) that:

$$dP_s = |F(\phi_i, \phi)|^2 S_i d\phi \tag{9}$$

and integrating over all angles:

$$P_s = \int_0^{2\pi} dP_s = \int_0^{2\pi} |F(\phi_i, \phi)|^2 S_i d\phi = S_i \sigma_s$$

where the scattering cross width σ_s is defined as:

$$\sigma_s = \frac{P_s}{S_i} = \int_0^{2\pi} |F(\phi_i, \phi)|^2 d\phi \tag{10}$$

This quantity represents the equivalent width that would produce the amount P_s of scattered power if illuminated by a wave with power density S_i.

The geometric cross width σ_g [m] of a particle is its geometric width projected onto a plane that is perpendicular to the direction of the incident wave t_i. The relationship between the geometric and scattering widths can be investigated in two regimes. If the size of the object D (maximum distance between two points inside the object) is much smaller than the wavelength λ, it follows that:

$$\frac{\sigma_s}{\sigma_g} << 1 \quad (D << \lambda)$$

According to Rayleigh scattering theory [19], it means meaning that in this regime the power scattered by the particle is much smaller than the product of geometric cross width and the amplitude of the Poynting vector. Besides, in the high frequency regime $D >> \lambda$:

$$\frac{\sigma_s}{\sigma_g} \to 1$$

which is known as geometric optics limit.

Similarly to (10), the absorption cross width can be defined as the ratio between the absorbed power P_a and the incident Poynting vector S_i. From Ohm's law [20]:

$$P_a = \frac{1}{2}\omega \int_A d\rho \varepsilon_r'' \mid E_{int}(\rho) \mid^2 \tag{11}$$

where $E_{int}(\varrho)$ is the internal field within the particle and A is the particle's area.

2.2. Extinction cross width and albedo

The extinction cross-section σ_{ext} of a particle is defined as:

$$\sigma_{ext} = \sigma_s + \sigma_a \tag{12}$$

and it represents the total power loss from the incident wave due to scattering and absorption. The fraction of scattering over the extinction cross width is defined as albedo α:

$$\alpha = \frac{\sigma_s}{\sigma_{ext}} \leq 1 \tag{13}$$

The computations of the extinction cross width (12) can be carried out from the knowledge of the scattering matrix, as stated by the forward scattering theorem (also known as optical theorem):

$$\sigma_{ext} = -4\sqrt{\frac{\pi}{2k_0}} Re(F(\phi_i, \phi))$$

This result can be proved either undertaking the explicit computation of (11) [15] or computing the received power over a given width and relating this quantity to the geometric dimensions [19]. Besides, it can be shown that for the high frequency regime [9]:

$$\frac{\sigma_a}{\sigma_g} \to 1 \tag{14}$$

$$\frac{\sigma_{ext}}{\sigma_g} \to 2 \tag{15}$$

Equation (15) is also known as extinction paradox.

2.3. Example: scattering from a circular cylinder

Figure 2 shows the scattering, the absorption and the extinction cross widths are computed for a circular cylinder of radius a [21]. In figure 2 these cross width values are normalized over the geometric cross width σ_g=2a and they are plotted versus the radius a normalized over the wavelength λ. The particle is characterized by ε_r'=4 and σ=10^{-3}. One could notice that as the size of the particle increases the absorption and scattering cross widths tend to σ_g

width; whereas the extinction cross width tends to $2\sigma_g$ (extinction paradox, (15)). Also, Figure 2 distinguishes three different regions: Rayleigh scattering ($a \ll \lambda$), Mie scattering ($a \approx \lambda$) and optical region ($a \gg \lambda$).

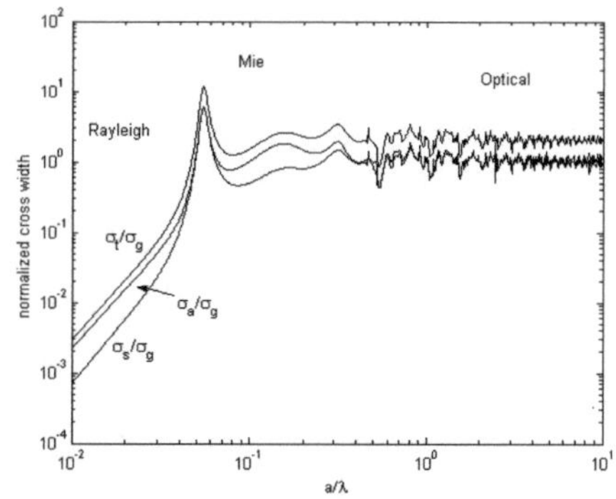

Figure 2. Scattering, absorption and extinction cross widths normalized over the geometric cross width $\sigma_g = 2a$ versus a/λ ($\varepsilon_r{}' = 4$ and $\sigma = 10^{-3}$).

2.4. Specific intensity

Transport theory deals with the propagation of energy in a medium containing randomly placed particles. For a given point ρ and a given direction specified by vector $\hat{t} = cos\phi \hat{x} + sin\phi \hat{y}$ (or equivalently by the azimuth angle ϕ), the power flux density within a unit frequency band centered at frequency f within a unit angle is defined as specific intensity and denoted by $I(\rho,\phi)$ [W m^{-1}rad^{-1}Hz^{-1}]. Hence, the amount of power dP flowing along direction ϕ within an angle $d\phi$ through an elementary width dl with normal that forms an angle $\Delta\phi$ with ϕ (see Figure 3) in a frequency interval $(f, f+df)$ is:

$$dP = I(\rho, \phi) cos\Delta\phi \, dl \, d\phi \, df \tag{16}$$

The specific intensity $I(\rho,\phi)$ as it appears in (16) could be related either to the power emitted from a surface or to the power received by the unit width. As far as the single particle of § 2.3 is concerned, the specific intensity carried by the incident plane wave (2) is:

$$I_i(\rho, \phi) = S_i \delta(\phi - \phi) \tag{17}$$

in which $\delta(.)$ is the Dirac Delta function, whereas, recalling (9), the specific intensity for the scattered wave can be written as:

$$I_s(\rho, \phi) = | F(\phi_i, \phi) |^2 S_i = | F(\phi_i, \phi) |^2 I_i(\rho, \phi_i) \tag{18}$$

Hence, the square modulus of the scattering function $F(\phi_i,\phi)$ relates the incident specific intensity with the scattered specific intensity. In a random medium, the specific intensity is computed as the ensemble average of the power per unit angle, frequency and area over the distribution of the random scatterers.

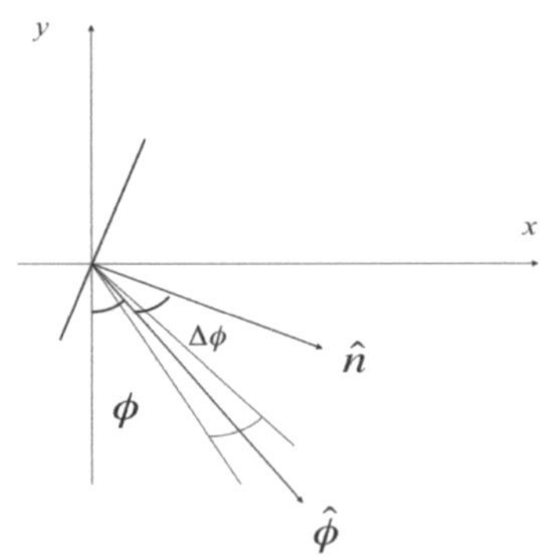

Figure 3. Geometry used for the definition of specific intensity: the amount of power *dP* flows along direction ^ within an angle *d* through an elementary width *dl* whose normal $\hat{n}$ forms an angle *Δ* with the direction of propagation ^.

3. Radiative transfer equation

The radiative transfer equation is an integro-differential equation that governs the propagation of specific intensity within a random medium. Let us assume that the random medium is a made of uniform slabs in the x-direction (see Figure 4). Therefore, specific intensity is a function of position in space only through y. In the following the derivation of the radiative transfer equation is briefly reminded. The reader is referred to [15, 17] for further details.

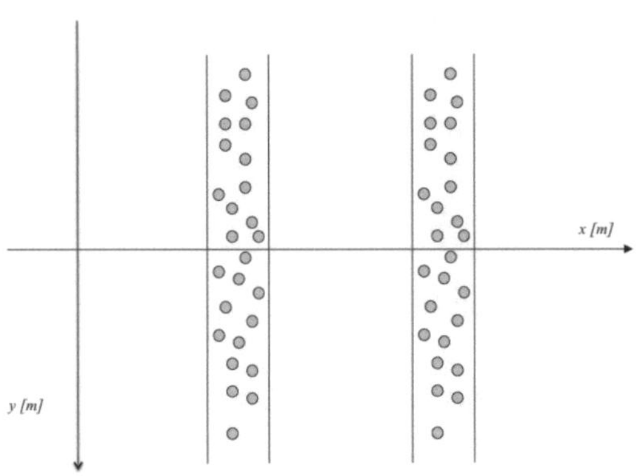

Figure 4. Geometry of the propagation environment.

http://dx.doi.org/10.5772/50871

3.1. Reduced intensity

Let us consider an area with side 1 in the x-direction and dy in the y-direction The area dy contains Ndy particles, where N is the number of particle per unit area. The particles are uniformly distributed in space. Each particle absorbs the power $P_a = \sigma_a I$ and scatters the power $P_s = \sigma_s I$, so that the decrease of specific intensity due to area dy is (recall (12)):

$$sin\phi \cdot dI(y, \phi) = -dy N \sigma_t I(y, \phi) = m - dy \cdot k_{ext} I(y, \phi) \quad (19)$$

where $k_{est} = N\sigma_{ext}$ [m^{-1}] is the extinction coefficient. Notice that if there are different kinds of particles (say m), each with possibly different orientation, density N_j and extinction cross width $\sigma_{ext,j}$ [9]:

$$k_{ext} = \sum_{j=1}^{m} N_j < \sigma_{ext,j} > \quad (20)$$

where < > represents the ensemble average over the distribution of particles orientations. Equation (19) defines the so-called reduced intensity since it only takes into account the extinction of the incident wave [9].

3.2. Independent scattering and limits of the transfer theory

Scattering of waves impinging on the area dy from all directions ϕ_i increases the intensity along direction ϕ according to (18). As the scatterers are assumed to be independent, the specific intensities due to different particles can be added. This is strictly true under some conditions that are discussed in the following as this point entails the limits of applicability of the radiative transfer theory [15]. As the Maxwell equations are linear, the total scattered field E can be written as the sum of the E_j fields scattered from each particle:

$$\boldsymbol{E} = \sum_j \boldsymbol{E}_j \quad (21)$$

The specific intensity is proportional to the square modulus of the electric field averaged over the distribution of scatterers:

$$< | E |^2 > = \sum_j < | E_j |^2 > + \sum_j \sum_{l \neq j} < E_j E_l^* >$$

Now, let $E_j = | E_j | e^{j\alpha_j}$, we get that:

$$< E_j E_l^* > = < | E_j | | E_l | e^{j(\alpha_j - \alpha_l)}$$

The phase difference $(\alpha_j - \alpha_l)$ depends on the distance between the particles d_{jl} through the product k_{0djl}. If the distribution of particle separation is not much smaller than the wavelength, i.e. the standard deviation (S.D.) of d_{jl} satisfies:

$$S.D.(d_{jl}) \geq \frac{\lambda}{4} \tag{22}$$

then $(\alpha_j - \alpha_l)$ is approximately uniformly distributed within $[0, 2\pi]$ so that:

$$< E_j E_l^* > = 0$$

From the discussion above, the assumption underlying the radiative transfer equation of independent scattering limits the applicability of the transfer theory to cases where the distance between particles is large enough (see (22)) so as to make negligible the near far interactions between particles. Experimental studies confirm this conclusion: for the radiative transfer to be applicable the spacing between scatterers must be larger than $\lambda/3$ and $0.4D$ where λ is the wavelength of the host medium and D the diameter of the scatterers [10]. Theoretical studies on the relationship between the radiative transfer approach and the wave approach using Maxwell's equations can be found in [22-23]. Moreover, the limits of the transfer theory are investigated for a two-dimensional problem similar to the one considered here in [16] through comparison with the wave approach.

Assuming independent scattering, the increase on the specific intensity along direction ϕ due to scattering within the area dy is:

$$dy \int_0^{2\pi} N \mid F(\phi_i, \phi) \mid^2 I(\rho, \phi_i) d\phi_i = dy \int_0^{2\pi} p(\phi_i; \phi) I(\rho, \phi_i) d\phi_i \tag{23}$$

where $p(\phi_i; \phi) = N F(\phi_i, \phi)$ is the phase function. Notice that if the random medium contains particles of different kinds, the overall phase function is defined as [9]:

$$p(\phi_i, \phi) = \sum_{j=1}^{m} N_j < p_j(\phi_i, \phi) > \tag{24}$$

3.3. Progressive and regressive intensity

The radiative transfer equation is obtained by combining (19) and (23):

$$sin\phi \frac{\partial I(y, \phi)}{\partial y} = -k_{ext} I(y, \phi) + \int_0^{2\pi} p(y, \phi_i, \phi) I(y, \phi_i) d\phi_i \tag{25}$$

In this formulation, the extinction coefficient and the phase function are considered function of the position y. A first step toward the solution of scalar radiative transfer problem is converting equation (25) into two coupled integro-differential equations by introducing the progressive intensity I^+, that corresponds to propagating directions $0 < \phi < \pi$, and the regressive intensity I^-, that accounts for the propagating directions $\pi < \phi < 2\pi$. The two specific intensities are defined as:

$$I^{+}(y,\phi)=I(y,\phi)$$
$$I^{-}(y,\phi)=I(y,\phi+\pi)$$

where the azimuth direction ϕ ranges within $0<\phi<\pi$. The scalar radiative transfer equation (25) can be equivalently stated as:

$$\begin{aligned} sin\phi\frac{\partial I^{+}(y,\phi)}{\partial y} &= -k_{ext}I^{+}(y,\phi)+\int_{0}^{\pi}p(y,\phi_i,\phi)I^{+}(y,\phi_i)d\phi_i+ \\ &+\int_{0}^{\pi}p(y,\phi_i+\pi,\phi)I^{-}(y,\phi_i)d\phi \\ -sin\phi\frac{\partial I^{-}(y,\phi)}{\partial y} &= -k_{ext}I^{-}(y,\phi)+\int_{0}^{\pi}p(y,\phi_i,\phi+\pi)I^{+}(y,\phi_i)d\phi_i+ \\ &+\int_{0}^{\pi}p(y,\phi_i+\pi,\phi+\pi)I^{-}(y,\phi_i)d\phi \end{aligned} \tag{26}$$

This equivalent formulation of (25) makes it easier to set the boundary conditions as explained in §3.3.1 and §3.3.2. For a uniform distribution of the particles over the random medium and circular cylindrical particles, the phase matrix becomes a function ϕ-ϕ_i only and the extinction matrix becomes independent on ϕ [17]. Therefore, in this case the scalar radiative transfer equations (26a-26b)can be written as:

$$\begin{aligned} sin\phi\frac{\partial I^{+}(y,\phi)}{\partial y} &= -k_{ext}I^{+}(y,\phi)+\int_{0}^{\pi}p^{+}(y,\phi-\phi_i)I^{+}(y,\phi_i)d\phi_i+ \\ &+\int_{0}^{\pi}p^{-}(y,\phi-\phi_i)I^{-}(y,\phi_i)d\phi \end{aligned} \tag{27}$$

$$\begin{aligned} -sin\phi\frac{\partial I^{-}(y,\phi)}{\partial y} &= -k_{ext}I^{-}(y,\phi)+\int_{0}^{\pi}p^{-}(y,\phi-\phi_i)I^{+}(y,\phi_i)d\phi_i+ \\ &+\int_{0}^{\pi}p^{+}(y,\phi-\phi_i)I^{-}(y,\phi_i)d\phi \end{aligned} \tag{28}$$

where $p^{+}(y,\phi-\phi_i)=p(y,\phi_i,\phi)$ and $p^{-}(y,\phi-\phi_i)=p(y,\phi_i,\phi+\pi)=p(y,\phi+\pi_i,\phi)$, based on the 2π periodicity of both p^{+} and p^{-} [17].

3.3.1. Boundary conditions on the specific intensity

The radiative transfer equation has to be solved by imposing appropriate boundary conditions. Here, the boundary conditions that the specific intensity must satisfy on a plane boundary between two media with indices of refraction n_1 and n_2 are considered. The expression for the transmission T and reflection coefficient R for the two polarizations are known in literature [20]. The reflected specific intensity I_r is related to the incident specific intensity I_i as:

$$I_r = | R |^2 I_i$$

As far as the transmitted specific intensity is concerned, we can write the conservation of power on a segment dl belonging to the boundary (recall (16)):

$$I_i cos\phi_1 d\phi_1 dl = I_r cos\phi_1 d\phi_1 dl + I_t cos\phi_2 d\phi_2 dl \tag{29}$$

As by the Snell's law$n_1 sin\phi_1 = n_2 dsin\phi_2$, it ends:

$$\frac{I_t}{I_i} = \frac{n_2}{n_1}(1 - | R |^2)$$

4. Solution of the radiative transfer equation through numerical quadrature

The radiative transfer equation is an integro-differential equation whose solution in analytical form is very difficult, if not impossible. However, efficient numerical solutions can be devised. A comprehensive treatment of the main techniques can be found in [15, 24].

The case under study concerns a random medium where relevant scattering occurs. An approximate solution can be obtained by computing the integrals in (26) by numerical quadrature as firstly proposed in [15]. The continuum of propagation directions ϕ is discretized into a set of n directions $\phi_{il} = (i-1)\,\Delta\phi$, where $\Delta\phi = \pi/n$ and $i = 1,\ldots, n$, and the corresponding n×1 vectors gathering the progressive and regressive intensities:

$$\boldsymbol{I}^+(r) = \begin{bmatrix} I^+(h, \phi_1) \\ I^+(h, \phi_2) \\ \vdots \\ I^+(h, \phi_n) \end{bmatrix} \tag{30}$$

$$\boldsymbol{I}^-(r) = \begin{bmatrix} I^-(h, \phi_1) \\ I^-(h, \phi_2) \\ \vdots \\ I^-(h, \phi_n) \end{bmatrix} \tag{31}$$

The two integro-differential equations (27) can be approximate as follows:

$$\frac{d\boldsymbol{I}^{+}(y)}{dy} = -\boldsymbol{K}(y)\boldsymbol{I}^{+}(y) + \boldsymbol{P}^{+}(y)\boldsymbol{I}^{+}(y) + \boldsymbol{P}^{-}(y)\boldsymbol{I}^{-}(y) \tag{32}$$

$$\frac{d\boldsymbol{I}^{-}(y)}{dy} = -\boldsymbol{K}(y)\boldsymbol{I}^{-}(y) - \boldsymbol{P}^{-}(y)\boldsymbol{I}^{+}(y) - \boldsymbol{P}^{+}(y)\boldsymbol{I}^{-}(y) \tag{33}$$

where then ×n matrices P^{+} and P^{-} are defined as:

$$[\boldsymbol{P}^{\pm}(y)]_{i,j} = \frac{1}{sin\phi_i} p^{\pm}(y,\ \phi_i - \phi_j)\Delta\phi$$

so that:

$$\frac{1}{sin\phi_i}\int_0^{\pi} p^{\pm}(y,\ \phi-\phi')I^{\pm}(y,\ \phi')d\phi' \approx \sum_{j=1}^{n}[P^{\pm}(y)]_{i,j} I^{\pm}(y,\ \phi_j) \tag{34}$$

The n×n K(y) matrix is defined as:

$$\begin{aligned}[\boldsymbol{K}(y)]_{i,j} &= \frac{1}{sin\phi_i} k_{ext}(y) i = j \\ &= 0 i \neq j\end{aligned} \tag{35}$$

Now, defining the 2n×1 vector$\boldsymbol{I}(y) = [\boldsymbol{I}^{+}(y)^{T} \boldsymbol{I}^{-}(y)^{T}]$, one obtains the system of first order linear equation:

$$\frac{d\boldsymbol{I}(y)}{dy} = \boldsymbol{G}(y)\boldsymbol{I}(y) \tag{36}$$

where

$$\boldsymbol{G}(y) = \begin{bmatrix} -\boldsymbol{K}(y) + \boldsymbol{P}^{+}(y) & \boldsymbol{P}^{-}(y) \\ -\boldsymbol{P}^{-}(y) & \boldsymbol{K}(y) - \boldsymbol{P}^{+}(y) \end{bmatrix}$$

4.1. Discrete ordinate Eigen analysis

If the space contains slabs of homogeneous random (and non-random) media, the linear differential system (36) within each slab (say the l th) can be written as

$$\frac{d\boldsymbol{I}(y)}{dy} = \boldsymbol{G}_l \boldsymbol{I}(y) \tag{37}$$

being matrix G_l a constant. This follows from the fact that, the extinction coefficient and the phase function within each slab are constant. In this case, the system (37) can be solved by the discrete ordinate analysis method as follows. Let $I(y)=\beta exp(-\lambda g)$be a tentative solution where β is a 2n×1, one get by substitution in (36) - dropping l for simplicity:

$(\lambda \boldsymbol{U} + \boldsymbol{G})\beta = 0$

As a result, (β, λ) represent pair of eigenvectors/eigenvalues of the matrix –G. Notice that given the symmetry relations:

$[\boldsymbol{G}_l]_{(n+i))n+i)} = -[\boldsymbol{G}_l]_{i,i}$

$[\boldsymbol{G}_l]_{(n+i))n+j)} = -[\boldsymbol{G}_l]_{iji} i, \ j=1, \cdots, n$

by ordering the eigenvalues in increasing order $(\lambda_i \geq \lambda_{i-1})$ it follows:

$\lambda_{(n+i)} = -\lambda_i i=1, \cdots, n$

and the corresponding eigenvectors satisfy the condition:

$[\beta_{(n+i)}]_{(n+)} = [\beta_i]_j i=1, \cdots, n$

The solution of (36) can be written as the linear combination

$$\boldsymbol{I}(y) = \sum_{i=1}^{2n} \beta_i e^{(-\lambda_i y) c_i} = \boldsymbol{B}\boldsymbol{D}(y)\boldsymbol{c} \tag{38}$$

where$\boldsymbol{B} = [\beta_1 \cdots \beta_{2n}]$, $\boldsymbol{c} = [c_1 \cdots c_{2n}]^T$, and D(y) is a diagonal matrix with elements $[e^{(-\lambda_1 y)} \cdots e^{(-\lambda_{2n} y)}]$.

The vector of unknown constant c can be computed from the knowledge of I(y) for one value of y, say$\bar{y}$, as:

$$\boldsymbol{c} = (\boldsymbol{B}\boldsymbol{D}(\bar{y}))^{-1} \boldsymbol{I}(\bar{y}) \tag{39}$$

4.1.1. Setting the boundary conditions for a single slab

Given geometry depicted in figure 4, with only one slab of random medium (ranging within $0 \leq y \leq d$), the semi-infinite medium along the positive direction of y is a dielectric one with given dielectric constant. The boundary conditions read:

$$\boldsymbol{I}^+(0) = \bar{\boldsymbol{I}}_0^+ \tag{40}$$

$$\boldsymbol{I}^-(d) = \boldsymbol{A}_R \bar{\boldsymbol{I}}^+(d) \tag{41}$$

where $\bar{\boldsymbol{I}}_0^+$is the incident specific intensity and A_R is the n×n reflection matrix relative to the interface at $y = d$ between the random medium and the last slab. This can be easily computed from the results in §2.4. The value of the unknown constant vector (39) for the specific intensity within the random medium is obtained through the computation of the specific intensity$\boldsymbol{I}(\bar{y})$. To do this, as first step one need to calculate from (38) and (39):

$$\mathbf{I}(d) = \underbrace{\mathbf{B}\mathbf{D}(d)\mathbf{B}^{-1}}_{\mathbf{T}}\mathbf{I}(\bar{y})\mathbf{I}(0) \tag{42}$$

Then, using (40) and (41), and partitioning the matrix T as:

$$\boldsymbol{T} = \begin{bmatrix} \boldsymbol{T}_{11} & \boldsymbol{T}_{12} \\ \boldsymbol{T}_{21} & \boldsymbol{T}_{22} \end{bmatrix} \tag{43}$$

in which T_{ij} is n×n, one gets:

$$\boldsymbol{I}(0)^{-} = (\boldsymbol{A}_R\boldsymbol{T}_{12} - \boldsymbol{T}_{22})^{-1}(\boldsymbol{T}_{21} - \boldsymbol{A}_R\boldsymbol{T}_{11})\bar{\boldsymbol{I}}_0^{+}$$

The substitution of this finding in (39) is used for the computation of the vector c.

4.1.2. Setting the boundary conditions for a multi-layer medium

In the multi-layered geometry of figure 4, each slab has depth *dl* and the specific intensity is (38). The unknown is the constant vector c_l and the subscript l runs over the different slabs, $l = 1,..., L$. Notice that slab can contain either a random medium characterized by its matrix G_l or vacuum (matrix G_l=0 for vacuum). The semi-infinite medium along the positive direction of y is a general dielectric with given dielectric constant. The boundary conditions (40) and (41) still hold with$d = \sum_{l=1}^{L} d_l$.

The specific intensity $I_{l,2}$ - at the interface between the l and the $(l+1)^{th}$ slab - is related to the specific intensity $I_{l,1}$ at the interface between the $(l-1)$ and l^{th} slab as:

$$\mathbf{I}_{l,2} = \underbrace{\mathbf{B}_l\mathbf{D}(d_l)\mathbf{B}_l^{-1}}_{\mathbf{T}_l}\mathbf{I}(\bar{y})\mathbf{I}_{l,1} \tag{44}$$

Therefore, similarly to (42) we can write:

$$\mathbf{I}(d) = \underbrace{\prod \mathbf{T}_l}_{\mathbf{T}}\mathbf{I}(0)$$

Proceeding as in the previous Section, the specific intensity I(0) is now calculated and used in (39) to obtain the constant vector c_1 for the first slab. Then, equation (44) can be iteratively applied so as to compute the specific intensity at the beginning of each slab. The latter allows through (39) the computation of c_l for each slab.

5. Numerical results

In this section two examples illustrate how radiative transfer theory could be employed to study the beam broadening and the corresponding spatial correlation for an indoor environment within the context of a communication system.

The main assumption is that the propagation environment can be modeled as a layered parallel plane medium. This situation is a first approximation of an open space office made of a succession of tables where each table is modeled as random medium containing randomly placed scatterers. These are modeled as circular cylinder so as to simplify the solution of the radiative transfer equation.

5.1. Case study I: one table

In this example, the geometry under study is depicted in Figure 5. The whole xy plane is characterized by the vacuum dielectric constant ε_0. One slab ranging within $1 \le y \le 1.8$ contains N [m^{-2}] uniform randomly distributed circular cylinders with radius a = 6 cm. Where not stated otherwise, frequency of operation is f = 5.2 GHz, according to standard wireless local area networks (WLAN) such as IEEE802.11x and Hyperlan/2.

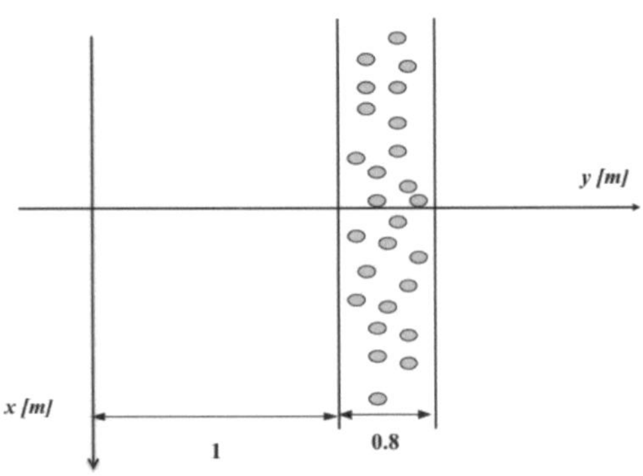

Figure 5. Geometry of the propagation environment (case study I).

5.1.1. Beam broadening

Figure 6 shows the specific intensity $I^{+}(y,\phi)$ for N=10, ε'_r =4, σ=10^{-3} Sm^{-1} and vertical polarization. Specifically, the image represents the beam broadening of a plane wave propagating in the direction ϕ_i = 90° with specific intensity I_0^{+}= 1 [W $m^{-1}rad^{-1}$Hz].

The specific intensities are assumed to be normalized with respect to $I_0^{+}$and thus are shown in dB. Right after the entrance into the table region the progressive beam $I^{+}(y, \phi)$ broadens since the energy is scattered in all directions by the cylinders. As expected the regressive beam $I^{-}(y, \phi)$ is zero on the right side of the table and almost uniform in the $y\phi$ plane (not shown).

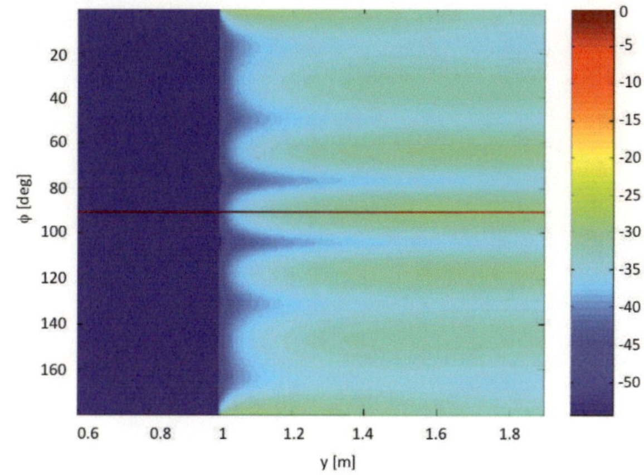

Figure 6. Specific intensity $I^{+}(y,)$ versus y (vertical polarization, $\varepsilon_r^{'} = 4$, $\sigma = 10^{-3}$ Sm^{-1}, $N = 10$).

In order to get a quantitative insight into the beam broadening discussed above, Figure 7 shows the specific intensity $I^{+}(y, \phi)$ for $y = 1.9$ m (on the right of the table) for different values of the density N=2, 10 (vertical polarization). The reduced or line of sight (LOS) contribution as shown in the box decreases for increasing object densities. However, a larger density of scatterers entails a more relevant contribution of the diffuse energy (i.e., $I^{+}(y, \phi)$ for $\phi \neq 90°$).

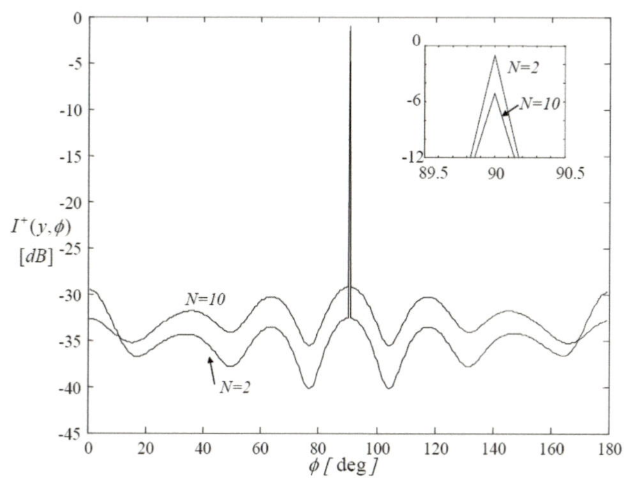

Figure 7. Specific intensity $I^{+}(y,)$ for $y = 1.9$ m (on the right of the table) for different values of the density $N = 2, 10$ (vertical polarization, $\varepsilon_r^{'} = 4$, σ=10^{-3} Sm^{-1}).

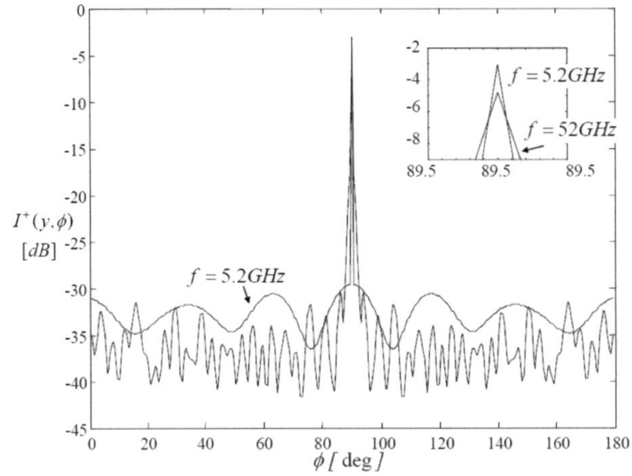

Figure 8. Specific intensity $I^+(y,)$ for $y = 1.9$ m (on the right of the table) for different values of the frequency $f = 5.2$, 52 GHz (vertical polarization, $N = 6$, $\varepsilon'_r = 4$, $\sigma = 10^{-3}$ Sm^{-1}).

The effect of an increase in the carrier frequency, envisioned for next generation wireless LAN, is shown in Figure 8. A larger carrier frequency (f = 52 GHz) yields a less consistent contribution of the LOS direction (see box) and a greater number of interference fringes. Employing the horizontal polarization instead of the vertical one yields qualitatively similar results as shown in Figure 9 for N=6. Therefore, in the following only the vertical polarization is considered.

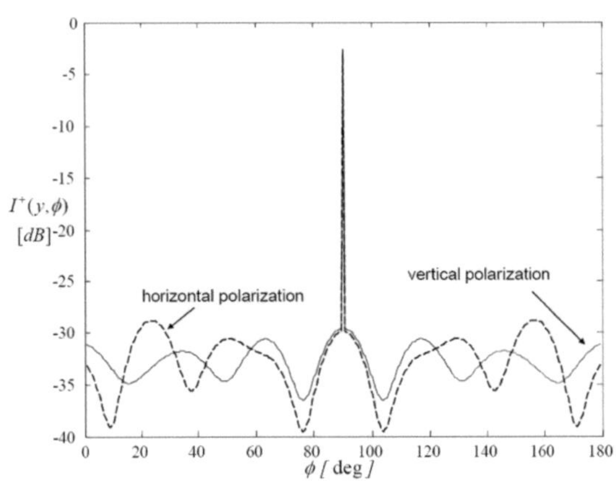

Figure 9. Specific intensity $I^+(y,)$ for $y = 1.9$ m (on the right of the table) for horizontal and vertical polarization frequency ($f = 5.2$ GHz, $N = 6$, $\varepsilon'_r = 4$, $\sigma = 10^{-3}$ Sm^{-1}).

http://dx.doi.org/10.5772/50871

5.1.2. Spatial correlation

In a communication link, it is of great interest to assess the degree of correlation between the signals received by different antennas as a function of their inter-spacing Δ [m]. In fact, wireless links can capitalize on the uncorrelation between the received signals to increase the degree of diversity of the system that in turn rules the asymptotic performance of the link in terms of probability of error [25].

The correlation $r_y(\Delta)$ of the signals received by antennas separated by Δ can be expressed in terms of the probability density function $p_y(\phi)$ of the direction of arrivals of the waves impinging on the receivers. This function can be obtained by interpreting (after appropriate scaling) the power received over a certain direction ϕ as a measure of the probability that a signal is received through direction ϕ:

$$p_y(\phi)=\frac{I^+(y,\phi)}{\int I^+(y,\phi)d\phi} \tag{45}$$

Then, recalling that the ratio between signals received at two points separated by Δ in the direction x is$exp(-j2\pi/\lambda cos\phi\Delta)$:

$$r_y(\Delta)=\int_0^{\pi} p_y(\phi)e^{\left(-\frac{j2\pi}{\lambda}cos\phi\Delta\right)}d\phi \tag{46}$$

The correlation (46) is evaluated for the example at hand under the same assumptions as in Figure 7.

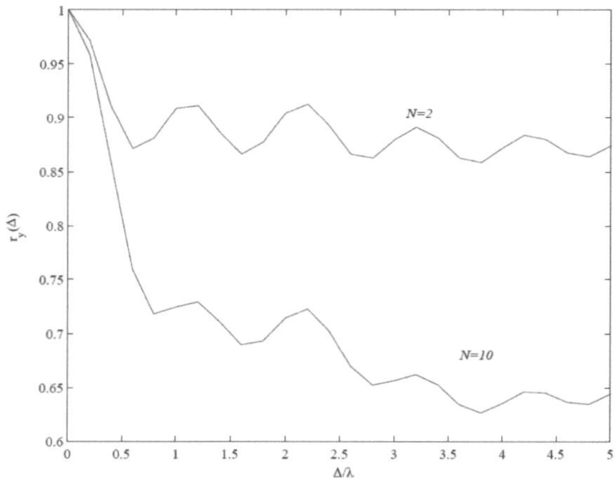

Figure 10. Correlation $r_y(\Delta)$ versus Δ/λ for $y = 1.9$ m (on the right of the table) for different values of the density $N = 2$, 10 (vertical polarization, $\varepsilon_r' = 4$, $\sigma = 10^{-3}$ Sm^{-1}).

The results are shown in figure 10. The correlation decreases with increasing object density. Thus, from the perspective of communication system performance, increasing the object density is beneficial in terms of degree of diversity. For instance, for $\Delta = \lambda$ the correlation decreases from around 0.9 to around 0.7.

5.2. Case study II: two tables with interface

In this example the environment has a more complicated geometry, where two tables are followed by a semi-infinite dielectric slab (dielectric constant $\varepsilon_{r,I}$) as shown in figure 11. The simulation parameters follow the setting described for Case study I, in particular the radius of the scatterers is a=6 cm and the carrier frequency is f=5.2 GHz. The dielectric constant of the slab is chosen as $\varepsilon_{r,I}$ =3.

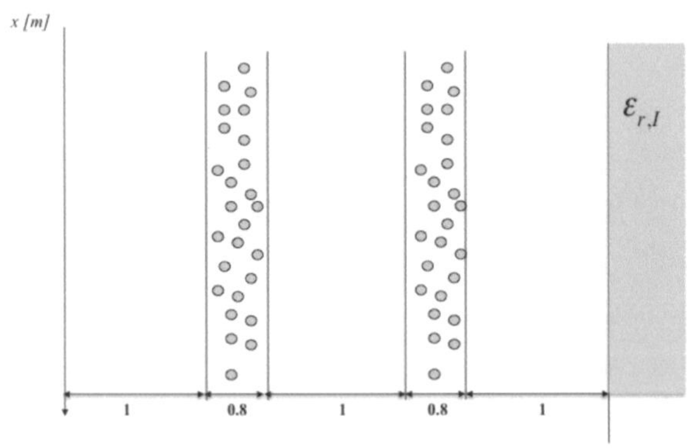

Figure 11. Geometry of the propagation environment (case study II).

5.2.1. *Beam broadening*

Figure 12 shows the specific intensity $I^{+}(y, \phi)$ for $N = 6$, $\varepsilon'_r = 4$, $\sigma = 10^{-3}$ Sm^{-1} and ϕ_i=90° and vertical polarization. Again, the specific intensity is normalized with respect to I_0^+ and thus are shown in dB. As discussed for case study I, right after the entrance into the table regions the progressive beam $I^{+}(y, \phi)$ broadens since the energy is scattered in all directions by the cylinders. Moreover, the refraction over the interface at y=4.6 m focuses the beam within the semi-infinite dielectric, hence reducing its angular spread. The refraction and the reflection of the reduced (or LOS) component are apparent in Figures 12 and 13, respectively. The incidence angle is chosen equal to 45°. Refracted and reflected angles satisfy the Snell's law.

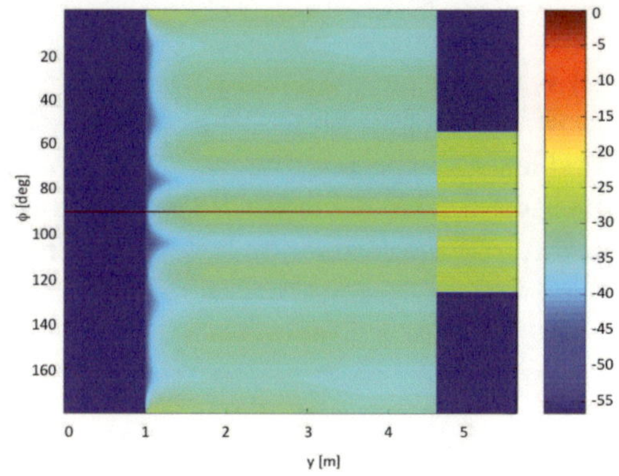

Figure 12. Specific intensity $I^{+}(y,)$ versus y (vertical polarization, $\varepsilon_r' = 4$, $\sigma = 10^{-3}$ Sm^{-1}, $N = 6$, $_i$=90°).

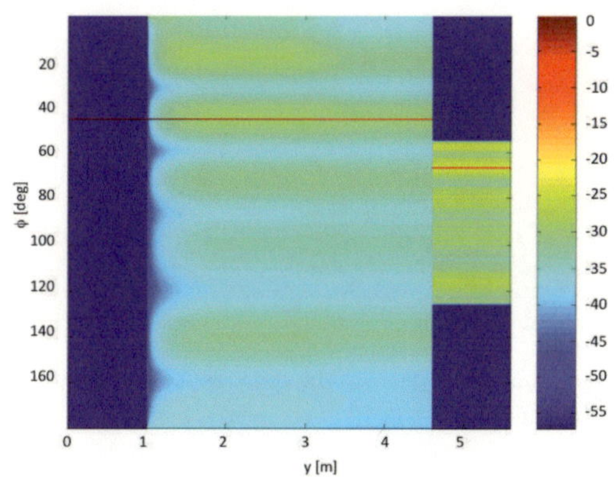

Figure 13. Specific intensity $I^{+}(y,)$ versus y (vertical polarization, $\varepsilon_r' = 4$, $\sigma = 10^{-3}$ Sm^{-1}, $N = 6$, $_i$=45°).

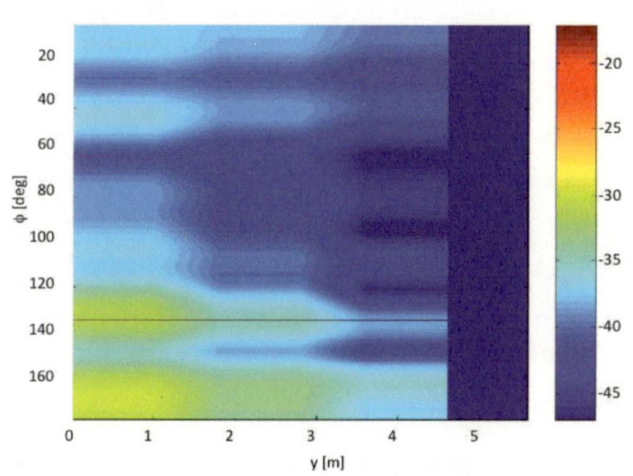

Figure 14. Specific intensity $I^{-}(y,)$ versus y (vertical polarization, $\varepsilon_r' = 4$, $\sigma = 10^{-3}$ Sm^{-1}, $N = 6$, $_i$=45°).

5.2.2. *Spatial correlation*

The spatial correlation $r_y(\Delta)$ is shown for the same parameters of the previous example in Figure 15. It is computed for $y = 1.9$ m (just on the right of the first table) and $y = 3.7$ m (just on the right of the second table). As expected, the increased scattering contribution due to the random scatterers of the second table decreases the spatial correlation. In particular, for inter-antenna spacing around $\Delta = 5\lambda$ the correlation is decreased from 0.75 to around 0.61.

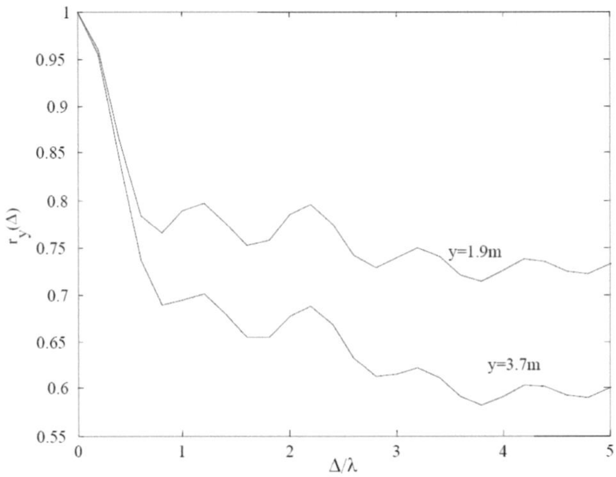

Figure 15. Correlation $r_y(\Delta)$ versus Δ/λ for $y = 1.9$ m (on the right of the first table) and $y = 3.7$ m (on the right of the second table) for $N = 6$, vertical polarization, $\varepsilon'_r = 4$, $\sigma = 10^{-3}$ Sm^{-1}, $_i$=90°.

5.3. Limits and validity of radiative transfer predictions

The validation of numerical results presented in §5.1 and §5.2 relies on a discussion, as there is no empirical evidence (measurements) to compare with. The two chosen cases are not realistic but they represent a possible benchmark to be employed against the numerical results that issue from analytical methods. Among these, the T-matrix approach [26, 27] seems to be the best suited to compute scattering from a random distribution of cylinders and to compare the radiated fields on a realization-by-realization basis with ray tracing. At this stage of development, the only comparison that the author is able to provide to assess the validity of a radiative transfer approach is against a ray tracing technique based upon the beam tracing method [26]. The numerical code used for the simulations was developed at the Politecnico di Milano [29] and it has already been used for different purposes related to the indoor propagation [30].

To make the beam tracing procedure suitable for the study of electromagnetic propagation, it is necessary to include the reflection coefficient associated with the interactions of the path with the environment. In particular, the signal received from each path has to be scaled by the product of the reflection coefficients corresponding to the bounces each path goes through when propagating from the transmitter to the receiver. To compare the outcome of the ray tracing simulation with the results of radiative transfer theory some approximations were made. Case I was reproduced in such a way to replicate the same geometrical and radio electrical conditions. The reader is referred to [31] for further details about the numerical computation and the comparison results, while here the author recalls the general approach and the basic information. The transmitted plane wave was approximated by building a linear antenna array of N_T elements with inter-element spacing Δ_T lying parallel to the y axis. To fulfill the requirement of far field regime, the transmitting array was placed at a range distance $R > (N \Delta_T)^2 / \lambda$ [9].

The circular scatterers were approximated by polygons of N_P sides. The quantities of interest were averaged over N_I realizations of the random medium. The signal is received at the desired points by a linear antenna array of N_R elements with inter-element spacing Δ_R lying parallel to the y axis. Figure 16 illustrates the geometry together with the transmitting and the receiving antenna arrays.

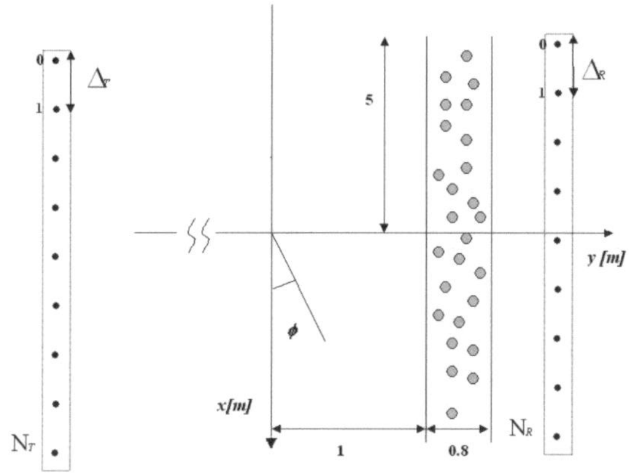

Figure 16. Geometry of the experiment studied by means of ray tracing.

The specific numerical values for the ray-tracing algorithm were selected after careful empirical investigation to yield negligible approximation errors. The transmitting array has N_T= 100 elements and inter-element distance Δ_T=λ/2, while the receiving array has N_R= 80 elements with inter-element spacing $\Delta_R = \lambda/4$, thus resulting in an angular resolution of approximately $\Delta\phi$=6°. The key parameter that discriminates the reliability of the radiative transfer approach was identified as the fraction of area that is effectively occupied by the scatterers. Let $A_{ext} = \pi(\sigma_{ext} / 2)^2$ be the effective area occupied by each particle. The fraction of effective area that is occupied by the scatterers is $\eta = N A_{ext}$. The numerical investigation proved that for scatterers that yield a relatively low value of η the radiative transfer provides a solution that closely matches the second order statistics (i.e. power fluctuations) as given by the ray tracing results.

Figure 17 shows the progressive specific intensity $I^{+}(y,\phi)$ computed according to radiative theory compared with $\hat{I}(y,\phi)$ from the ray tracing procedure. To compare the two approaches, the effect of limited angular resolution is accounted for by processing the outcome of the radiative transfer $I^{+}(y,\phi)$ and the number of averaging iterations for ray tracing is N_I = 100 [31]. By increasing the density N, and consequently the fraction η of effective area occupied by the scatterers, the difference between the prediction of radiative transfer and those from ray tracing increases.

Figure 18 shows the comparison between the spatial correlation $r_y(\Delta)$ computed by the radiative transfer theory and the same quantity $\hat{r}_y(\Delta)$from ray tracing versus Δ/λ for y = 1.9 m (on the right of the table) and two values of the density N=6,10.

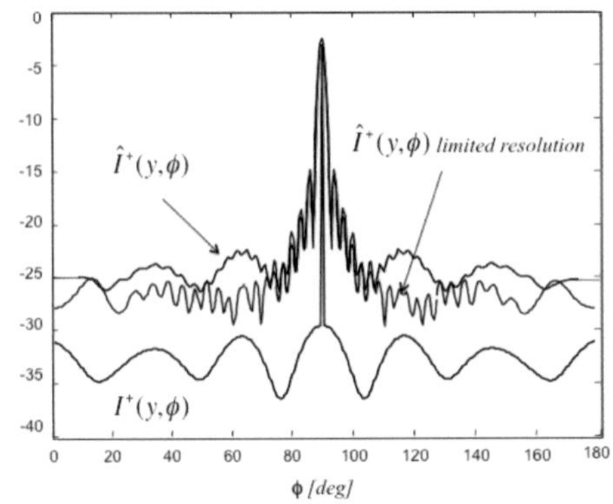

Figure 17. Specific intensity $I^{+}(y,)$ computed according to the radiative theory with and without limited resolution and $\hat{I}(y,\ )$ from the ray tracing procedure for $\varepsilon'_r = 4$, $\sigma = 10^{-3}$ Sm^{-1}, y = 1.9 m (on the right of the random medium), N = 10 m $^{-2}$ and vertical polarization.

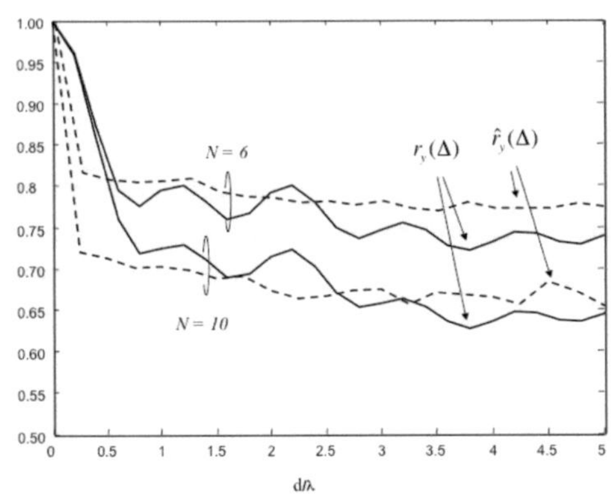

Figure 18. Spatial correlation $r_y(\Delta)$ computed by the radiative transfer theory and $\hat{r}_y(\Delta)$ from ray tracing versus Δ/λ for y = 1.9 (on the right of the random medium) and different values of the density N = 6, 10 m^{-2} (vertical polarization and $\varepsilon'_r = 4$, $\sigma = 10^{-3}$ Sm^{-1}).

http://dx.doi.org/10.5772/50871

6. Conclusions

In this work, the use of radiative transfer theory to study the propagation in an indoor environment was reported. A particular focus was made on two specific aspects related to the channel performances, such as the beam broadening and the spatial correlation, as they are parameters of interest when deploying a network. The pretty good agreement, although under well-controlled conditions, encourages proceeding along two distinct directions. From one side, an improvement in the modeling and in the sketching of actual environments is mandatory if one wants to use the radiative transfer technique as a possible approach to analyze indoor channel performances in practical scenarios. From the other side, validation should be performed either against measurements or - at least - against appropriate analytical solution without limitations or approximations in the frequency domain (or wavelength scale) of interest.

Acknowledgements

The author would like to gratefully acknowledge the work of Osvaldo Simeone (now Assistant Professor at the New Jersey Institute of Technology), who contributed with ideas, numerical developments and simulations to this research during his doctoral studies in Milano at the Politecnico, Faculty of Telecommunication Engineering.

Author details

Ada Vittoria Bosisio*

Address all correspondence to: bosisio@elet.polimi.it

National Research Council of Italy, Istituto di Elettronica ed Ingegneria dell'Informazione e delle Telecomunicazioni, Italy

References

[1] Falsafi, A., Pahlavan, K., & Ganning, Y. (1996). Transmission techniques for radio LAN's-a comparative performance evaluation using ray tracing. *Journal on Selected Areas in Communications*, 14(3), 477-491.

[2] Ji, Z., Li, B., Wang, Chem, H., & Sarkar, T. K. (2001). Efficient ray-tracing methods for propagation H. prediction for indoor wireless communications. *Antennas and Propagation Magazine*, 43(2), 41-49.

[3] Stuber, G. (2001). Principles of Mobile Communication. Dordrecht (The Netherlands), Kluwer Academic Publishers.

[4] Ullmo, D., & Baranger, H. U. (1999). Wireless propagation in buildings: A statistical scattering approach. *IEEE Transactions on Vehicular Technology*, 48(3), 947-955.

[5] Iskander, M. F., & Yun, Z. (2002). Propagation prediction models for wireless communication systems. *IEEE trans. Microwave Theory and Techniques*, 50(3), 662-673.

[6] Valenzuela, R. A. (1994). Ray tracing prediction for indoor radio propagation in: Proceedings of 5th IEEE International Symposium on Personal, Indoor and Mobile Radio Communication, PIMRC1994, 18-23 September 1994, The Hague, The Netherlands, 1(3), 140-144.

[7] Muller, A. (1994). Monte-Carlo multipath simulation of ray tracing channel models in: Proceedings of IEEE Global Telecommunications Conference GLOBECOM'94, 28 November - 2 December 1994, San Francisco (CA), 3, 1446-1450.

[8] Hassan-Ali, M., & Pahlavan, K. (2002). A new statistical model for site-specific indoor radio propagation prediction based on geometric optics and geometric probability. *IEEE Trans on Wireless Communications*, 1(1), 112-124.

[9] Ishimaru, A. (1978). Wave propagation and scattering in random media. Chap 7 and 11, New York, Academic Press, 1.

[10] Chandrasekhar, S. (1960). Radiative transfer theory, New York, Dover Publications.

[11] Kim, A. D., & Ishimaru, A. (1998). Optical and millimeter wave pulse propagation through fog and rain layers in Proceedings of International Geoscience and Remote Sensing Symposium IGARSS98, 6-10 July 1998, Seattle (WA), 1, 33-35.

[12] Caldeirinha, R. F. S., & Al-Nuaimi, M. O. (2002). Propagation modeling of bistatic scattering of isolated trees for micro- and millimetre wave urban microcells in Proceedings of 13th IEEE International Symposium on Personal, Indoor and Mobile Radio Communication PIMRC'02, 15-18 September 2002, Lisbon, 1(3), 135-139.

[13] Al-Nuaimi, M. O., & Hammoudeh, A. M. (1994). Measurements and predictions of attenuation and scatter of microwave signal by trees. *IEEE Proceedings on Microwave, Antennas and Propagation*, 141(2), 70-76.

[14] Hawley, M. S., Conway, J., & Anderson, A. P. (1998). The influence of tissue layering on microwave thermographic measurements. *International Journal of Hyperthermia*, 4(4), 427-435.

[15] Ulaby, F. T., Moore, R. K., & Fung, A. K. (1986). Microwave Remote Sensing. Active and Passive, Vol. III: From Theory to Applications. Norwood (MA), Artech House.

[16] Roux, L., Mareschal, P., Vukadinovic, N., Thibaud, J. B., & Greffet, J. J. (2001). Scattering by a slab containing randomly located cylinders: comparison between radiative transfer and electromagnetic simulation. *Journal of the Optical Society of America A*, 18(2), 374-384.

[17] Sarabandi, K., Tavakoli, A., & Ulaby, F. T. (1992). Propagation in a two-dimensional periodic random medium with inhomogeneous particle distribution. *IEEE Transactions on Antennas and Propagation*, 40(10), 1175-1186.

[18] Ishimaru, A., Jaruwatanadilok, S., Ritcey, J. A., & Kuga, Y. (2010). A MIMO Propagation Channel Model in a Random Medium. *IEEE Transactions on Antenna and Propagation*, 58(1), 178-186.

[19] Van de Hulst, H. C. (1957). Light scattering by small particles, place, J. Wiley & Sons.

[20] Ramo, S., Whinnery, J. R., & Van Duzer, T. (1994). Fields and waves in communication electronics, place John Wiley and sons.

[21] Karam, M. A., Fung, A. K., & Antar, Y. H. M. (1998). Electromagnetic wave scattering from some vegetation samples. *IEEE Transactions on Geoscience and Remote Sensing*, 26(6), 799-808.

[22] Fante, R. L. (1981). Relationship between radiative transfer theory and Maxwell's equations in dielectric media. *Journal of the Optical Society of America*, 71(4), 460-468.

[23] Ishimaru, A., & Kuga, Y. (1982). Attenuation constant of a coherent field in a dense distribution of particles. *Journal of the Optical Society of America*, 72(10), 1317-1320.

[24] Tsang, L., Kong, J. A., & Shin, R. (1984). Radiative transfer theory for active remote sensing of a layer of nonspherical particles. *Radio Science*, 19(2), 629-642.

[25] Acampora, A., & Winters, J. (1987). System applications for wireless indoor communications. *IEEE Communications Magazine*, 25(8), 11-20.

[26] Siqueira, P. R., & Sarabandi, K. (2000). T-Matrix Determination of Effective Permittivity for Three-Dimensional Dense Random Media. *IEEE Transactions on Antennas and Propagation*, 48(2), 317-327.

[27] Chew, W. C., Wang, Y-M., Otto, G., Wagner, R. L., & Liu, Q. H. (1992). A Generalized Recursive Algorithm for Wave-Scattering Solutions in Two Dimensions. *IEEE Transactions on Microwave Theory and Techniques*, 40(4), 716-723.

[28] Funkhouser, T., Carlbom, I., Elko, G., Pingali, G., Sondhi, M., & West, J. (1998). A beam approach to acoustic modeling for interactive virtual environment in ACM Computer Graphics Proceedings of SIGGRAPH'98; Orlando (F), 19-24 July, 21-32.

[29] Foco, M., Polotti, P., Sarti, A., & Tubaro, S. (2003). Sound spatialization based on fast beam tracing in the dual space in Proceedings of the International Conference DAFx-03, London (UK), 8-11 September.

[30] Bosisio, A. V. (2011). Performances of an RSSI-based positioning and tracking algorithm in Proceedings of International Conference on Indoor Positioning and indoor Navigation IPIN2011, Guimaraes (P), 21-23 September, doi:10.1109/IPIN.2011.6071952.

[31] Bosisio, A. V., & Simeone, O. (2007). Comparison between radiative transfer and ray tracing for indoor propagation in Proceedings of International Conference on Electromagnetics in Advanced Applications ICEAA07, Torino (I), 17-21 Sept., 395-398.

Chapter 10

Electromagnetic Wave Packets in the Theory of Bremsstrahlung and Transition Radiation by High-Energy Electrons

Nikolai Fyodorovich Shul'ga and
Sergii Valeriyovich Trofymenko

Additional information is available at the end of the chapter

http://dx.doi.org/10.5772/50866

1. Introduction

A lot of high-energy physical processes develop within large domains of space along the direction of particle motion (see, for example monographs [1-3] and references in them). In the case of electromagnetic processes the size of these domains can substantially exceed sometimes not only interatomic distances of substance but the size of experimental facility (detectors) as well [1,2,4-13]. Essential in this case is the fact that interaction of particles with atoms and experimental facility situated within such domains and outside them can substantially differ. Such situation arises, for example, when considering long-wave radiation in processes of bremsstrahlung and transition radiation by ultra relativistic electrons. Therefore, it is necessary to know what happens within such regions and what the peculiarities of evolution of such processes in space and time are. The present chapter is dedicated to the consideration of different aspects of this problem, which concern the behavior of high-energy wave packets, which take place in processes of bremsstrahlung and transition radiation.

We begin with consideration of the behavior of localized high-energy wave packets of a scalar massive particle in wave mechanics [2,14], of Gaussian packets of free electromagnetic waves and of packets, which take place in the equivalent photon method [15]. It is shown that in all cases high energies make the stabilizing effect upon the packet motion. Some peculiarities of dispersion of such packets and their reconstruction into the packets of diverging waves are considered as well.

INTECH
open science | open minds

© 2012 Shul'ga and Trofymenko; licensee InTech. This is an open access article distributed under the terms of the Creative Commons Attribution License (http://creativecommons.org/licenses/by/3.0), which permits unrestricted use, distribution, and reproduction in any medium, provided the original work is properly cited.

Further we show that the discussed wave packets naturally arise in processes of bremsstrahlung and transition radiation by high-energy electrons. For this purpose, firstly, the process of bremsstrahlung at an ultra relativistic electron instantaneous scattering to a large angle is considered [16,17]. The consideration is made on the basis of classical electrodynamics. In this case the moving electron is considered as a charge with its own coulomb field moving together with it. At the instantaneous scattering the perturbation of this field occurs. This perturbation is treated here as appearance of a packet of free plane electromagnetic waves, which reconstructs then into a packet of diverging waves. For ultra relativistic particles, however, this does not happen at once. The length within which this process develops has a name of the coherence length of the radiation process [1,2]. It is $2\gamma^2$ times larger than the length λ of the considered radiated waves ($\gamma$is here the electron's Lorentz-factor). We show that within this length the field around the electron substantially differs from the coulomb one. This leads, in particular, to the fact that the bremsstrahlung characteristics in this case substantially depend on both the detector's size and its position relative to the scattering point.

It is shown further that analogous effects take place in the process of transition radiation by an ultra relativistic electron during its traverse of thin ideally conducting plate as well [16-18]. The picture of evolution in space and time of the electromagnetic field, which arises before and after the electron traverse of thin metallic plate is considered here. The main attention is paid to effects in the process of backward transition radiation. In this case the wave packets of the field reflected from the plate are the packets of free waves, which reconstruct into the field of transition radiation. We show that the structure of these packets is in many respects analogous to the structure of the packets, which take place at instantaneous scattering of the particle to a large angle. This fact explains the presence of analogous effects in transition radiation and bremsstrahlung in the considered cases.

The special attention is drawn to the transition radiation by a scattered electron, which own field is not totally reconstructed after the scattering [17-19]. During a long period of time in this case the electron is in 'half-bare' state, which is the state in which some Fourier harmonics in the field around the electron are suppressed compared to the equilibrium coulomb field. The large values of distances, which the electron covers in this state allow us to place the plate within these distances and to consider the transition radiation by such electron on this plate. We show that in this case characteristics of backward transition radiation substantially differ from transition radiation characteristics in the case when the target is situated on large distances from the scattering point. The effect of transition radiation suppression and the effect of oscillatory dependence of transition radiation characteristics on the distance between the plate and the scattering point take place in this case. The causes of such effects are discussed.

2. High energy wave packets

2.1. Dispersion of relativistic wave packets

The general solution of the wave equation can be presented in the form of a wave packet, which spatially disperses in course of time. In semiclassical approximation such packet does not disperse. It moves according to the laws of classical mechanics (see, for example [2, 20]). It is going beyond the semiclassical approximation that leads to the packet dispersion. The high-energy wave packets are of special interest because the speed of their dispersion decreases with the increase of their energy. Let us pay attention to some peculiarities of dispersion of such packets. Significant here is the fact that characteristic features of this dispersion are similar for all fields. Therefore it is sufficient to consider just scalar field.

The general solution of the wave equation

$$\left(\frac{\partial^2}{\partial t^2}-\nabla^2-m^2\right)\phi(\vec{r},\, t)=0 \tag{1}$$

for a scalar particle with the mass m can be written in the following form of the expansion of the field $\phi(\vec{r},\, t)$over plane waves:

$$\phi(\vec{r},\, t)=\int\frac{d^3\kappa}{(2\pi)^3}e^{i(\vec{\kappa}\vec{r}-\omega t)}C_{\vec{\kappa}} \tag{2}$$

where $\omega=\sqrt{\vec{\kappa}^2+m^2}$ and $C_{\vec{\kappa}}$ - are the expansion coefficients. Here and further we will use the system of units in which the speed of light c and the Plank constant $\hbar$ equal unit.

Let us consider the dispersion of the wave packet, which at the initial moment of time coincides with the Gaussian packet modulated by the plane wave with large value of the momentum $\vec{p}$ [2,14]. Moreover we will assume that the initial widths of the packet $a_{||}$ and $a_{\perp}$ parallel and perpendicular to the particle momentum $\vec{p}$ are different. For such packet at the initial moment of time the field $\phi(\vec{r},\, t)$ has the following form:

$$\phi(\vec{r},\, t)=e^{i\vec{p}\vec{r}-\frac{z^2}{2a_{||}^2}-\frac{\vec{\rho}^2}{2a_{\perp}^2}} \tag{3}$$

where z and $\vec{\rho}$ are the coordinates parallel and orthogonal to$\vec{p}$. At the moment of time t this packet will be defined by the relation (2) with

$$C_{\vec{\kappa}}=(2\pi)^{3/2}a_{||}a_{\perp}^2e^{-\frac{(p-\kappa_z)^2a_{||}^2}{2}-\frac{\kappa_{\perp}^2a_{\perp}^2}{2}} \tag{4}$$

We can write the obtained expression for the field $\phi(\vec{r},\, t)$ in the form

$$\phi(\vec{r},\, t) = Ae^{i(\vec{p}\vec{r} - \varepsilon\, t)} I(\vec{r},\, t) \tag{5}$$

in which $A = a_{||} a_{\perp}^2$, $\varepsilon = \sqrt{\vec{p}^2 + m^2}$and

$$I(\vec{r},\, t) = (2\pi)^{-3/2} \int d^3\kappa e^{i(\vec{\kappa} - \vec{p})\vec{r} - \frac{(p-\kappa_z)^2 a_{||}^2}{2} - \frac{\kappa_{\perp}^2 a_{\perp}^2}{2} - i(\omega_k - \varepsilon)t} \tag{6}$$

Having made in this expression the variable substitution $\vec{\kappa} = \vec{p} + \vec{q}$ we find that

$$I(\vec{r},\, t) = (2\pi)^{-3/2} \int d^3 q e^{i\vec{q}\vec{r} - \frac{q_z^2 a_{||}^2}{2} - \frac{q_{\perp}^2 a_{\perp}^2}{2} - i(\omega_{\vec{p}+\vec{q}} - \varepsilon)t} \tag{7}$$

In the case of large energies it is possible to make the expansion over $|\vec{q}| / p$ in the quantity $(\omega_{\vec{p}+\vec{q}} - \varepsilon)$ in (7). Having preserved the quadratic terms of expansion we obtain

$$\omega_{\vec{p}+\vec{q}} - \varepsilon \approx v q_z + \frac{q_z^2}{2\varepsilon\gamma^2} + \frac{q_{\perp}^2}{2\varepsilon} \tag{8}$$

where $v = p / \varepsilon$ and$\gamma = (1 - v^2)^{-1/2}$. Substituting this expression into (6) after simple calculations we obtain

$$I(\vec{r},\, t) = \frac{1}{\sqrt{a_{||}^2 + i\frac{t}{\varepsilon\gamma^2}}} \frac{1}{a_{\perp}^2 + i\frac{t}{\varepsilon}} \exp\left\{ -\frac{(z - vt)^2}{2\left(a_{||}^2 + i\frac{t}{\varepsilon\gamma^2}\right)} - \frac{\rho^2}{2\left(a_{\perp}^2 + i\frac{t}{\varepsilon}\right)} \right\} \tag{9}$$

The formula (9) can be written in the following form as well:

$$I(\vec{r},\, t) = A(t) \exp\left\{ i\alpha(\vec{r},\, t) - \frac{(z - vt)^2}{2\Delta_{||}^2(t)} - \frac{\rho^2}{2\Delta_{\perp}^2(t)} \right\} \tag{10}$$

in which $A(t)$is a slowly changing quantity

$$A(t) = \frac{1}{\left(a_{\perp}^2 + i\frac{t}{\varepsilon}\right)\sqrt{a_{||}^2 + i\frac{t}{\varepsilon\gamma^2}}} \tag{11}$$

$\alpha(\vec{r},\, t)$is the real phase

$$\alpha(\vec{r},\ t)=\frac{(z-vt)^2}{2}\frac{t/\varepsilon\gamma^2}{a_{||}^4+(t/\varepsilon\gamma^2)^2}+\frac{\rho^2}{2}\frac{t/\varepsilon}{a_{\perp}^4+(t/\varepsilon)^2} \tag{12}$$

$\Delta_{||}(t)$ and $\Delta_{\perp}(t)$ are the longitudinal and transverse widths of the packet at the moment of time t

$$\Delta_{||}^2(t)=a_{||}^2+\left(\frac{t}{a_{||}\varepsilon\gamma^2}\right)^2,\ \Delta_{\perp}^2(t)=a_{\perp}^2+\left(\frac{t}{a_{\perp}\varepsilon}\right)^2 \tag{13}$$

In the case of $a_{||}=a_{\perp}$ the obtained above formulae coincide with the corresponding result of the paper [14].

The formulae (13) show that in longitudinal and transverse directions the squares of the widths of the packet $\Delta_{||}^2(t)$ and $\Delta_{\perp}^2(t)$ grow with time proportionally to t^2m^4/ε^6 andt^2/ε^2. In nonrelativistic case these quantities do not depend on the particle energy ($t^2/\varepsilon^2=t^2/m^2$). In relativistic case the quantities t^2m^4/ε^6 and t^2/ε^2 are substantially smaller than the corresponding values for nonreativistic particles. Let us note that the additional factor m^4/ε^4 exists for longitudinal direction in$\Delta_{||}^2(t)$. It leads to the substantial decrease of the speed of the packet dispersion in this direction compared to the speed of the packet dispersion in transverse direction. Thus the relativistic effects do the stabilizing influence upon the wave packets.

In conclusion let us note that while deriving the formula (9) we neglected the terms proportional to tq^4/ε^3 in the exponential factor in (7). Therefore the formula (9) is valid during the interval of time, which satisfies the condition

$$tq_{eff}^4/\varepsilon^3\ll 1 \tag{14}$$

where q_{eff} are the characteristic values of the variable $\vec{q}$ in (7), which make contribution to this integral. For $a_{||}\approx a_{\perp}$ the inequality (14) can be written in the form

$$\frac{t}{a_{\perp}^2\varepsilon}\frac{1}{(a_{\perp}\varepsilon)^2}\ll 1 \tag{15}$$

Thus the formula (9) is valid for the time interval t, which satisfies the inequality (15). For ultra relativistic particles this interval rapidly grows with the increase of the particle's energy.

2.2. Dispersion of a high-energy packet of electromagnetic waves

Now let us consider high energy packets of free electromagnetic waves. Scalar and vector potentials of such packets are the solutions of the wave equation (1) with $m=0$. Therefore in order to analyze the peculiarities of dispersion of such packets we can use the formulae of the previous section assuming that all the terms in them containing the Lorentz-factor γ equal zero. In this case for scalar potential we find that

$$\phi(\vec{r}, t) = Ae^{i(\vec{k}\vec{r}-\omega t)} I(\vec{r}, t) \tag{16}$$

where $\vec{k}$ and ω are the wave vector and the frequency of the electromagnetic wave and

$$I(\vec{r}, t) = A(t)\exp\left\{i\alpha_k(\vec{r}, t) - \frac{(z-t)^2}{2\Delta_{||}^2(t)} - \frac{\rho^2}{2\Delta_{\perp}^2(t)}\right\} \tag{17}$$

Here

$$A(t) = \frac{1}{a_{||}\left(a_{\perp}^2 + it/\omega\right)}, \quad \alpha(\vec{r}, t) = \frac{\rho^2}{2}\frac{t/\omega}{a_{\perp}^4 + (t/\omega)^2}$$

and

$$\Delta_{||}^2(t) = a_{||}^2, \quad \Delta_{\perp}^2(t) = a_{\perp}^2 + (t/a_{\perp}\omega)^2 \tag{18}$$

The obtained formulae show that the initially Gaussian packet does not disperse in the direction parallel to the $\vec{k}$ vector. In transverse direction the square of the packet widths grows proportionally to$(t/\omega)^2$. Thus the speed of the packet dispersion decreases with the increase of the wave frequencyω.

When considering a process of radiation by relativistic electrons it is often necessary to deal with packets, which are constructed of plane waves with wave vectors, which directions are close to the direction of a given vector$k$. Such wave packets differ somehow from the ones considered above. Let us consider some peculiarities of dispersion of such packets assuming for simplicity that at the initial moment of tome $t=0$ the distribution of the waves over the wave vectors is Gaussian relative to the given vector $\vec{k}$ [15]. For such distribution in the initial moment of time the scalar potential $\phi_k(\vec{r},0)$ has the following form:

$$\phi_k(\vec{r},0) = \frac{1}{\pi\bar{\Delta}_{\vartheta}^2}\int d^2\vartheta\, e^{-\vartheta^2/\bar{\Delta}_{\vartheta}^2} e^{i\vec{k}\vec{r}} \tag{19}$$

where ϑ is the angle between the packet wave vector and the wave vector$\vec{k}$, $\bar{\Delta}_{\vartheta}^2$is the average value of the square of the angleϑ,$\bar{\Delta}_{\vartheta}^2 \ll 1$.

The coefficients $C_{\vec{q}}$ of the Fourier expansion (2) for such initial packet have the following form

$$C_{\vec{q}}=(2\pi)^3\int\frac{d^2\vartheta}{\pi\bar{\Delta}_\vartheta^2}e^{-\vartheta^2/\bar{\Delta}_\vartheta^2}\delta(\vec{k}-\vec{q}) \tag{20}$$

in which $\delta(\vec{k}-\vec{q})$ is the delta-function. As a result we come to the following expression for the scalar potential

$$\phi_k(\vec{r},\, t)=\frac{1}{1+ikz\bar{\Delta}_\vartheta^2/2}e^{ik(z-t)-\frac{(k\rho/2)^2\bar{\Delta}_\vartheta^2}{1+ikz\bar{\Delta}_\vartheta^2/2}} \tag{21}$$

where z and $\vec{\rho}$ are the coordinates parallel and orthogonal to$\vec{k}$.

Let us note that the considered case corresponds to the wave packet, which consists of the plane waves the directions of the wave vectors of which have some small scatter around the z axis. If the initial packet has some small variation in frequencyω, the formula (21) should be averaged over this variation of the waves in frequency. Assuming for simplicity that the distribution of the waves over frequency has the Gaussian form with the average value of the square of the frequency variation $\bar{\Delta}_\omega^2$ we find that

$$\langle\phi_k(\vec{r},\, t)\rangle=\frac{1}{1+i\omega z\bar{\Delta}_\vartheta^2/2}e^{i\omega(z-t)-\frac{(z-t)^2\bar{\Delta}_\omega^2}{4}-\frac{(\omega\rho/2)^2\bar{\Delta}_\vartheta^2}{1+i\omega z\bar{\Delta}_\vartheta^2/2}} \tag{22}$$

The given expression for the wave packet has the same structure as the corresponding expression for the packet (16). If the substitutions $(t/a_\perp^2\varepsilon)\to(\omega z\bar{\Delta}_\vartheta^2/2)$ and $a_{||}^2\to 4\bar{\Delta}_\omega^2$ are made in the latter expression the both formulae for the wave packet will become identical.

The formula (22) shows that for $\omega z\bar{\Delta}_\vartheta^2/2\ll 1$

$$\langle\phi_k(\vec{r},\, t)\rangle\approx\exp\left\{i\omega(z-t)-\frac{(z-t)^2\bar{\Delta}_\omega^2}{4}-\left(\frac{\omega\rho}{2}\right)^2\bar{\Delta}_\vartheta^2\right\} \tag{23}$$

and for $\omega z\bar{\Delta}_\vartheta^2/2\gg 1$

$$\langle\phi_k(\vec{r},\, t)\rangle\approx-\frac{2i}{\omega z\bar{\Delta}_\vartheta^2}\exp\left\{i\omega(z-t)-\frac{(z-t)^2\bar{\Delta}_\omega^2}{4}+i\omega\frac{\rho^2}{2z}-\frac{\rho^2}{z^2\bar{\Delta}_\vartheta^2}\right\} \tag{24}$$

For $z\gg\rho$ the latter formula can be written in the form of a diverging wave

$$\langle\phi_k(\vec{r},\, t)\rangle\approx-\frac{2i}{\omega r\bar{\Delta}_\vartheta^2}\exp\left\{i\omega(r-t)-\frac{(z-t)^2\bar{\Delta}_\omega^2}{4}-\frac{\rho^2}{z^2\bar{\Delta}_\vartheta^2}\right\} \tag{25}$$

where $r = \sqrt{z^2 + \rho^2} \approx z + \rho^2 / 2z$. Thus on distances z from the center of the initial packet, which satisfy the condition

$$\omega z \bar{\Delta}_{\vartheta}^2 / 2 \ll 1 \tag{26}$$

the form of the packet (22) coincides with the form of the packet at $t=0$. Only on the distances, which satisfy the condition

$$\omega z \bar{\Delta}_{\vartheta}^2 / 2 \gg 1 \tag{27}$$

the transformation of the packet (22) to the packet of spherical diverging waves occurs.

Let us note that in the theory of radiation of electromagnetic waves by a moving electron the spatial region in which the formation of spherical diverging waves occurs has a name of the wave zone (see for example [21]). In particular, for nonrelativistic charged particles the wave zone begins on distances from the radiation region, which exceed the length of the radiated wave λ. However, the condition (27) shows that for $\bar{\Delta}_{\vartheta}^2 \ll 1$ the wave zone formation occurs not on distances $z \gg \lambda$ as in the case of a nonrelativistic particle but on distances

$$z \gg 2\lambda / \bar{\Delta}_{\vartheta}^2 \tag{28}$$

which are much larger than the wave length $\lambda = 1/\omega$. For sufficiently small values of $\bar{\Delta}_{\vartheta}^2$ the length $z = 2\lambda / \bar{\Delta}_{\vartheta}^2$ can reach macroscopic size.

2.3. Wave packets in the equivalent photon method

The problem of dispersion of wave packets naturally arises in the equivalent photon method (or the method of virtual photons) in which at the certain moment of time ($t=0$) the electron's coulomb field is substituted by a packet of free electromagnetic waves (see [22] and references there). Let us consider some peculiarities of wave packets behavior in this method [15].

For this purpose we write the scalar potential of the coulomb field of the electron moving along the z axis with the velocity $\vec{v}$ in the form of the following Fourier-expansion:

$$\phi_C(\vec{r}, t) = \mathrm{Re} \int \frac{d^3k}{(2\pi)^3} e^{i(\vec{k}\vec{r} - \vec{k}\vec{v}t)} C_k^C \tag{29}$$

in which

$$C_k^C = \frac{8\pi e \theta(k_z)}{k_\perp^2 + k_z^2 / v^2 \gamma^2} \tag{30}$$

Here $e$is the electron's charge, $\gamma$is the Lorentz-factor, $k_z$and $k_\perp$ are the components of vector $\vec{k}$ parallel and orthogonal to z axis, $\theta(k_z)$is the Heaviside step function.

In the equivalent photon method it is assumed that for $t=0$ the packet (2), which consists of free electromagnetic waves coincides with the packet (29). It corresponds to the Fourier expansion (2) with the coefficients$C_{\vec{\kappa}} = C_k^C$. Let us consider the behavior of such packet in ultra relativistic case in which$\gamma \gg 1$.

For $\gamma \gg 1$ the main contribution to (2) are made by the values of $\vec{\kappa}$ with directions close to the direction of the electron's velocity$\vec{v}$. Taking this fact into account we can present the packet (2) in the following form:

$$\phi(\vec{r}, t) = \mathrm{Re} \int_0^\infty dk \phi_k(\vec{r}, t) \tag{31}$$

where

$$\phi_k(\vec{r}, t) = \frac{2}{\pi} e^{ik(z-t)} \int_0^\infty \frac{\vartheta d\vartheta}{\vartheta^2 + \gamma^{-2}} J_0(k\rho\vartheta) e^{-ikz\vartheta^2/2} \tag{32}$$

Here ϑ is the angle between $\vec{k}$ and $\vec{v}$ ($\vartheta \ll 1$) and $J_0(x)$ is the Bessel function.

The function $\phi_k(\vec{r}, t)$ has the same structure as the function (21) corresponding to the Gaussian distribution of vectors $\vec{k}$ over the anglesϑ. Namely, if$kz\vartheta^2/2 \ll 1$, the main contribution to the integral (32) is made by the values $\vartheta \approx \gamma^{-1}$ and

$$\phi_k(\vec{r}, t) \approx \frac{2}{\pi} K_0(k\rho / \gamma) e^{ik(z-t)} \tag{33}$$

where $K_0(x)$ is the modified Hankel function. In this case after integration over $\vec{k}$ in (31) we find that

$$\phi(\vec{r}, t) = \frac{e}{\sqrt{\rho^2 \gamma^{-2} + (z-t)^2}} \tag{34}$$

The main contribution to (31) is made by the values$k \approx \gamma / \rho$, hence expression (34) is valid in the range of coordinates ρ and z that satisfy the condition$z < \gamma\rho$. In this range of coordinates the packet under consideration moves with the velocity of light in the z axis direction.

So on distances $z < 2\gamma^2\lambda$ the considered wave packet practically coincides with the initial one (at$t = 0$). Substantial transformation of the packet would happen only on distances

$$z > 2\gamma^2\lambda \tag{35}$$

In this case for the evaluation of the integral in (32) over ϑ one could apply the method of stationary phase. As a result of using of this method we find that

$$\phi_k(\vec{r}, t) = -\frac{2i}{\pi}\frac{1}{\theta_0^2 + \gamma^{-2}}\frac{1}{kr}e^{ik(r-t)} \tag{36}$$

where $r \approx z + \rho^2 / 2z$ and $\vartheta_0 = \rho / z$ is the point of stationary phase of the integral (32). We see that the components (36) of our packet have in the case under consideration the form of diverging spherical waves. Under this condition the angle ϑ_0 corresponds to the direction of radiation, and the function before the diverging wave describes the angular distribution of the radiation. So, the condition (35) draws out the wave zone in application to given problem.

The value $2\gamma^2\lambda$ presenting in the condition (35) is known in the theory of radiation by ultrarelativistic particles as the formation length or the coherence length [1, 2].

3. The bremsstrahlung at an electron instantaneous scattering

3.1. The electromagnetic field structure at an electron instantaneous scattering. The 'half-bare' electron

The electromagnetic wave packets similar to the ones considered above arise, for example, in the processes of bremsstrahlung by relativistic electron at its instantaneous scattering to a large angle and in the process of transition radiation during an electron traverse of thin metallic plate in vacuum. The present section is dedicated to the analysis of evolution of electromagnetic wave packets and peculiarities of formation of radiation by relativistic electron in the wave and the pre wave zones in the first process, while the next one – to the analogous questions concerning the second one.

Let a relativistic electron move along the z axis with the velocity $\vec{v}$ and at the moment of time $t = 0$ in the point $z = 0$ change it abruptly from $\vec{v}$ to $\vec{v}\,'$ (Figure 1).

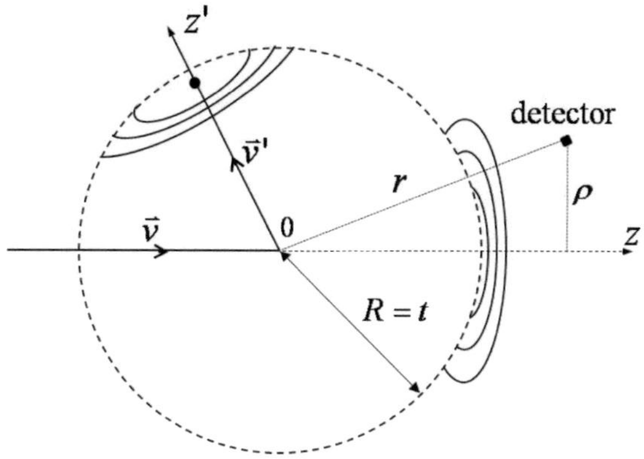

Figure 1. The picture of total field after instantaneous scattering of an electron to a large angle.

Scalar and vector potentials of the total electromagnetic field, which takes place in such process can be defined from inhomogeneous Maxwell equations

$$\begin{aligned} \Delta\phi - \frac{\partial^2 \phi}{\partial t^2} &= -4\pi\rho(\vec{r}, t), \\ \Delta\vec{A} - \frac{\partial^2 \vec{A}}{\partial t^2} &= -4\pi\vec{j}(\vec{r}, t) \end{aligned} \tag{37}$$

in which ρ and $\vec{j}$ are respectively the charge and the current density of the particle, which in the case of a single point particle motion should be taken as

$$\begin{aligned} \rho(\vec{r}, t) &= e\delta(\vec{r} - \vec{r}(t)) \\ \vec{j}(\vec{r}, t) &= e\vec{v}(t)\delta(\vec{r} - \vec{r}(t)) \end{aligned} \tag{38}$$

where $\delta(\vec{r})$is the delta-function and $\vec{r}(t)$and $\vec{v}(t)$ are respectively the particle's trajectory and its velocity. The solution of the set of equations (37) has the physical sense if it has the form of the retarded potentials. In our case of a point particle the required form of the solution of (37) and (38) is the following [21]:

$$\{\vec{A}, \phi\} = e\int_{-\infty}^{+\infty} \frac{dt'}{|\vec{r} - \vec{r}(t')|}\{\vec{v}(t'), 1\}\delta(t' - t + |\vec{r} - \vec{r}(t')|) \tag{39}$$

In the case of a uniform particle motion with the velocity $\vec{v}$along the z axis (39) leads to the following explicit expressions for the potentials:

$$\vec{A}_v(\vec{r}, t) = \frac{e\vec{v}}{\sqrt{\rho^2\gamma^{-2} + (z - vt)^2}}, \quad \phi_v(\vec{r}, t) = \frac{e}{\sqrt{\rho^2\gamma^{-2} + (z - vt)^2}} \tag{40}$$

which are the coulomb potentials of the moving particle. Here $\gamma = E/m$ is the particle's Lorentz-factor and ρ is the absolute value of the coordinate $\vec{\rho}$ of the observation point in the plane orthogonal to the z axis. The formula (40) for $\phi(\vec{r}, t)$shows that the equipotential surfaces of the field of the particle, which moves uniformly straightforward are the ellipsoids flattened in the direction of particle motion.

In order to obtain the solution of (37) for the considered case of the particle instantaneous scattering it is convenient to express the potentials in the form of Fourier-integrals. Let us consider, for example, the vector potential:

$$\vec{A}(\vec{r}, t) = \frac{1}{(2\pi)^3} \int \vec{A}_{\vec{k}}(t) e^{i\vec{k}\vec{r}} d^3k \tag{41}$$

In order to obtain the expansion in the form of the retarded potential we should calculate the Fourier-component $\vec{A}_{\vec{k}}(t)$ with the use of the inverse Fourier-transform of the expression (39) for the vector potential:

$$\vec{A}_{\vec{k}}(t) = \int \vec{A}(\vec{r}, t) e^{-i\vec{k}\vec{r}} d^3r =$$

$$= e \int_{-\infty}^{+\infty} dt' \int d^3r e^{-i\vec{k}\vec{r}} \frac{\vec{v}(t')}{|\vec{r} - \vec{r}(t')|} \delta(t' - t + |\vec{r} - \vec{r}(t')|)$$

Making here the substitution $\vec{R} = \vec{r} - \vec{r}(t')$ and integrating the expression over d^3R it is possible finally to present the Fourier-expansion (41) of the vector potential in the following form:

$$\vec{A}(\vec{r}, t) = -\frac{e}{2\pi^2} \mathrm{Im} \int \frac{d^3k}{k} e^{-i(kt - \vec{k}\vec{r})} \int_{-\infty}^{t} dt' \vec{v}(t') e^{i(kt' - \vec{k}\vec{r}(t'))} \tag{42}$$

Substituting into (42)$\vec{v}(t') = \vec{v}\theta(-t') + \vec{v}'\theta(t')$, where $\theta(x)$is the Heaviside theta function, we have for $t < 0$ (which is before the scattering moment):

$$\vec{A}(\vec{r}, t) = \frac{e}{2\pi^2} \mathrm{Re} \int \frac{d^3k}{k} \frac{\vec{v}}{ck - \vec{k}\vec{v}} e^{i(\vec{k}\vec{r} - \vec{k}\vec{v}t)} \tag{43}$$

As integration over $\vec{k}$may show, (43) is nothing else than the Fourier-expansion of the own coulomb field (40) of a uniformly moving particle.

The different and rather interesting situation takes place after the scattering moment (for $t>0$). In this case from (42) we obtain:

$$\vec{A}(\vec{r},\ t)=\frac{e}{2\pi^2}\mathrm{Re}\int\frac{d^3k}{k}\left\{\frac{\vec{v}\,'}{k-\vec{k}\vec{v}\,'}\left[1-e^{-i(k-\vec{k}\vec{v}\,')t}\right]e^{-i\vec{k}\vec{v}'t}+\frac{\vec{v}}{k-\vec{k}\vec{v}}e^{-ikt}\right\}e^{i\vec{k}\vec{r}} \tag{44}$$

The integration over $\vec{k}$ gives in this case:

$$\vec{A}(\vec{r},\ t)=\vec{A}_{\vec{v}'}(\vec{r},\ t)\theta(t-r)+\vec{A}_{\vec{v}}(\vec{r},\ t)\theta(r-t) \tag{45}$$

where $\vec{A}_{\vec{v}}$ and $\vec{A}_{\vec{v}'}$ are respectively the coulomb potentials of the particles, which uniformly move in the directions of z and z' axes. The analogous expressions for the scalar potential can be easily obtained by the same method.

The expression (43) shows that before the scattering moment the total field around the electron coincides with its own coulomb field, which moves with the velocity $\vec{v}$ together with the electron. After the scattering, according to (44) and (45), the total field around the electron breaks into two parts represented by two items in the considered expressions.

The first item in braces in (44) corresponds to the nonequilibrium field, which the scattered electron has already managed to rebuild around itself by the moment of time t. This field consists of the own coulomb field of the electron, which moves along the z' axis with the velocity $\vec{v}\,'$ (the first item in square brackets in (44) corresponds to it) and the packet of free waves, which moves in the same direction (it is described by the second item in square brackets). There is substantial interference between these fields due to which (as the first item in (45) shows) the total field vanishes in the region $r>t$, which the signal about the electron scattering at the moment of time $t=0$ has not yet reached. Inside the sphere of radius $r=t$ the total field coincides with the equilibrium coulomb field of the electron.

The second item in braces in (44) describes the field, which as though `tears away' from the electron at the scattering moment. It is a packet of free electromagnetic waves, which moves in the direction of the initial electron's velocity $\vec{v}$ and gradually transforms into bremsstrahlung. This field is different from zero outside the sphere $r=t$ and vanishes inside it.

The equipotential surfaces of the scalar potential of the field around the electron after its scattering to a large angle are presented on Figure 1.

The behavior of certain Fourier-components of the nonequilibrium field of the electron after its scattering and of the field 'torn away' from the electron at its scattering is of special interest. According to (44), the values of $\vec{k}$, which make contribution to each item of the nonequilibrium field of the electron has directions close to the direction of the electron's final velocity $\vec{v}\,'$. During the period of time $t<(k-\vec{k}\vec{v}\,')^{-1}$, however, both of these items substantially cancel each other due to the interference. Therefore the coulomb field, which corresponds to the scattered electron, does not appear instantly. Namely, during the period of time $t\leq(k-\vec{k}\vec{v}\,')^{-1}$ the Fourier components with wave vector k are nearly absent in the field around the scattered electron. As the main contribution to the coulomb field is made by $\vec{k}$ with directions close to the direction of the velocity $\vec{v}\,'$ this period of time is defined by the relation$t \leq 2\gamma^2 / k$. Figuratively speaking we can say that after the scattering during such period of time the electron is in 'half-bare' state, which means without considerable part of its field. Such state of electron with nonequilibrium field manifests itself during further scatterings of the electron, which causes different effects of bremsstrahlung suppression during an electron's motion in substance, such as Landau-Pomeranchuk-Migdal effect [23-25], the effect of radiation suppression in thin layer of substance (TSF-effect [26-28]), etc. Recently the series of detailed experimental investigations of these effects at ultra high energies was performed on SLAC [29,30] and CERN[31,32] accelerators, which confirmed the main theoretical predictions.

The notion of a 'half-bare' electron was introduced in the papers of E.L. Feinberg [33,34] who studied the time evolution of the state vector of the system 'electron + photon' after the scattering of a fast electron to a large angle on atom. The classical theory of this effect was given in [2, 27, 35].

Let us note that during the period of time $t = 2\gamma^2 / \omega$ the electron covers the distance

$$l_C = 2\gamma^2 v / \omega \tag{46}$$

In the theory of radiation by relativistic electrons the length l_C has a name of the coherence length of the radiation process [1,2]. Within this length the interference effects in radiation are significant. For ultra relativistic electrons in the region of small frequencies of radiated waves the length l_C can have macroscopic size, which exceeds not only interatomic distance of the substance but the size of the experimental facility as well. Indeed, in the millimeter range of the waves radiated by electrons with the energy of $50Mev$ the coherence length is $l_C = 20m$. It means that within such length after scattering the electron is in 'half-bare' state, which means that the Fourier-components of the corresponding wavelengths are absent in the field around the electron.

The field, which 'tears away' from the electron at its scattering (the second item in braces in (44)) has the structure similar to the one, which has the packet of free waves considered above in the equivalent photons method. Therefore the main peculiarities of the reconstruction of the 'torn away' field to the field of radiation will be the same as the considered above peculiarities of the wave packets evolution. Let us consider this process in detail.

3.2. The problem of measurement of bremsstrahlung characteristics

The results presented above show that for ultra relativistic electrons the radiation formation process develops on large distances along the initial and final directions of the electron motion, which can be of macroscopic size. In this case a detector, which registers the radiation characteristics can be situated both in the wave zone (which means on large distances $r \to \infty$ from the scattering point) and in the pre wave zone (which means on within the coherence length l_C of the radiation process). Let us show that the results of measurements can substantially differ in these cases. For this purpose let us consider the peculiarities of the formation of radiation from the 'torn-away' field on different distances from the scattering point [17]. The consideration will be made for a point detector situated in the point with coordinates$\vec{r}=(\vec{\rho}, z)$. Here the z axis is parallel to the vector of the electron's initial velocity and $\vec{\rho}$ are the coordinates in the orthogonal plane. In this case by the point detector we mean the detector, which measures the characteristics of electromagnetic waves (wave packet), which fall on a small element of surface situated in the point with the coordinates $\vec{r}=(\vec{\rho}, z)$ and seen at solid angle *do* from the scattering point.

Making in the second item in (44) the variable substitution $k_z \to k$by $k_z=\sqrt{k^2-q^2}$($k_z$and $q$are respectively the components of the wave vector $\vec{k}$ along the z axis and orthogonal to it) and denoting$k=\omega$, it is possible to present the Fourier-expansion of the `torn-away' field scalar potential in the following form:

$$\phi(\vec{r}, t)=\frac{e}{\pi v^2}\int_{-\infty}^{+\infty} d\omega e^{-i\omega t}\int_0^{|\omega|} dq\frac{qJ_0(q\rho)}{q^2+\omega^2/v^2\gamma^2}Q(z) \tag{47}$$

where

$$Q(z)=\frac{\omega\cos\left(\sqrt{\omega^2-q^2}z\right)}{\sqrt{\omega^2-q^2}}+iv\sin\left(\sqrt{\omega^2-q^2}z\right) \tag{48}$$

and $\rho=|\vec{\rho}|$.

In (47) and (48) the square root $\sqrt{\omega^2-q^2}$ is considered to be a single-valued branch of the analytical function, which is equal to $|\sqrt{\omega^2-q^2}|$ for $\omega>q$ and $-|\sqrt{\omega^2-q^2}|$ for$\omega<-q$.

In ultra relativistic case ($\gamma\gg 1$) the range of$q$, which make the main contribution to the integral (47) is $q\leq\omega/\gamma\ll\omega$ and it is possible to expand the square roots $\sqrt{\omega^2-q^2}$ in (48) in the small factorq/ω. Let us leave the items proportional to the second power of q/ωin the arguments of sine and cosine, while in the other parts of the expression (48) neglect them. Moreover the integration over $q$can be extended to the region$0<q<\infty$. This leads to the following expression for the `torn-away' field potential in ultra relativistic case:

$$\phi(\vec{r}, t)=\frac{e}{\pi}\int_{-\infty}^{+\infty} d\omega\int_0^{+\infty} dq\frac{qJ_0(q\rho)}{q^2+\omega^2/v^2\gamma^2}e^{i\omega(z-t)-iq^2z/2\omega} \tag{49}$$

The equations (37) are presented in Lorentz gauge

$$div\vec{A} + \frac{\partial\phi}{\partial t} = 0 \tag{50}$$

If we know $\phi(\vec{r}, t)$, we can derive the vector potential component A_z from this equation ($\vec{A}_\perp = 0$ in this case). In the considered case the vector potential equals to the scalar one up to the neglected items of the order of $(q/\omega)^2$.

In order to determine the total energy radiated in the direction of a small area $d\vec{s} = \vec{n}r^2 do$ situated in the point $\vec{r} = (\vec{\rho}, z)$ we can calculate the flux of the Poynting vector through this area during the whole time of the particle motion

$$d\varepsilon = \frac{r^2 do}{4\pi}\int dt(\vec{E}\times\vec{H})\vec{n}$$

where do is the element of solid angle in the direction of radiation $\vec{n} = \vec{r}/r$. The fields $\vec{E}$ and $\vec{H}$ in this expression are the fields of the packet of free electromagnetic waves (the field 'torn-away' from the electron at its scattering). They are related to scalar ϕ and vector $\vec{A}$ potentials of the radiation field by

$$\vec{E} = -\frac{\partial\vec{A}}{\partial t} - \nabla\phi,\ \vec{H} = \nabla\times\vec{A} \tag{51}$$

Proceeding to the Fourier-expansions of the fields ϕ and $\vec{A}$ over frequency ω we obtain the following expression for the radiation spectral-angular density:

$$\frac{d\varepsilon}{d\omega do} = \frac{r^2}{4\pi^2}[\vec{E}_\omega(\vec{r})\times\vec{H}_{-\omega}(\vec{r})]\vec{n} \tag{52}$$

in which $\omega \geq 0$.

With the use of the Maxwell equation $\vec{H}_\omega = (-i/\omega)\nabla\times\vec{E}_\omega$, the radiation spectral-angular density can be expressed in the terms of Fourier-component of the electric field alone:

$$\frac{d\varepsilon}{d\omega do} = \frac{ir^2}{\omega}[\vec{E}_\omega\times(\nabla\times\vec{E}_{-\omega}(\vec{r}))]\cdot\vec{n} \tag{53}$$

Let us note that the formulae (52) and (53) are valid for arbitrary distances from the scattering point. Therefore they can be used for radiation consideration both in the wave and the pre wave zones. In the wave zone (which means in the region $r\to\infty$) they can be considerably simplified. Therefore, firstly, let us dwell on the consideration of this case.

On large distances from the scattering point ($r\to\infty$) a Fourier-component of the field of the radiation waves $\vec{E}_\omega(\vec{r})$ reconstructs into a packet of diverging waves, which amplitude is proportional to the factor$r^{-1}\exp(i\omega r)$. The action of the Hamilton operator ∇ in (53) upon $\vec{E}_{-\omega}$ in this case can be applied only to the factor$\exp(i\omega r)$:

$$\nabla \times \vec{E}_{-\omega} = -i\omega \vec{n} \times \vec{E}_{-\omega} \tag{54}$$

As a result we obtain the following expression for the radiation spectral-angular density for $r \to \infty$:

$$\frac{d\varepsilon}{d\omega do} = \frac{r^2}{4\pi^2} \mid \vec{E}_\omega(\vec{r}) \mid^2 \tag{55}$$

Let us note that the formula (55) is valid on large distances from the scattering point ($r \to \infty$). While proceeding from (53) to (55) we took into account that$\vec{n}\vec{E}_\omega=0$. This relation directly comes from the expression of Fourier-components of electric and magnetic fields in terms of potentials $\phi_\omega(\vec{r})$ and $\vec{A}_\omega(\vec{r})$

$$\vec{E}_\omega = -\nabla \phi_\omega(\vec{r}) + i\omega \vec{A}_\omega(\vec{r}),\ \vec{H}_\omega = \nabla \times \vec{A}_\omega(\vec{r}) \tag{56}$$

where $\phi_\omega(\vec{r})$ and $\vec{A}_\omega(\vec{r})$ are defined from the expression (49).

In the case of ultrarelativistic particle in the region of characteristic for this process small radiation angles $\vartheta \ll 1$ the 'torn-away' electric field can be considered as transverse having only $E_\perp$ component orthogonal to z axis. In this case the formula (55) for the radiation spectral-angular density has the following form:

$$\frac{d\varepsilon}{d\omega do} = \frac{r^2}{4\pi^2} \mid E_{\omega\perp}(\vec{r}) \mid^2 \tag{57}$$

Using (49) we can derive the electric field Fourier-component orthogonal to z axis. Substituting it into (57) for the spectral-angular distribution of bremsstrahlung we achieve:

$$\frac{d\varepsilon}{d\omega do} = \left(\frac{ez}{\pi}\right)^2 \left| \int_0^\infty dq \frac{q^2 J_1(q\rho)}{q^2 + \omega^2/v^2\gamma^2} e^{-i\frac{q^2 z}{2\omega}} \right|^2 \tag{58}$$

For large distances from the scattering point, namely in the wave zone of the radiation process ($z \gg 2\gamma^2/\omega$) the integral in (58) can be calculated with the use of stationary phase method [36]. It leads to the well known expression for radiation distribution from the bremsstrahlung theory [1,4,37]:

$$\frac{d\varepsilon}{d\omega do} = \frac{e^2}{\pi^2} \frac{\vartheta^2}{(\vartheta^2 + \gamma^{-2})^2} \tag{59}$$

where $\vartheta = \rho / z$ is the angle between the direction of radiation and the z axis. As we can see from (59) in the wave zone the radiation is mainly concentrated within characteristic angles $\vartheta \approx 1/\gamma$. The schematic form of the bremsstrahlung angular distribution in the wave zone is presented on the Figure 2 by the solid curve.

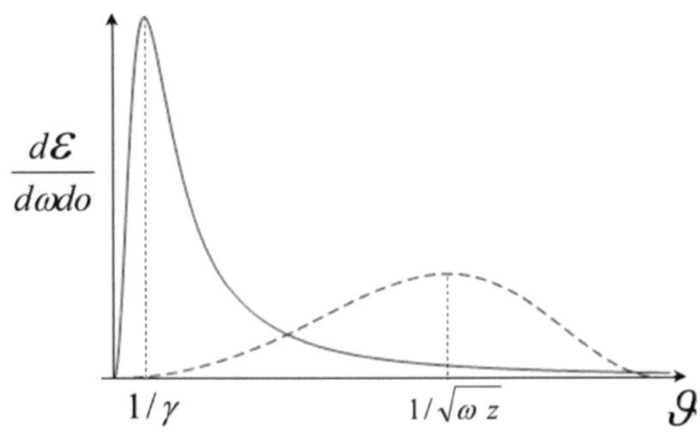

Figure 2. Bremsstrahlung angular distribution given by a point detector in the wave zone (solid curve) and in the pre-wave zone (dashed curve) for a certain frequency ω

For ultra high energies of the radiating particle in the region of characteristic small angles of radiation the expression (57) (and hence the expression (58)) is valid for the description of radiation spectral-angular density on small distances from the scattering point as well (in particular, in the pre-wave zone ($z \ll 2\gamma^2 / \omega$) of the radiation process). Indeed, as the electric field can be considered transverse in this case than $\vec{E}_\omega = -\nabla \phi_\omega = -\partial\phi_\omega / \partial\vec{\rho}$. The equality of $\phi_\omega(\vec{r})$ and $A_\omega(\vec{r})$ leads to the following expression for the magnetic field:

$$\vec{H}_\omega = \nabla \times \vec{A}_\omega = \nabla \phi_\omega \times \vec{e}_z$$

where $\vec{e}_z$ is a unit vector in the direction of z axis. Taking into account the orthogonality of $\nabla\phi_\omega$ and $\vec{e}_z$ for the absolute value of the magnetic field we obtain $|\vec{H}_\omega| = |\nabla\phi_\omega|$, which is equal to the absolute value of the electric field. Moreover, these fields are orthogonal $(\vec{E}_\omega \vec{H}_\omega = \nabla\phi_\omega(\nabla\phi_\omega \times \vec{e}_z) = 0)$. Taking into account the fact that the magnetic field $\vec{H}(\vec{r}, t)$ is the real function, which leads to the relation $\vec{H}_{-\omega} = \vec{H}^*_\omega$, we can present the general expression (52) in the following form:

$$\frac{d\varepsilon}{d\omega do} = \frac{r^2}{4\pi^2}[E_{\omega\perp} E^*_{\omega\perp}(\vec{r})]\vec{e}_z\vec{n}$$

which coincides with (57) at small angles between $\vec{e}_z$ and $\vec{n}$. Now let us consider radiation in the pre-wave zone on the basis of this formula.

http://dx.doi.org/10.5772/50866

In the pre-wave zone ($z \ll 2\gamma^2 / \omega$) of the radiation process it is not possible to use the stationary phase method for the analysis of radiation characteristics. Here, making the substitutions $q = \omega x / \gamma$ and $\rho = z\vartheta$, we can present the integral (58) in the form:

$$\frac{d\varepsilon}{d\omega do} = \left(\frac{e\omega z}{\pi\gamma}\right)^2 | I_1 - I_2 |^2 \tag{60}$$

where

$$I_1 = \int_0^{+\infty} dx J_1(\omega z \gamma^{-1} x \vartheta) e^{-i\frac{\omega z}{2\gamma^2} x^2}$$

$$I_2 = \int_0^{+\infty} dx \frac{J_1(\omega z \gamma^{-1} x \vartheta)}{x^2 + 1} e^{-i\frac{\omega z}{2\gamma^2} x^2}$$

In the case $\gamma \gg 1$ the absolute value of the integral I_2 is negligibly small comparing to the corresponding value of I_1 and for spectral-angular density of bremsstrahlung in the pre-wave zone we obtain:

$$\frac{d\varepsilon}{d\omega do} = \left(\frac{e\omega z}{\pi}\right)^2 | I_1 |^2 = \frac{4e^2}{\pi^2} \frac{1}{\vartheta^2} \sin^2\left(\frac{\omega z \vartheta^2}{4}\right) \tag{61}$$

From (61) we can conclude that in the pre-wave zone the radiation is mainly concentrated within angles $\vartheta \approx 2 / \sqrt{\omega z}$, which exceed the characteristic angles $\vartheta \approx 1 / \gamma$ of the wave zone. Therefore in the pre-wave zone ($z \ll 2\gamma^2 / \omega$) the point detector gives broader angular distribution of radiation (dashed curve on Figure 2) than in the wave zone ($z \gg 2\gamma^2 / \omega$). Moreover this distribution depends on the frequency ω of the radiated waves.

By the point detector we mean here the detector of the smaller size $\delta\rho$ than the transversal radiation length of the process $l_T \approx \gamma / \omega$, which is the characteristic transversal distance on which at the moment of time $t = 0$ the Fourier harmonics of frequency ω are concentrated in the wave packet (49). Such detector registers the radiation of frequency ω, which falls on a small domain of space, where the detector is situated.

The measurements, however, can be made by the extended detector of the larger size than the characteristic transversal length of the radiation process, so that $\delta\rho \gg l_T$. Such detector registers not only the waves of frequency ω, which fall on the small element of surface with coordinates $\vec{\rho}$ and z, as the point detector does, but all the electromagnetic waves of frequency ω, which propagate in the direction of wave vector $\vec{k}(|\vec{k}| = \omega)$. In order to calculate the bremsstrahlung spectral-angular distribution, which is registered by an extensive detector, which is a plate of large size, we need to integrate the expression (58) over the entire considered plate and express the obtained result in the form of an integral over the directions of wave vectors of radiated waves. The integrand in this case will be nothing else than the required distribution. In our case after performing the procedures described above we can present the expression (58) in the following form:

$$\frac{d\varepsilon}{d\omega do_{\gamma}} = \frac{e^2}{\pi^2} \frac{\vartheta_{\gamma}{}^2}{\left(\vartheta_{\gamma}{}^2 + \gamma^{-2}\right)^2} \tag{62}$$

where $\vartheta_{\gamma} = q / \omega$ is the angle between the direction of the wave vector $\vec{k}$ and the z axis. Hence the bremsstrahlung spectral-angular distribution obtained by the extended detector coincides with the one (59) obtained by the point detector in the wave zone. But unlike the case with point detector this distribution does not depend on the distance from the scattering point and is the same both in the wave and the pre-wave zones.

4. Transition radiation by relativistic electron on thin metallic plate

4.1. Scalar and vector potentials of transition radiation field

The electromagnetic wave packets of the structure analogous to the one considered in the process of an electron instantaneous scattering take place also in the process of relativistic electron traverse of thin conducting plate. Let us consider a problem about transition radiation that arises during normal traverse of thin ideally conducting plate, situated in the plane $z=0$, by an electron, which moves along the z axis from $z=-\infty$ to $z=+\infty$ (Figure 3). Let us investigate the structure of electromagnetic fields that take place before and after the electron's traverse of the plate in vacuum [16-18].

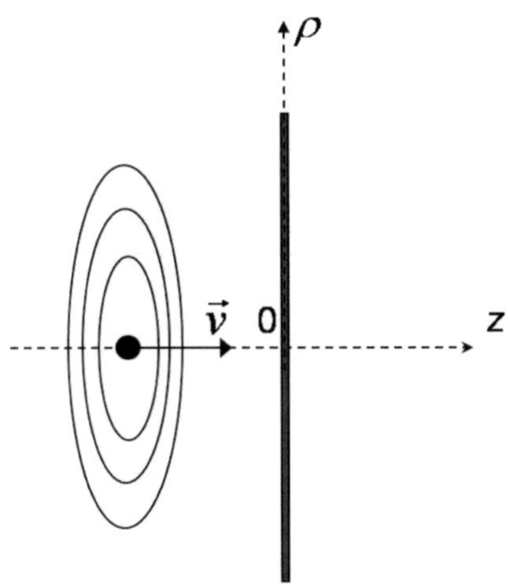

Figure 3. The electron normal traverse of thin metallic plate.

Scalar and vector potentials of the electromagnetic field, which is generated by an electron moving in vacuum are the solutions of inhomogeneous Maxwell equations (37). The equations set (37) in the considered problem should be supplemented by a boundary condition, which corresponds to the fact that on the plate's surface the tangential component of the total electric field equals zero. The general solutions of the equations (37) for the electron, which moves uniformly with the velocity $\vec{v}$ can be represented in the form of the following Fourier-expansions of the potentials $\phi(\vec{r}, t)$and$\vec{A}(\vec{r}, t)$:

$$\begin{aligned}\phi(\vec{r}, t)&=\int\frac{d^3kd\omega}{(2\pi)^4}e^{i(\vec{k}\vec{r}-\omega t)}\left[\phi^C_{\vec{k},\omega}\delta(\omega-\vec{k}\vec{v})+\phi^f_{\vec{k},\omega}\delta(k^2-\omega^2)\right]\\ \vec{A}(\vec{r}, t)&=\int\frac{d^3kd\omega}{(2\pi)^4}e^{i(\vec{k}\vec{r}-\omega t)}\left[\vec{A}^C_{\vec{k},\omega}\delta(\omega-\vec{k}\vec{v})+\vec{A}^f_{\vec{k},\omega}\delta(k^2-\omega^2)\right]\end{aligned} \tag{63}$$

The first items in (63) are the Fourier-expansion of the electron's Coulomb field, for which

$$\phi^C_{\vec{k},\omega}=-\frac{8\pi^2e}{\omega^2-k^2},\ \vec{A}^C_{\vec{k},\omega}=\vec{v}\phi^C_{\vec{k},\omega} \tag{64}$$

The second items in (63) are the Fourier-expansion of the field of induced surface currents on the plate (we will name it the free field), for which in vacuum $|\vec{k}|=|\omega|$. On large distances from the region in which the transformation of the surrounding electron field takes place this items form the transition radiation field. It can be derived from the boundary condition for the total electric field $\vec{E}=\vec{E}^C+\vec{E}^f$ on the surface of the plate:

$$\vec{E}^C_\perp(\vec{\rho}, z=0,t)+\vec{E}^f_\perp(\vec{\rho}, z=0,t)=0 \tag{65}$$

Here $\vec{E}^C$is the electron's Coulomb field and $\vec{E}^f$ is the field of induced surface currents on the plate. Moreover it is required that the free field produced by the plate propagates on the left and on the right of the plate respectively in the negative and positive directions of z axis. The scalar and the vector potentials define electric and magnetic fields by the relation (51). Due to the symmetry of the problem the vector potential is directed along the particle velocity$\vec{v}$:

$$\vec{A}(\vec{r}, t)=\vec{v}A(\vec{r}, t) \tag{66}$$

and the relation

$$A^C(\vec{r}, t)=\phi^C(\vec{r}, t) \tag{67}$$

for the particle's field in vacuum is valid. In general case such relation between potentials is not valid.

Using Fourier expansion (63) it is possible to derive the potential $\phi(\vec{r},\, t)$ from the condition (65). For this let us perform the integration over the component $k_z$in (63). Taking also into account the relation between potentials and fields (51) we obtain the following expression for transversal component of the electric field:

$$\vec{E}_{\perp}(\vec{r},\, t) = -\frac{\partial}{\partial \vec{\rho}} \int \frac{d^2 k_{\perp} d\omega}{(2\pi)^4} e^{i(\vec{k}_{\perp}\vec{\rho} + k_z z - \omega t)} \left[\phi^{C}_{\vec{k},\omega} \Big|_{k_z = \frac{\omega}{v}} + \frac{1}{2\left|\sqrt{\omega^2 - k_{\perp}^2}\right|} \left(\phi^{f}_{\vec{k},\omega} \Big|_{k_z = \sqrt{\omega^2 - k_{\perp}^2}} + \right. \right. \tag{68}$$

$$\left. \left. + \phi^{f}_{\vec{k},\omega} \Big|_{k_z = -\sqrt{\omega^2 - k_{\perp}^2}} \right) \right].$$

For$z = 0$, according to (65) this component should equal zero. From this we find that

$$-\phi^{C}_{\vec{k},\omega} \Big|_{k_z = \omega/v} = \frac{1}{2\left|\sqrt{\omega^2 - k_{\perp}^2}\right|} \left(\phi^{f}_{\vec{k},\omega} \Big|_{k_z = \sqrt{\omega^2 - k_{\perp}^2}} + \phi^{f}_{\vec{k},\omega} \Big|_{k_z = -\sqrt{\omega^2 - k_{\perp}^2}} \right) \tag{69}$$

The values $k_z = \pm\sqrt{\omega^2 - k_{\perp}^2}$ satisfy the dispersion relation$\omega^2 = k_z^2 + k_{\perp}^2$, which is defined by the respective δ- function in (63). The sign before the square root$\sqrt{\omega^2 - k_{\perp}^2}$determines the direction of propagation of plane waves (Fourier components) with given values of$\omega$and $|k_{\perp}|$ [1, 4]. Indeed, the equation of a plane wave constant phase along the $z$axis is$k_z z - \omega t = const$. The plane waves, which the free field produced by the plate consists of, should propagate away from the plate. Hence for $\omega > 0$ on the right of the plate (which means$z > 0$) it is necessary to take into account only Fourier-components with positive sign before the root $\sqrt{\omega^2 - k_{\perp}^2}$ in (62), while for $\omega < 0$ and $z > 0$- only Fourier-components with negative sign before this root in (68). In the region $z < 0$ for $\omega > 0$ and $\omega < 0$ in (68) we should take into account the items with opposite signs before the root $\sqrt{\omega^2 - k_{\perp}^2}$ relatively to the case for$z > 0$. The value of the square root itself is considered either positive or to belong to the upper complex half plane.

Thus, taking into account all requirements mentioned above we can write the scalar potential of the free field in the following form:

$$\phi^f(\vec{r},\, t) = -\frac{e}{2\pi^2 v} \int d^2 k_{\perp} \int_{-\infty}^{\infty} d\omega \frac{1}{k_{\perp}^2 + \omega^2 / p^2} e^{i\left(z\omega\sqrt{1 - k_{\perp}^2/\omega^2} - \omega t + \vec{k}_{\perp}\vec{\rho}\right)} \tag{70}$$

where $p = v\gamma$ (γ– electron's Lorentz-factor). It is a packet of free electromagnetic waves, which gradually turns into the field of transition radiation in such way that each harmonic with frequency ω reconstructs into diverging spherical wave on distance$z > l_C \approx 2\gamma^2 / \omega$, which is the formation length of the radiation process.

Let us note that the value of $|k_\perp|$ in (63) is arbitrary. Therefore it is necessary to perform the integration in (63) not only over travelling waves$k_\perp^2 < \omega^2$, but over surface ones $k_\perp^2 > \omega^2$ as well.

Using (70) and the Lorentz gauge (50) in which the equations (37) are presented we can derive the vector potential:

$$A^f(\vec{r}, t) = -\frac{e}{2\pi^2}\int d^2k_\perp \int_{-\infty}^{\infty} d\omega \frac{1}{k_\perp^2 + \omega^2/p^2} \frac{1}{\sqrt{1-k_\perp^2/\omega^2}} e^{i\left(z\omega\sqrt{1-k_\perp^2/\omega^2} - \omega t + \vec{k}_\perp \vec{\rho}\right)} \tag{71}$$

Making in (70) the substitution $|k_\perp| = |\omega| x$ and separating the contributions to the potential by the free field of travelling and surface waves, we can write the potential of this field in the following form:

$$\phi^f(\vec{r}, t) = \Phi_1(\vec{r}, t) + \Phi_2(\vec{r}, t) \tag{72}$$

where

$$\Phi_1(\vec{r}, t) = -\frac{2e}{\pi v}\int_0^1 \frac{x dx}{x^2 + p^{-2}} \int_0^\infty d\omega\, J_0(\omega x \rho) \cos\left[\omega\left(|z|\sqrt{1-x^2} - t\right)\right] \tag{73}$$

$$\Phi_2(\vec{r}, t) = -\frac{2e}{\pi v}\int_1^\infty \frac{x dx}{x^2 + p^{-2}} \int_0^\infty d\omega\, J_0(\omega x \rho) \cos(\omega t) e^{-|z|\omega\sqrt{x^2-1}} \tag{74}$$

Deriving (72), we performed in (70) integration over azimuth angle between $k_\perp$ and $\rho$and proceeded from integration over ω along the interval $-\infty < \omega < +\infty$ to integration over only positive values of this variable.

The corresponding expressions for vector potential$A^f = A_1 + A_2$, according to (71), differ from (73) and (74) only by additional factor $1/\sqrt{1-x^2}$ in the integrands.

4.2. The structure of transition radiation field

Let us discuss the structure of the fields that arise during the electron traverse of thin ideally conducting plate. Firstly, let us consider the structure of this field along the z axis for $\rho = 0$ [16]. This case is interesting for the fact that the calculation of all the integrals in (73) and (74) essentially simplifies for$\rho = 0$. Nevertheless, in this case all the main peculiarities of free waves formation in the considered process remain intact.

As a result of rather simple calculations (see Appendix) we obtain the following expression for the field produced by the plate on the z axis for $\rho = 0$:

$$\phi^f(z, t) = -\frac{e}{||z| - vt|}\theta(|z| - t) - \frac{e}{||z| + vt|}\theta(t - |z|) \tag{75}$$

The total field produced by the electron and the plate can be obtained by addition of the expression (75) and the electron's own coulomb field on the z axis:

$\phi^C(z, t) = e / |z - vt|$

The obtained results show that for $t < 0$ the total field produced by the electron and the plate in the region $z < 0$ is the electron's own coulomb field, which moves towards the plate and the field, which coincides with the field of electron's image inside the plate. By the field of electron's image we assume the field, which is created by an imaginary particle with a charge of the opposite sign, which is situated on the opposite side of the plate and moves symmetrically to the electron relatively to the plate. In the region $z > 0$ the total field equals zero for $t < 0$.

For $t > 0$ the picture of the total field distribution is the following. In the region $z < 0$ for $|z| > t$ the total field is defined by the electron's field in this coordinate region (the electron in this case is situated on the right of the plate) and the field of its 'image', which moves in the direction opposite to the z axis. In the region $z < 0$ for $|z| < t$ the total field equals zero. In the region of positive z values for $z > t$ the total field equals zero while for $z < t$ this field is the sum of the field of the electron, which is situated in this coordinate region for $t > 0$ and the field of its 'image' situated on the left of the plate.

The integrals in (73) and (74) can be analytically calculated as well. After rather long calculations we finally obtain the following expression for $\phi^f(\vec{r}, t)$:

$$\phi^f(\vec{r}, t) = -\frac{e}{\sqrt{\rho^2\gamma^{-2} + (|z| - vt)^2}}\theta(r - t) - \frac{e}{\sqrt{\rho^2\gamma^{-2} + (|z| + vt)^2}}\theta(t - r) \tag{76}$$

The electron's own coulomb field has the following form:

$$\phi^C(\vec{r}, t) = \frac{e}{\sqrt{\rho^2\gamma^{-2} + (z - vt)^2}} \tag{77}$$

The structure of the expressions for scalar potential is the same as the structure of these expressions for $\rho = 0$. If $t < 0$ then, according to (76) and (77), the total field in the left half-space is equal to the sum of the electron's coulomb field in this region and the field of its image:

$$\phi(\vec{r}, t) = \frac{e}{\sqrt{\rho^2\gamma^{-2} + (z - vt)^2}} - \frac{e}{\sqrt{\rho^2\gamma^{-2} + (|z| - vt)^2}}$$

In the right half-space the total field equals zero for $t < 0$.

After electron's traverse of the plate, which means for $t>0$, according to (76) and (77), the total field in the left half-space is defined by the formula

$$\phi(\vec{r},\, t)=\left[\frac{e}{\sqrt{\rho^2\gamma^{-2}+(z-vt)^2}}-\frac{e}{\sqrt{\rho^2\gamma^{-2}+(|z|-vt)^2}}\right]\theta(r-t) \tag{78}$$

In the right half-space, where the electron is situated after the traverse of the plate, the total field has the following form:

$$\phi(\vec{r},\, t)=\left[\frac{e}{\sqrt{\rho^2\gamma^{-2}+(z-vt)^2}}-\frac{e}{\sqrt{\rho^2\gamma^{-2}+(z+vt)^2}}\right]\theta(t-r) \tag{79}$$

Thus for $t>0$ the picture of the total field, which is created by the electron-plate system is as following. In the left half-space in the coordinate region $r>t$ it is a sum of the electron's coulomb field of the opposite sign reflected from the plate and the own field of electron, which is situated on the right of the plate. The reflected field in this case moves with velocity $-\vec{v}$ in the direction opposite to the direction of electron's motion. In the coordinate region $r<t$, which the signal about the electron's traverse of the plate at $t=0$ has already reached, the total field equals zero. In the right half-space for $t>0$ and $r<t$ the total field equals the sum of the fields of the electron and its `image' on the left of the plate. For $r>t$ the total field equals zero.

The analogous expressions can be obtained for vector potential as well. Namely, for $t>0$ the total field vector potential (the sum of particle's coulomb field and radiation potentials) has the following form:

$$\vec{A}(\vec{r},\, t)=\vec{v}\left[\frac{e}{\sqrt{\rho^2\gamma^{-2}+(z-vt)^2}}+\frac{e}{\sqrt{\rho^2\gamma^{-2}+(|z|-vt)^2}}\right]\theta(r-t) \tag{80}$$

The expression in square brackets in (80) differs from the same expression for scalar potential (78) only by the sign of second item. The reason of this can be understood from the following reasoning. For $z<0$ in the region $r>t$ the field (78) is the difference between two coulomb fields, the sources of which are the electron and its image. The vector potential of the coulomb field is related to its scalar potential by $\vec{A}=\vec{v}\phi$, where $\vec{v}$ is the velocity of the field source. As electron moves with velocity $\vec{v}$ and its image - with velocity $-\vec{v}$, their vector potentials respectively equal $\vec{A}_e=\vec{v}\phi_e$ and $\vec{A}_i=-\vec{v}\phi_i$. It is the presence of the `minus' sign in the expression for $\vec{A}_i$ that causes the discussed difference in the signs of the items in square brackets in (78) and (80).

In the region $z>0$ for $t>0$ the vector potential has the following form:

$$\vec{A}(\vec{r},\, t)=\vec{v}\left[\frac{e}{\sqrt{\rho^2\gamma^{-2}+(z-vt)^2}}+\frac{e}{\sqrt{\rho^2\gamma^{-2}+(z+vt)^2}}\right]\theta(t-r) \tag{81}$$

The obtained results are valid for arbitrary electron velocities. The case of an ultra relativistic particle is of special interest because for such particles the reconstruction of the total field, created by the plate and the electron after its traverse of the plate, into the field of radiation occurs on large distances. The results obtained in this case are illustrated by Figure 4. Here the equipotential surfaces of the scalar potential of the field reflected to the left half-space and the field around the electron on the right of the plate are presented for $t > 0$.

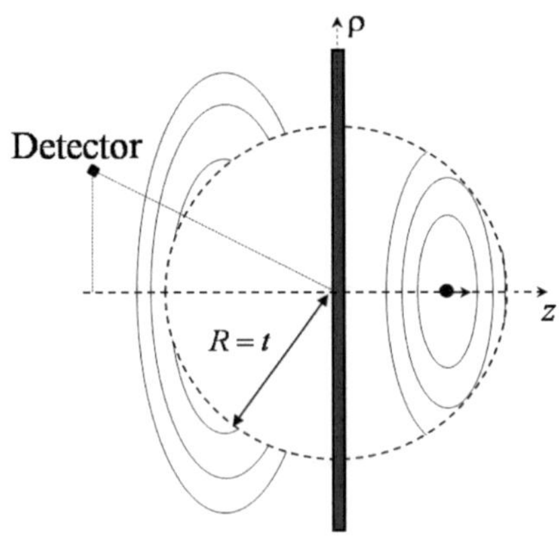

Figure 4. The total field in ultra relativistic case for $t > 0$.

In order to understand what occurs with the electric field on the surface of the sphere of radius $r = t$ with the center in the point of the electron traverse of the plate (θ - sphere) let us consider the structure of the force lines of the total field for $t > 0$ and, for example, $z < 0$.

While building the force lines it is necessary to take into account that they should originate or end either on charges or in the infinity and not to cross each other. In the case of an infinite plate all the field lines originate and end either on the surface charges of the plate or on the electron, which traverses it.

Thus each force line of the total field (78), which originates on a surface charge of the plate somewhere in the area $\rho > t$ and stretches through the space region $r > t$ to the θ - sphere, should be refracted and stretch further along the surface of the sphere, ending on another surface charge of the plate at $\rho = t$ (Figure 5). It is the force lines, which overlap each other on the θ - sphere are the force lines of the transition radiation field. Indeed, the field on the θ - sphere propagates in the radial direction with the speed of light and is perpendicular to this direction. Moreover, as will be shown in further discussion, this field decreases with the distance as $1/r$.

It is necessary to note that the given picture of the force lines indicates the necessity of the existence of the field (78) outside the θ - sphere along with the radiation field on it. It is only in this case that the force lines of the radiation field, which originate (end) on the surface charges of the plate and stretch along the θ - sphere can proceed to the region $r > t$ forming the lines of the field (78) and not intersect or break on the θ - sphere.

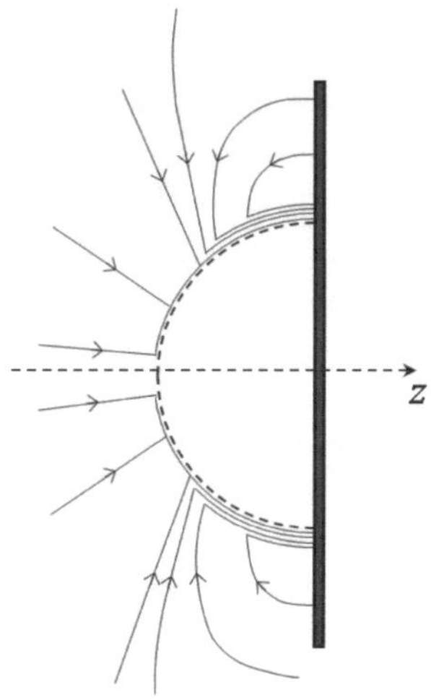

Figure 5. The picture of the total field force lines for $t>0, z<0$.

4.3. The transition radiation field and its intensity

In ultra relativistic case ($\gamma \gg 1$) the range of $k_{\perp}$, which make the main contribution to the integral (70), which represents the transition radiation field is $k_{\perp} \le \omega / \gamma \ll \omega$ and it is possible to make in (70) the same expansion of the square root $\sqrt{\omega^2 - k_{\perp}^2}$ in the small factor $k_{\perp} / \omega$ as we did in (47) and (48) for the bremsstrahlung field. Such expansion represents the fact that transition radiation is considered at small angles around the particle's trajectory $\vartheta = k_{\perp} / \omega \approx 1 / \gamma$ within which the most part of the radiation is concentrated in ultra relativistic case. In the result of the expansion of (70) we obtain the expression for the transition radiation field, which coincides with the analogous expression (49) for the bremsstrahlung field. This means that the packets of free electromagnetic waves, which arise in the processes of electron instantaneous scattering and the particle traverse of thin metallic plate have similar (but not totally identical) structure and the radiation effects, which take place in these processes should be analogous.

Let us consider the radiation, which arises during an electron normal traverse of thin ideally conducting plate [16-18]. In this case by radiation we mean the part of the electromagnetic energy, which belongs to the frequency interval (ω, $\omega + d\omega$) and falls into the small detector, which is placed on different distances r from the point of electron traverse of the plate. The detector's position relative to this point is defined by the transverse $\vec{\rho}$ and longitudinal z coordinates (Figure 4). The radiation direction ϑ is then defined by the ratio of these coordinates:

$$\vartheta = arctg \frac{\rho}{|z|} \tag{82}$$

In further discussion we will mainly concentrate our attention on the peculiarities of the radiation formation process in the left half-space, which is the region of negative z. Naturally, the general expression (55) for the radiation spectral-angular density on large distances from the target is valid in this case as well. The discussed structural similarity of the fields in the processes of electron instantaneous scattering and electron traverse of thin metallic plate also gives us the possibility to use the expression (57) for the description of the transition radiation field in the range of small angles $\vartheta \approx 1/\gamma$ characteristic to this process on arbitrary distances z from the plate. Firstly, let us consider the transition radiation in the wave zone.

Let us show that in the region of large distances from the target ($r \to \infty$) the formula (55) gives the well known result from the theory of transition radiation by electron on metallic plate [1, 4, 37]. Indeed, in differentiating of (78) and (80) the items proportional to derivatives of the square brackets will give the total field outside the θ- sphere, which decreases with the distance r faster thanr^{-1}. The item proportional to the derivative of the θ- function is on large distances the diverging wave, which amplitude is proportional tor^{-1}. This item defines all the transition radiation characteristics.

The scalar ϕ^{f} and vector $\vec{A}^{f}$ potentials on the θ- sphere can be written as

$$\phi = -2v\frac{e}{r}\frac{\cos\vartheta}{1-v^2\cos^2\vartheta}\theta(r-t), \ \vec{A} = -\vec{v}\frac{\phi}{\cos\vartheta} \tag{83}$$

where ϑ is the angle between r and$-\vec{v}$. So the transition radiation electric field is totally defined by the spatial derivative of the scalar potential:

$$\vec{E}^{f}(\vec{r}, \vartheta, t) = 2v\frac{e}{r}\frac{\delta(r-t)}{1-v^2\cos^2\vartheta}\left(\frac{\vec{v}}{v} + \vec{n}\cos\vartheta\right) \tag{84}$$

The Fourier-component of this expression has the following form:

$$\vec{E}^{f}_{\omega}(\vec{r}, \vartheta) = 2v\frac{e}{r}\frac{1}{1-v^2\cos^2\vartheta}\left(\frac{\vec{v}}{v} + \vec{n}\cos\vartheta\right)e^{i\omega r} \tag{85}$$

Let us note that the relation $\vec{n}\vec{E}_{\omega}=0$ is valid for (81) as$\vec{n}\vec{v} = -v\cos\vartheta$.

Substituting (85) into (55) we obtain the well known expression for spectral-angular density of backward transition radiation [1, 4, 5]:

$$\frac{d\varepsilon}{d\omega do} = \frac{e^2v^2}{\pi^2}\frac{\sin^2\vartheta}{(1-v^2\cos^2\vartheta)^2} \tag{86}$$

The presented derivation of the formula (86) for the transition radiation spectral-angular density is based on the analysis of the electron's field reflected from the plate on large distances $r \to \infty$ from the plate. In this case by large distances we mean the distances from the target, which substantially exceed the coherence length of the radiation process [1, 2]

$$l_C(\vartheta) = \frac{\omega^{-1}}{1 - v\cos\vartheta} \tag{87}$$

For ultrarelativistic electrons the transition radiation is mainly concentrated inside a narrow cone with the opening angle$\vartheta \approx 1/\gamma$. In this region of angles ϑ the coherence length (87) quickly increases with the increase of the electron's Lotentz-factor:

$$l_C(\vartheta) = \frac{2\gamma^2/\omega}{1 + \gamma^2\vartheta^2} \tag{88}$$

For large γ and small ω this length can have ultra large values, which allow, for example, a detector to be situated within this length. The analysis of the transition radiation process in this case requires the development of the methods, which allow to consider the radiation process not only on large distances from the target comparing to the coherence length, l_C, but on the small ones as well. One of such methods is based on the application in the considered problem the Fourier-expansions of scalar, ϕ^f, and vector, $\vec{A}^f$, potentials of the free field, which is generated by the plate during its traverse by the electron. Primarily, it is necessary to note that these fields are the wave packets, which consist of plane electromagnetic waves. At large values of the time interval after the interaction of the electron with the target these wave packets reconstruct into spherical diverging waves of radiation. On small distances from the target the spatial structure of these packets is close to the structure of the packet formed by the electron's own field, which is reflected from the plate.

Firstly, let us consider on the base of this method the transition radiation on large distances from the plate $|z| \gg 2\gamma^2/\omega$. In this case the integrals over $x$in (72) can be calculated with the use of the stationary phase method. Using for this purpose the asymptotic expression for the Bessel function $J_0(\omega x\rho)$ for large values of its argument we can write $\phi^f_\omega(\vec{r})$ in the next way:

$$\phi^f_\omega(\vec{r}) = -\frac{2e}{v}\mathrm{Re}\int_0^1 \frac{x dx}{x^2 + p^{-2}}\sqrt{\frac{2}{\pi\omega\rho x}}\left\{e^{i\left(\omega\rho x - \frac{\pi}{4} + |z|\omega\sqrt{1-x^2}\right)} + e^{i\left(\omega\rho x - \frac{\pi}{4} - |z|\omega\sqrt{1-x^2}\right)}\right\} \tag{89}$$

The first item in the braces gives the stationary phase point

$$x_0 = \frac{\rho/|z|}{\sqrt{\rho^2/z^2 + 1}} = \sin\vartheta \tag{90}$$

where the ratio $\rho / |z|$ is defined by the relation (82). As stationary phase point is absent in the second item in (89) in the considered region of the variable x, we can neglect the contribution of this item to the integral over x. For the same reason we neglect the contribution to $\phi_\omega^f(\vec{r})$ of the item in (72), which contains the integration over the values $x > 1$. In the result for $\phi_\omega^f(\vec{r})$ we obtain:

$$\phi_\omega^f(\vec{r}) = -\frac{2e}{\omega v r}\frac{\cos^2\vartheta}{\sin^2\vartheta + p^{-2}} e^{i(\omega r - \pi/2)} \tag{91}$$

We took into account the fact that in spherical coordinate system $\rho = r\sin\vartheta$ and $|z| = r\cos\vartheta$.

With the use of (56) the expression $(\vec{E}_\omega \times \vec{H}_{-\omega})\vec{n}$ in (52) can be written in the form:

$$(\vec{E}_\omega \times \vec{H}_{-\omega})\vec{n} = (-\partial_z \phi_\omega^f + i\omega \vec{A}_\omega^f)\vec{n}_\perp \cdot \nabla_\perp A_{-\omega}^f - \cos\nabla_\perp \phi_\omega^f \cdot \nabla_\perp A_{-\omega}^f \tag{92}$$

In the point of stationary phase for $r \to \infty$ $A_\omega^f = -\phi_\omega^f / \cos\vartheta$. Taking into account $\frac{\partial}{\partial\rho}\phi_\omega = i\omega\sin\vartheta\phi_\omega$ and $\frac{\partial}{\partial z}\phi_\omega = i\omega\cos\vartheta\phi_\omega$ we find that

$$(\vec{E}_\omega \times \vec{H}_{-\omega})\vec{n} = \omega^2 tg^2\vartheta \phi_\omega^f \phi_{-\omega}^f \tag{93}$$

Substituting the asymptotic (91) for the potential ϕ_ω into this expression we obtain the formula (86) for the radiation spectral-angular density.

For ultra relativistic particles characteristic values of the radiation angles $\vartheta \approx \gamma^{-1}$ are much less than unit. The radiation spectral-angular density (86) in this case has the following form:

$$\frac{d\varepsilon}{d\omega do} = \frac{e^2}{\pi^2}\frac{\vartheta^2}{(\gamma^{-2} + \vartheta^2)^2} \tag{94}$$

In this region of radiation angles the characteristic values of the variable x, which make contribution to the integral over x in (72) are small $x_{eff} \approx \gamma^{-1}$. The stationary phase method is valid for calculation of this integral if

$$\frac{|z|\omega}{2\gamma^2} \gg 1$$

In other words, it is required that the distance between the target and the detector should substantially exceed the coherence length of the radiation process, which means the radiation should be considered in the wave zone.

As it was pointed out, the length l_C can have macroscopic values and the detector can be situated in the pre-wave zone, which is on smaller distances that the coherence length of the radiation process. The stationary phase method is not applicable for calculation of the integral over x in (72) in this case. For this purpose the expansion of the phase over small values of x in the exponential factor in (72) can be made. Neglecting the items proportional to higher powers than x^2 we find that

$$\phi_\omega^f = -\frac{2e}{\pi v}\mathrm{Re}\int_0^\infty \frac{xdx}{x^2+p^{-2}} J_0(\omega x\rho) e^{i(\omega|z|-\omega|z|x^2/2)} \tag{95}$$

In the region of small radiation angles for which $\vartheta^2|z|\omega \le 1$, the convergence of the integral (95) is determined by the values $x \approx 1/\sqrt{|z|\omega}$, therefore the items proportional to x^2 should be preserved in the exponent phase. With the same accuracy in the region of small angles the vector potential A_ω^f is related ϕ_ω^f by

$$A_\omega^f = -\phi_\omega^f$$

In this case $(\vec{E}_\omega \times \vec{H}_{-\omega})\vec{n} = \nabla_\perp \phi_\omega^f \cdot \nabla_\perp \phi_{-\omega}^f$ and the radiation spectral-angular density has the following form:

$$\frac{d\varepsilon}{d\omega do} = \left(\frac{2e^2}{\pi v}\right)^2 |B(z)|^2 \tag{96}$$

where

$$B(z) = \omega|z| \int_0^\infty x^2 dx \frac{J_1(x\omega|z|\vartheta)}{x^2+\gamma^{-2}} e^{-i\omega|z|x^2/2} \tag{97}$$

Let us note that during the derivation of the expression (97) we only took into account the fact that the consideration of the radiation process was made in the region of small angles of radiation. Therefore the formula (97) is valid both for large ($z \gg l_C$) and small ($z \ll l_C$) distances between the detector and the target. If $|z| \gg l_C$, the calculation of the integral $B(z)$ can be made with the use of the stationary phase method. As a result formula (97) gives the corresponding result of the transition radiation theory (see formula (94)). In the region $|z| < l_C$ the formula (96) leads to the broader radiation angular distribution than the one defined by the expression (94). The possibility of existence of such effect was considered in the paper [10]. Its theory was elaborated in the paper [8]. According to it, in particular, in the region of small angles on distances $|z| < l_C$ the radiation spectral-angular density is defined by the formula:

$$\frac{d\varepsilon}{d\omega do} = \frac{4e^2}{\pi^2} \frac{1}{\vartheta^2} \sin^2\left(\frac{\omega \mid z \mid \vartheta^2}{4}\right) \tag{98}$$

Such modification of the radiation angular distribution for $\mid z \mid < l_C$ comparing to the case of $\mid z \mid \gg l_C$ is caused by the fact that in the region $\mid z \mid < l_C$ only part of the waves reflected from the plate fall into the point detector (see [8]). It is necessary to note that for $\mid z \mid < l_C$ the radiation angular distribution depends on the frequency of the radiated electromagnetic wave. If the measurements are performed by the detector of the larger size than the characteristic transversal length of the radiation process the radiation spectral-angular density is defined by the expression (94) on arbitrary distances from the plate and the discussed above pre-wave zone effects disappear.

The obtained results show that the effects analogous to the ones which take place in the process of bremsstrahlung at the instantaneous scattering of the electron to a large angle (the broadening of the radiation angular distribution and its dependence on the frequency of the registered photon in the measurements performed by a point detector on small distances from the scattering point) take place also for backward transition radiation in the process of an electron traverse of metallic plate. For ultra relativistic particles, according to (94) and (98), the radiation is mainly concentrated in the region of small angles $\vartheta \ll 1$: $\vartheta_{eff} \approx 1/\gamma$ in the wave zone, in the pre-wave zone $\vartheta_{eff} \approx 2/\sqrt{\omega \mid z \mid}$.

5. Transition radiation by 'half-bare' electron

The wave packets, which arise at ultra relativistic electron instantaneous scattering to a large angle, reconstruct into radiation field on distances along the initial and final directions of the electron's velocity, which are of the order of the coherence length of the radiation process. For large energies of the electron and low frequencies of the radiated waves, as was stated above, this length can be of macroscopic size. In this case the possibility of investigating of the evolution of such wave packets in space and time by macroscopic devices appears. In [15] one of such possibilities, which concerns the reflection of wave packets from an ideally conducting plate situated on different distances from the scattering point, was discussed. In this case if the plate is situated in the direction of motion of the scattered electron perpendicular to its velocity the reflected field is the backward transition radiation. However, the peculiarity of this process lies in the fact that unlike the ordinary backward transition radiation the considered one is the radiation by the particle with non-eqiulibrium field. The plate in this case can be considered as an element of the radiation detector. Let us obtain the formulae, which describe the given process and discuss some of its peculiarities on their basis [17].

http://dx.doi.org/10.5772/50866

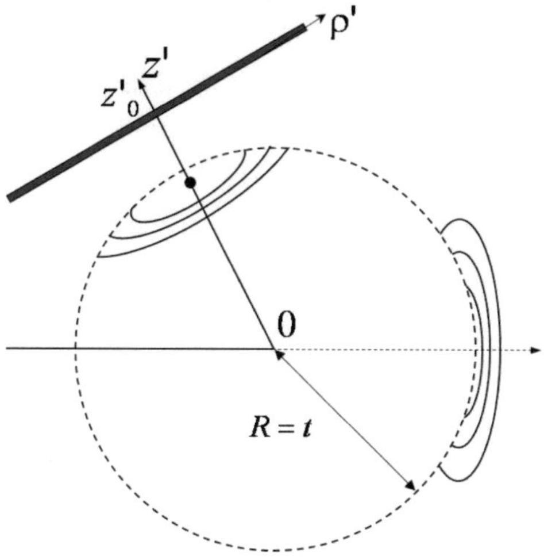

Figure 6. Normal incidence of the scattered 'half-bare' electron on thin metallic plate.

Let the ideally conducting plate be situated in the plane $z'=z'_0$(Figure 6). The Fourier-expansion of the field around the scattered electron

$$\phi(\vec{r},\, t)=\frac{e}{2\pi^2}\mathrm{Re}\int\frac{d^3k}{k}\left[\frac{e^{i\vec{k}(\vec{r}-\vec{v}'t)}}{k-\vec{k}\vec{v}'}-\frac{e^{i\vec{k}\vec{r}-ikt}}{ck-\vec{k}\vec{v}'}\right] \tag{99}$$

consists of two parts, the first of which describes the equilibrium coulomb field of the electron, which moves with the velocity $\vec{v}'$ along the direction of scattering, while the second part is the nonequilibrium field, which is structurally equal to the 'torn away' field (it is equal to equilibrium coulomb field outside the θ- sphere and vanishes inside it). Hence, the second part of the field (99) can be presented in the form (47) with a mere substitution$\vec{v}\rightarrow\vec{v}'$. The first part of the field (99) can be presented in the analogous form by making the substitution $k_z\rightarrow k$ from $k=\sqrt{k_z^2+q^2}$ and denoting$k_z v'=\omega$. From the expression for scalar potential obtained by the considered transformations we can derive the expression for the Fourier-component of the electric field perpendicular to z axis, which in ultra relativistic case is:

$$E_{\perp}(\vec{r},\, \omega)=2e\int_{-\infty}^{+\infty}d\omega\, e^{i\frac{\omega z'}{v'}}\int_0^{\infty}dq\frac{q^2 J_1(q\rho)}{q^2+\omega^2/v'^2\gamma^2}\left[1-e^{-i\frac{\omega z'}{2v'^2}\left(\gamma^{-2}+q^2 v'^2/\omega^2\right)}\right] \tag{100}$$

From (100) it follows that the rebuilding of the field around the electron occurs in such way that each Fourier-harmonic of frequency ω_0 totally reconstructs and becomes the harmonic of equilibrium coulomb field on the distance from the scattering point, which coincides with radiation formation length ($|z'|\approx 2\gamma^2/\omega_0$) for this$\omega_0$. It is possible to place the plate quite close to the scattering point so that at the moment of electron's traverse of the plate the Fourier-harmonics of certain frequencies $\omega<\omega_0$ will have not yet reconstructed. In other words it is possible to place the plate in the pre-wave zone for these frequencies. In this case the incident electron will be 'half-bare' and its transition radiation should differ from such radiation by electron with equilibrium field.

The total field of the electron-plate system consists of the field of 'half-bare' electron $E_\perp$ and the field $E_\perp^f$ of currents induced on the surface of the plate. Applying the boundary condition for electric field on the surface of the plate

$$E_\perp\big|_{z'=z'_0}+E_\perp^f\big|_{z'=z'_0}=0$$

we can find the expression for the Fourier-harmonic of the field of induced surface currents:

$$E_\perp^f(\vec{r},\omega)=2e\frac{e^{i\omega R}}{R}\frac{\vartheta}{\vartheta^2+\gamma^{-2}}[F_\omega(\vec{r})-1] \tag{101}$$

where $F_\omega(\vec{r})=\frac{1}{v}\frac{\rho'^2+\gamma^{-2}(z'-2z'_0)^2}{\rho'^2+\gamma^{-2}(z'-z'_0)^2}\exp\left\{\frac{i\omega z'_0}{2}\left[\frac{1}{v^2\gamma^2}+\frac{\rho'^2}{(z'-z'_0)(z'-2z'_0)}\right]\right\}$,

R is the distance between the point of the electron's traverse of the plate and the point where the field is considered, $R\approx z'_0-z'+\rho'^2/2(z'-z'_0)$ and ϑ is counted from the direction of $-\vec{v}'$. This field gradually transforms into backward transition radiation.

The expression (101) can be simplified for $-z'\gg 2\gamma^2/\omega$. In this case:

$$E_\perp^f(\vec{r},\omega)=2e\frac{e^{i\omega R}}{R}\frac{\vartheta}{\vartheta^2+\gamma^{-2}}\left[\frac{1}{v}e^{i\frac{z'_0\omega}{2\gamma^2}(1+\gamma^2\vartheta^2)}-1\right] \tag{102}$$

and using (57) for spectral-angular density of transition radiation by 'half-bare' electron we obtain:

$$\frac{d\varepsilon}{d\omega do}=\frac{e^2}{\pi^2}\frac{\vartheta^2}{(\vartheta^2+\gamma^{-2})^2}2\left\{1-\cos\left[\frac{\omega z'_0}{2}(\gamma^{-2}+\vartheta^2)\right]\right\} \tag{103}$$

The expression (103) differs from the corresponding expression for transition radiation by electron with equilibrium field by the interference factor inside the braces and the coefficient two in front of them. As we can see from (103), when the distance z'_0 between the scattering point and the plate is much less than the radiation formation length ($l_C\approx 2\gamma^2/\omega$) the radiation is highly suppressed. For larger values of z'_0 the dependence of the radiation intensity on z'_0 has the oscillation type with the period of the order of the formation length:

http://dx.doi.org/10.5772/50866

$$\Lambda = \frac{4\pi}{\omega(\vartheta^2 + \gamma^{-2})} \tag{104}$$

Due to the nonzero frequency resolution $\Delta\omega$ of the detector it is possible to observe such oscillations only in the area limited by the condition

$$z'_0 < \frac{2\pi}{\Delta\omega(\vartheta^2 + \gamma^{-2})} \tag{105}$$

Also due to the nonzero size and, therefore, angular resolution of the detector the oscillations can be observed only inside the region

$$z'_0 < \frac{\pi}{\omega\vartheta\Delta\vartheta} \tag{106}$$

For large distances $z'_0 \gg \Lambda$ the considered oscillations disappear and the detector registers an incoherent sum of contributions to transition radiation by electron's own field reflected from the plate and by the field of bremsstrahlung in this direction.

6. Conclusion

The behavior of localized high-energy electromagnetic wave packets, which take place in processes of transition radiation and bremsstrahlung by relativistic electrons has been considered. It was shown that with the increase of the energy the stabilization of characteristics of motion of such packets takes place, which consists in substantial decrease of the speed of their dispersion. Essential here is the fact that at high energies the lengths, on which the reconstruction of the form of such packets into packets of diverging waves takes place, can reach macroscopic size, which can exceed the size of experimental facility. In this case both the size of the used detector and its position relative to the region of the wave packet formation become essential for measurements.

Such situation takes place, for example, after the sharp scattering of an electron to a large angle. It was shown that as a result of such scattering the electron's own coulomb field tears away from it and turns into a localized packet of free electromagnetic waves, which transforms into a packet of diverging waves on large distance from the scattering point. For ultra relativistic electrons such transformation of certain Fourier-harmonics of the packet field takes place within the coherence length of the radiation process, which substantially exceeds the length of the considered wave of radiation. In the case of low-frequency radiation this coherence length can be macroscopic. It gives birth to the problem of bremsstrahlung characteristics measurement by different detectors, which consists in the dependence of the results of measurement on the detector's size and its position relative to the scattering point.

In the final direction of the electron motion (after the scattering) certain Fourier-components of the field around it do not appear at once. The regeneration of these Fourier-components occurs within the coherence length of the radiation process$l_C \approx 2\gamma^2 / \omega$. In this case during a long period of time the ultra relativistic electron is in 'half-bare' state without a certain part of Fourier-components of its field. Such state of electron manifests itself, for example, during further collisions of the electron with atoms of a substance, which causes different effects of bremsstrahlung suppression (Landau-Pomeranchuk-Migdal effect, the effect of radiation suppression in thin layer of substance – TSF-effect, etc.). We have shown that the 'half-bare' state of the scattered electron should manifest itself in the process of further transition radiation by such electron as well. This manifestation consists in the fact that characteristics of transition radiation by such electron substantially depend on the position of the metallic plate relative to the scattering point. In this case both the transition radiation suppression and the oscillatory dependence of its characteristics on the distance between the plate and the scattering point take place.

The transition radiation formation process also develops within distances of the order of the coherence length of the radiation process,l_C. Such situation takes place both in the direction of the electron motion and in the opposite one. In the case of transition radiation, as in the case of bremsstrahlung, these distances can be macroscopic. The analysis of development of this process in space and time has shown that the packet of the waves reflected from the plate has the same structure as the wave packet torn away from the electron at its instantaneous scattering to a large angle. The structure of the field around the electron after its traverse of thin metallic plate is similar to the structure of the field, which forms around the electron after its scattering to a large angle. In other words, in this case the electron can be in the 'half-bare' state during a long period of time: a certain part of the Fourier components in the field around it is suppressed. The mentioned analogies cause the existence of the similar effects in transition radiation and bremsstrahlung. Thus the possibility of long existence of electron in the 'half-bare' state and of different manifestations of such state of electron can be investigated on the basis of the process of transition radiation by such electron.

Appendix

According to (72) the potential $\phi^f(\vec{r}, t)$ of the free field of electromagnetic waves is the sum of the contributions to it by the propagating $\Phi_1(\vec{r}, t)$ and the surface $\Phi_2(\vec{r}, t)$ waves. While deriving these functions we will pay special attention to the reasons of the step functions appearance in (75).

In order to calculate $\Phi_1(\vec{r}, t)$ we will present this function in the form:

$$\Phi_1(\vec{r}, t) = -\frac{2e}{\pi} \lim_{\eta \to 0} \int_0^1 \frac{x dx}{x^2 + p^{-2}} \int_0^\infty d\omega \cos\omega\left(|z|\sqrt{1-x^2} - t\right) e^{-\eta\omega} \quad (107)$$

where $\eta > 0$. Then after the variable substitution $y = \sqrt{1 - x^2}$ and integration over ω it is easy to obtain that

$$\Phi_1(\vec{r}, t) = -\frac{2e}{\pi v} \lim_{\eta \to 0} \int_0^1 \frac{y dy}{v^2 - y^{-2}} \frac{\eta}{\eta^2 + (y |z| - t)^2} \tag{108}$$

Integration over y in this expression leads to

$$\begin{aligned} \Phi_1(\vec{r}, t) = -\frac{2e}{\pi v} \lim_{\eta \to 0} \frac{\eta}{1 - v^2(\eta^2 + t^2 + z^2/v^2)^2 / (4z^2 t^2)} \Bigg\{ -\frac{v^2(\eta^2 + t^2 + z^2/v^2)}{4z^2 t^2} \ln\gamma \\ -\frac{v}{2zt} \ln\gamma(1+v) - \frac{v^2(\eta^2 + t^2 + z^2/v^2)}{4t^2} \left[\frac{1}{2z^2} \ln \frac{\eta^2 + (|z| - t)^2}{\eta^2 + t^2} \right. \\ \left. + \frac{t}{z^2 \eta} \left(arctg \frac{|z| - t}{\eta} + arctg \frac{t}{\eta} \right) \right] + \frac{v^2(\eta^2 + t^2)}{2zt} \frac{1}{z\eta} \left(arctg \frac{|z| - t}{\eta} + arctg \frac{t}{\eta} \right) \Bigg\} \end{aligned} \tag{109}$$

Taking into account that for $\eta \to 0$ only items proportional to $(arctg[(|z| - t)/\eta] + arctg[t/\eta])$ remain in (109) and that for these items

$$arctg \frac{|z| - t}{\eta} + arctg \frac{t}{\eta} = \frac{\pi}{2} [sign(|z| - t) + sign(t)] = \frac{\pi}{2} \theta(|z| - t) \theta(t)$$

we obtain the following expression for $\Phi_1(\vec{r}, t)$:

$$\Phi_1(\vec{r}, t) = \frac{e}{2} \left(\frac{1}{|z| + vt} - \frac{1}{|z| - vt} \right) \theta(|z| - t) \theta(t) \tag{110}$$

Thus the appearance of the step function $\theta(|z| - t)$ in $\Phi_1(\vec{r}, t)$ is connected with the limiting procedure $\eta \to 0$ in (109).

The calculation of $\Phi_2(\vec{r}, t)$ does not represent any difficulties as the integral over ω in it converges. As a result of the variable substitution $u = \sqrt{1 - x^2}$ and elementary integration we find that:

$$\begin{aligned} \Phi_2(\vec{r}, t) = \frac{e}{2} \left(\frac{1}{|z| - vt} - \frac{1}{|z| + vt} \right) sign(t) - \frac{e}{2} \left(\frac{1}{|z| - vt} + \frac{1}{|z| + vt} \right) = \\ = -\frac{1}{|z| - vt} \theta(-t) - \frac{1}{|z| + vt} \theta(t). \end{aligned} \tag{111}$$

Substituting the obtained expressions (110) and (111) for Φ_1 and Φ_2 into (72) we obtain the expression (75) for the potential of the free waves.

Author details

Nikolai Fyodorovich Shul'ga* and Sergii Valeriyovich Trofymenko

*Address all correspondence to: shulga@ kipt.kharkov.ua

Akhiezer Institute for Theoretical Physics of National Science Centre "Kharkov Institute of Physics and Technology" Kharkov, Ukraine

References

[1] Ter-Mikaelyan, M. L. (1972). High-Energy Electromagnetic Processes in media. New York: Wiley.

[2] Akhiezer, A. I., & Shul'ga, N. F. (1996). High Energy Electrodynamics in Matter. Amsterdam: Gordon and Breach Publ., 388 p.

[3] Dokshitzer, Yu. L., Khoze, V. A., Mueller, A. H., & Troyan, S. I. (1991). Basics of Perturbative QCD. Gif sur Yvette Cedex: Editions Frontiers. 274p.

[4] Garibyan, G. M., & Shi, Y. (1983). X-Ray transition radiation. Yerevan: Publ. of Acad. of Sc. of Arm. SSR, 320p. (in Russian)

[5] Ginzburg, V. L., & Tsytovich, V. N. (1984). Transition Radiation and Transition Scattering. Bristol: Adam Hilger.

[6] Rullhusen, P., Artru, X., & Dhez, P. (1998). Novel radiation sources using relativistic electrons. Singapore: World Scientific Publ. 202p.

[7] Shibata, Y., Hasebe, S., Ishiki, K., et al. (1995). Observation of coherent diffraction radiation from bunched electrons passing through a circular aperture in the millimeter- and submillimeter- wave length regions. *Phys. Rev.*, 52 E, 6737.

[8] Dobrovolsky, S. N., & Shul'ga, N. F. (2003). Transversal spatial distribution of transition radiation by relativistic electron in formation zone by the dotted detector. *Nucl. Instrum. Methods B*, 201, 123-132.

[9] Gorham, P., et al. (2000). Radio-frequency measurements of coherent transition and Cherenkov radiation: implication for high energy neutrino detection. *Phys. Rev.*, 62 E, 8590-8605.

[10] Verzilov, V. A. (2000). Transition radiation in the pre-wave zone. *Phys. Lett.*, 273 A, 135-140.

[11] Akhiezer, A. I., Shul'ga, N. F., & Fomin, S. P. (2005). Landau-Pomeranchuk-Migdal Effect. *Physics Reviews*, Edited by I. M. Khalatnikov. Cambridge Sci. Publ., Printed in UK, 22, 1-215.

[12] Potylitsyn, A. P. (2011). Electromagnetic Radiation of Electrons in Periodic Structures. Berlin: Springer. 213p. 10.1007/978-3-642-19248-7.

[13] Naumenko, G. N., Potylitsyn, A. P., Sukhih, L. G., et al. (2009). Macroscopic effect of shadow of electromagnetic field of relativistic electrons. *JETP Letters*, 90(2), 105-110.

[14] Blokhintsev, D. I. (1967). In the book: High-Energy Physics and Elementary Particles Theory. Kiev: Nauk. Dumka. p. 778.

[15] Shul'ga, N. F., Syshchenko, V. V., & Shul'ga, S. N. (2009). On the motion of high-energy wave packets and transition radiation by "half-bare" electron. *Phys. Lett.*, 374 A, 331-334.

[16] Shul'ga, N. F., Trofymenko, S. V., & Syshchenko, V. V. (2010). The Space-Time Evolution of the Process of Transition Radiation by Relativistic Electron. *Journal of Kharkiv National University*, No. 916, phys. series "Nuclei, Particles, Fields", 3(47), 23-41.

[17] Shul'ga, N. F., Trofymenko, S. V., & Syshchenko, V. V. (2011). On transition radiation and bremsstrahlung by relativistic electron with nonequilibrium field. *JETP Letters*, 93(1), 3-7.

[18] Shul'ga, N. F., Trofymenko, S. V., & Syshchenko, V. V. (2011). On transition radiation by relativistic electron. *Il Nuovo Cimento*, 34 C(4), 327-334.

[19] Shul'ga, N. F., Trofymenko, S. V., & Syshchenko, V. V. (2012). The prewave zone effect in transition radiation and bremsstrahlung by relativistic electron. *Problems of atomic science and technology*, No. 1 series: Nuclear Physics Investigations, 57, 134-138.

[20] Miller, W. (1974). Classical Limit Quantum Mechanics and the Theory of Molecular Collisions. New York: Wiley, 69-177.

[21] Landau, L. D., & Lifshitz, E. M. (1987). The Classical Theory of Fields. Oxford: Pergamon.

[22] Jackson, J. D. (1999). Classical Electrodynamics. New York: Wiley. 808p.

[23] Landau, L. D., & Pomeranchuk, I. Ya. (1953). Electron-cascade processes at ultra-high energies. *Dokl. Acad. Nauk. SSSR*, 92, 735.

[24] Migdal, A. B. (1956). Bremsstrahlung and pair production in condensed media at high energies. *Phys. Rev.*, 103, 1811.

[25] Akhiezer, A. I., & Shul'ga, N. F. (1987). Influence of multiple scattering on the radiation of relativistic particles in amorphous and crystalline media. *Sov. Phys. Usp.*, 30, 197-219.

[26] Ternovsky, F. F. (1961). On the theory of radiative processes in piece wise homogeneous media. *Sov. Phys. JETP*, 12, 123.

[27] Shul'ga, N. F., & Fomin, S. P. (1978). Suppression of radiation in an amorphous medium and a crystal. *JETP Letters*, 27, 117-120.

[28] Fomin, S. P., & Shul'ga, N. F. (1986). On space-time evolution of the process of ultra relativistic electron radiation in thin layer of substance. *Phys. Lett.*, 114 A(3), 148-152.

[29] Anthony, P., et al. (1997). Bremsstrahlung suppression due to LPM and dielectric effects in a variety of targets. *Phys. Rev.*, 12 D, 1286.

[30] Klein, S. (1998). Suppression of bremsstrahlung and pair production due to environmental factors. *Rev. of Mod. Phys.*, 71, 1501.

[31] Thomsen, H. D., Esberg, J., Kirsebom, K., et al. (2009). On the Macroscopic Formation Length for GeV Photons. *Phys. Lett.*, 672 B, 323-327.

[32] Thomsen, H. D., Esberg, J., Andersen, K. K., et al. (2010). Distorted Coulomb field of the scattered electron. *Phys. Rev.*, 81 D, 18.

[33] Feinberg, E. L. (1966). Consecutive interactions at high energies. *Sov Phys. JETP*, 23, 132.

[34] Feinberg, E. L. (1972). The problems of theoretical physics: paper collection dedicated to the memory of I.Y. Tamm. Moscow, *Nauka*, p. 248 (in Russian).

[35] Akhiezer, A. I., & Shul'ga, N. F. (1982). Radiation of relativistic particles in single crystals. *Sov. Phys. Usp.*, 25, 541-564.

[36] Bleistein, N., & Handelsman, R. (1986). Asymptotic Expansions of Integrals. New York: Holt, Dover Publications. 425p.

[37] Bolotovsky, B. M., & Serov, A. V. (2009). Peculiarities of transition radiation field. *Sov. Phys. Usp.*, 179, 517-524, (in Russian).

Chapter 11

The Effect of Weak Fields at Multiple Frequencies on the Scattering and Generation of Waves by Nonlinear Layered Media

Lutz Angermann and Vasyl V. Yatsyk

Additional information is available at the end of the chapter

http://dx.doi.org/10.5772/ 50679

1. Introduction

The chapter deals with the effects of weak fields at multiple frequencies on the scattering and generation of waves by an isotropic, nonmagnetic, linearly polarised (E-polarisation), layered, cubically polarisable, dielectric structure. In the domain of resonance frequencies we consider wave packets consisting of both strong electromagnetic fields at the excitation frequency of the nonlinear structure, leading to the generation of waves, and of weak fields at the multiple frequencies, which do not lead to the generation of harmonics but influence on the process of scattering and generation of waves by the nonlinear structure. The electromagnetic waves for a nonlinear layer with a cubic polarisability of the medium can be described by an infinite system of nonlinear boundary-value problems in the frequency domain. As has been shown in previous articles of the authors ([3], [4]), in the study of particular nonlinear effects it is possible to restrict this system to a finite number of equations. If the classical formulation of the problem is supplemented by the condition of phase synchronism, we arrive at a self-consistent formulation of a system of boundary-value problems with respect to the components of the scattered and generated fields. It is known that this system is equivalent to a system of nonlinear boundary-value problems of Sturm-Liouville type and also to a system of one-dimensional nonlinear Fredholm integral equations of the second kind.

The solution of the system of integral equations is approximated by the help of numerical algorithms. Those include the application of suitable quadrature rules and iterative procedures to solve the resulting nonlinear algebraic problems. Since in each iteration step the solution of linear algebraic systems is required, the approximate solution of the nonlinear problems is described by means of solutions of linear problems with an induced nonlinear permittivity.

In continuation of our results from previous works ([3], [4]), where we only considered one excitation field at the basic frequency, here results of calculations of characteristics of the

INTECH
open science | open minds

© 2012 Angermann and V. Yatsyk; licensee InTech. This is an open access article distributed under the terms of the Creative Commons Attribution License (http://creativecommons.org/licenses/by/3.0), which permits unrestricted use, distribution, and reproduction in any medium, provided the original work is properly cited.

scattered and generated fields of plane waves are presented, taking into account the influence of weak fields at multiple frequencies on the cubically polarisable layer. We restrict ourselves to the investigation of the third harmonic generated by layers with a positive value of the cubic susceptibility of the medium.

Within the framework of the self-consistent system which is given by a system of nonlinear integral equations, we show the following. The variation of the imaginary parts of the permittivities of the layer at the multiple frequencies can take both positive and negative values along the height of the nonlinear layer. It is induced by the nonlinear part of the permittivities and is caused by the loss of energy in the nonlinear medium which is spent for the generation of the electromagnetic fields. The magnitudes of these variations are determined by the amplitude and phase characteristics of the fields which are scattered and generated by the nonlinear layer.

Layers with negative and positive values of the coefficient of cubic susceptibility of the nonlinear medium have fundamentally different scattering and generation properties. In the case of negative values of the susceptibility, a decanalisation of the electromagnetic field can be detected. The maximal portion of the total energy generated in the third harmonic is observed in the direction normal to the structure and nearly amounts to 4% of the total dissipated energy. For a layer with a positive value of the susceptibility an effect of energy canalisation is observed (see [4]). Increasing intensities of the incidents fields lead to an increase of the angle of transparancy which increasingly deviates from the direction normal to the layer. In this case, the maximal portion of energy generated in the third harmonic is observed near the angle of transparency of the nonlinear layer. In the numerical experiments there have been reached intensities of the excitation field of the layer such that the relative portion of the total energy generated in the third harmonic is about 35% (see [4]).

In the chapter the effect of weak fields at multiple frequencies on the scattering and generation of waves is investigated numerically. The results indicate a possibility of designing a frequency multiplier and nonlinear dielectrics with controllable permittivity.

2. The mathematical model

We consider a layered nonlinear medium which is located in an infinite plate of thickness $4\pi\delta$, where $\delta > 0$ is a given parameter: $\{\mathbf{r} = (x, y, z)^\top \in \mathbb{R}^3 : |z| \le 2\pi\delta\}$.

It is assumed that the vector of the polarisation moment $\mathbf{P}$ can be expanded in terms of the electric field intensity $\mathbf{E}$ as follows:

$$\mathbf{P} = \boldsymbol{\chi}^{(1)}\mathbf{E} + (\boldsymbol{\chi}^{(2)}\mathbf{E})\mathbf{E} + ((\boldsymbol{\chi}^{(3)}\mathbf{E})\mathbf{E})\mathbf{E} + \dots , \tag{1}$$

where $\boldsymbol{\chi}^{(1)}, \boldsymbol{\chi}^{(2)}, \boldsymbol{\chi}^{(3)}$ are the media susceptibility tensors of rank two, three and four, with components $\{\chi_{ij}^{(1)}\}_{i,j=1}^{3}$, $\{\chi_{ijk}^{(2)}\}_{i,j,k=1}^{3}$ and $\{\chi_{ijkl}^{(3)}\}_{i,j,k,l=1}^{3}$, respectively (see [5]). In the case of isotropic media, the quadratic term disappears.

It is convenient to split $\mathbf{P}$ into its linear and nonlinear parts as $\mathbf{P} = \mathbf{P}^{(L)} + \mathbf{P}^{(NL)} := \boldsymbol{\chi}^{(1)}\mathbf{E} + \mathbf{P}^{(NL)}$. Similarly, with $\boldsymbol{\varepsilon} := \mathbf{I} + 4\pi\boldsymbol{\chi}^{(1)}$ and $\mathbf{D}^{(L)} := \boldsymbol{\varepsilon}\mathbf{E}$, where $\mathbf{I}$ denotes the identity in $\mathbb{C}^3$,

http://dx.doi.org/10.5772/ 50679

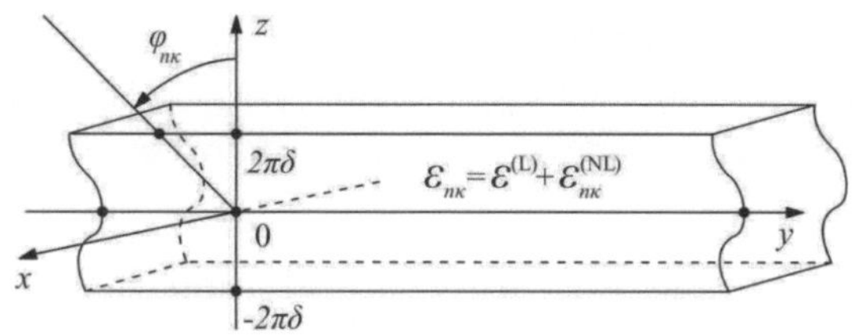

Figure 1. The nonlinear dielectric layered structure

the displacement field **D** can be decomposed as

$$\mathbf{D} = \mathbf{D}^{(L)} + 4\pi\mathbf{P}^{(NL)}. \tag{2}$$

ε is the linear term of the permittivity tensor. Furthermore, if the media under consideration are nonmagnetic, isotropic and transversely inhomogeneous w.r.t. z, i.e. $\varepsilon = \varepsilon^{(L)}\mathbf{I}$ with a scalar, possibly complex-valued function $\varepsilon^{(L)} = \varepsilon^{(L)}(z)$, if the wave is linearly E-polarised, i.e.

$$\mathbf{E} = (E_1, 0, 0)^\top, \quad \mathbf{H} = (0, H_2, H_3)^\top, \tag{3}$$

and if the electric field $\mathbf{E}$ is homogeneous w.r.t. the coordinate x, i.e. $\mathbf{E}(\mathbf{r}, t) = (E_1(t; y, z), 0, 0)^\top$, then the Maxwell's equations together with (2) are simplified to

$$\nabla^2\mathbf{E} - \frac{\varepsilon^{(L)}}{c^2}\frac{\partial^2}{\partial t^2}\mathbf{E} - \frac{4\pi}{c^2}\frac{\partial^2}{\partial t^2}\mathbf{P}^{(NL)} = 0, \tag{4}$$

where ∇^2 reduces to the Laplacian w.r.t. y and z, i.e. $\nabla^2 := \partial^2/\partial y^2 + \partial^2/\partial z^2$.

A stationary electromagnetic wave (with the oscillation frequency $\omega > 0$) propagating in a nonlinear dielectric structure gives rise to a field containing all frequency harmonics (see [20], [22], and other authors). Therefore, representing $\mathbf{E}$, $\mathbf{P}^{(NL)}$ via Fourier series

$$\mathbf{E}(\mathbf{r}, t) = \frac{1}{2}\sum_{s\in\mathbb{Z}} \mathbf{E}(\mathbf{r}, s\omega)e^{-is\omega t}, \quad \mathbf{P}^{(NL)}(\mathbf{r}, t) = \frac{1}{2}\sum_{s\in\mathbb{Z}} \mathbf{P}^{(NL)}(\mathbf{r}, s\omega)e^{-is\omega t}, \tag{5}$$

we obtain from (4) an infinite system of coupled equations w.r.t. the Fourier amplitudes. In the case of a three-component E-polarised electromagnetic field (3), this system reduces to a system of scalar equations w.r.t. E_1:

$$\Delta E_1(\mathbf{r}, s\omega) + \frac{\varepsilon^{(L)}(s\omega)^2}{c^2}E_1(\mathbf{r}, s\omega) + \frac{4\pi(s\omega)^2}{c^2}P_1^{(NL)}(\mathbf{r}, s\omega) = 0, \quad s \in \mathbb{N}. \tag{6}$$

In writing equation (6), the property $E_1(\mathbf{r}; j\omega) = \overline{E}_1(\mathbf{r}; -j\omega)$ of the Fourier coefficients and the lack of influence of the static electric field $E_1(\mathbf{r}, s\omega)|_{s=0} = 0$ on the nonlinear structure were taken into consideration.

We assume that the main contribution to the nonlinearity is introduced by the term $\mathbf{P}^{(NL)}(\mathbf{r}, s\omega)$ (cf. [25], [14], [8], [1], [24], [10], [19], [11]), and we take only the lowest-order terms in the Taylor series expansion of the nonlinear part $\mathbf{P}^{(NL)}(\mathbf{r}, s\omega) = \left(P_1^{(NL)}(\mathbf{r}, s\omega), 0, 0\right)^\top$ of the polarisation vector in the vicinity of the zero value of the electric field intensity, cf. (1). In this case, the only nontrivial component of the polarisation vector is determined by susceptibility tensor of the third order $\boldsymbol{\chi}^{(3)}$. In the time domain, this component can be represented in the form (cf. (1) and (5)):

$$P_1^{(NL)}(\mathbf{r}, t) \doteq \frac{1}{8} \sum_{\left\{\substack{n,m,p,s \in \mathbb{Z} \setminus \{0\} \\ n+m+p=s}\right.} \chi_{1111}^{(3)}(s\omega; n\omega, m\omega, p\omega) E_1(\mathbf{r}, n\omega) E_1(\mathbf{r}, m\omega) E_1(\mathbf{r}, p\omega) e^{-is\omega t}, \quad (7)$$

where the symbol $\doteq$ means that higher-order terms are neglected. Applying to (7) the Fourier transform w.r.t. time, we obtain an expansion in the frequency domain:

$$\begin{aligned} P_1^{(NL)}(\mathbf{r}, s\omega) = & \frac{1}{4} \sum_{j \in \mathbb{N}} 3\chi_{1111}^{(3)}(s\omega; j\omega, -j\omega, s\omega) |E_1(\mathbf{r}, j\omega)|^2 E_1(\mathbf{r}, s\omega) \\ & + \frac{1}{4} \sum_{\left\{\substack{n,m,p \in \mathbb{Z} \setminus \{0\} \\ n \neq -m,\ p=s \\ m \neq -p,\ n=s \\ n \neq -p,\ m=s \\ n+m+p=s}\right.} \chi_{1111}^{(3)}(s\omega; n\omega, m\omega, p\omega) E_1(\mathbf{r}, n\omega) E_1(\mathbf{r}, m\omega) E_1(\mathbf{r}, p\omega). \end{aligned} \quad (8)$$

We see that, under the above assumptions, the electromagnetic waves in a nonlinear medium with a cubic polarisability are described by an infinite system (6) and (8) of coupled nonlinear equations ([25], [14], [8], [2]).

Here we study nonlinear effects involving the waves at the first three frequency components of E_1 only. That is, we further neglect the influence of harmonics of order higher than 3. Then it is possible to restrict the examination of the system (6) and (8) to three equations, and also to leave particular terms in the representation of the polarisation coefficients. Taking into account only the nontrivial terms in the expansion of the polarisation coefficients and using the so-called *Kleinman's rule* (i.e. the equality of all the coefficients $\chi_{1111}^{(3)}$ at the multiple frequencies, [7], [9]), we arrive at the following system:

$$\begin{aligned} & \Delta E_1(\mathbf{r}, n\kappa) + (n\kappa)^2 \varepsilon_{n\kappa}(z, \alpha(z), E_1(\mathbf{r}, \kappa), E_1(\mathbf{r}, 2\kappa), E_1(\mathbf{r}, 3\kappa)) \\ = & -\delta_{n1} \kappa^2 \alpha(z) E_1^2(\mathbf{r}, 2\kappa) \overline{E}_1(\mathbf{r}, 3\kappa) \\ & - \delta_{n3} (3\kappa)^2 \alpha(z) \left\{ \frac{1}{3} E_1^3(\mathbf{r}, \kappa) + E_1^2(\mathbf{r}, 2\kappa) \overline{E}_1(\mathbf{r}, \kappa) \right\}, \quad n = 1, 2, 3, \end{aligned} \quad (9)$$

with $\kappa := \frac{\omega}{c} = \frac{2\pi}{\lambda}$, $\varepsilon_{n\kappa} := \begin{cases} 1, & |z| > 2\pi\delta, \\ \varepsilon^{(L)} + \varepsilon_{n\kappa}^{(NL)}, & |z| \le 2\pi\delta, \end{cases}$ and $\varepsilon^{(L)} := 1 + 4\pi\chi_{11}^{(1)}$,

$$\begin{aligned} \varepsilon_{n\kappa}^{(NL)} &= \alpha(z)\left[|E_1(\mathbf{r},\kappa)|^2 + |E_1(\mathbf{r},2\kappa)|^2 + |E_1(\mathbf{r},3\kappa)|^2 \right. \\ &\quad \left. + \delta_{n1}\frac{\left[\overline{E}_1(\mathbf{r},\kappa)\right]^2}{E_1(\mathbf{r},\kappa)}E_1(\mathbf{r},3\kappa) + \delta_{n2}\frac{\overline{E}_1(\mathbf{r},2\kappa)}{E_1(\mathbf{r},2\kappa)}E_1(\mathbf{r},\kappa)E_1(\mathbf{r},3\kappa)\right], \end{aligned} \tag{10}$$

where $\alpha(z) := 3\pi\chi_{1111}^{(3)}(z)$ is the so-called *function of cubic susceptibility* and δ_{nj} denotes Kronecker's symbol. For transversely inhomogeneous media (a layer or a layered structure), the linear part $\varepsilon^{(L)}$ of the permittivity is described by a piecewise smooth or even a piecewise constant function. Similarly, the function of the cubic susceptibility $\alpha = \alpha(z)$ is also a piecewise smooth or even a piecewise constant function. This assumption allows us to investigate the scattering and generation characteristics of a nonlinear layer and of a layered structure (consisting of a finite number of linear and nonlinear dielectric layers) within one and the same mathematical model.

Here and in what follows we use the following notation: $(\mathbf{r}, t)$ are dimensionless spatial-temporal coordinates such that the thickness of the layer is equal to $4\pi\delta$. The time-dependence is determined by the factors $\exp(-in\omega t)$, where $\omega = \kappa c$ is the dimensionless circular frequency and κ is a dimensionless frequency parameter. This parameter characterises the ratio of the true thickness h of the layer to the free-space wavelength λ, i.e. $h/\lambda = 2\kappa\delta$. $c = (\varepsilon_0\mu_0)^{-1/2}$ denotes a dimensionless parameter, equal to the absolute value of the speed of light in the medium containing the layer ($\mathfrak{Im}\, c = 0$). ε_0 and μ_0 are the material parameters of the medium. The absolute values of the true variables $\mathbf{r}', t', \omega'$ are given by the formulas $\mathbf{r}' = h\mathbf{r}/4\pi\delta$, $t' = th/4\pi\delta$, $\omega' = \omega 4\pi\delta/h$.

The scattered and generated field in a transversely inhomogeneous, nonlinear dielectric layer excited by a plane wave is *quasi-homogeneous* along the coordinate y, hence it can be represented as

(C1) $\quad E_1(\mathbf{r}, n\kappa) =: E_1(n\kappa; y, z) := U(n\kappa; z)\exp(i\phi_{n\kappa} y), \quad n = 1, 2, 3.$

Here $U(n\kappa; z)$ and $\phi_{n\kappa} := n\kappa \sin\varphi_{n\kappa}$ denote the complex-valued transverse component of the Fourier amplitude of the electric field and the value of the longitudinal propagation constant (longitudinal wave-number) at the frequency $n\kappa$, respectively, where $\varphi_{n\kappa}$ is the given angle of incidence of the exciting field of frequency $n\kappa$ (cf. Fig. 1).

Furthermore we require that the following *condition of the phase synchronism of waves* is satisfied:

(C2) $\quad \phi_{n\kappa} = n\phi_\kappa, \quad n = 1, 2, 3.$

Then the permittivity of the nonlinear layer can be expressed as

$$\begin{aligned} &\varepsilon_{n\kappa}(z, \alpha(z), E_1(\mathbf{r},\kappa), E_1(\mathbf{r},2\kappa), E_1(\mathbf{r},3\kappa)) = \varepsilon_{n\kappa}(z, \alpha(z), U(\kappa;z), U(2\kappa;z), U(3\kappa;z)) \\ &= \varepsilon^{(L)}(z) + \alpha(z)\left[|U(\kappa;z)|^2 + |U(2\kappa;z)|^2 + |U(3\kappa;z)|^2 \right. \\ &\quad + \delta_{n1}|U(\kappa;z)||U(3\kappa;z)|\exp\{i[-3\arg(U(\kappa;z)) + \arg(U(3\kappa;z))]\} \\ &\quad \left. + \delta_{n2}|U(\kappa;z)||U(3\kappa;z)|\exp\{i[-2\arg(U(2\kappa;z)) + \arg(U(\kappa;z)) + \arg(U(3\kappa;z))]\}\right], \\ &\qquad n = 1, 2, 3. \end{aligned} \tag{11}$$

A more detailed explanation of the condition (C2) can be found in [3, Sect. 3]. In the considered case of spatially quasi-homogeneous (along the coordinate y) electromagnetic fields (C1), the condition of the phase synchronism of waves (C2) reads as

$$\sin\varphi_{n\kappa} = \sin\varphi_{\kappa}, \quad n = 1,2,3.$$

Consequently, the given angle of incidence of a plane wave at the frequency κ coincides with the possible directions of the angles of incidence of plane waves at the multiple frequencies $n\kappa$, i.e. $\varphi_{n\kappa} = \varphi_{\kappa}$, $n = 1,2,3$. The angles of the wave scattered by the layer are equal to $\varphi^{\text{scat}}_{n\kappa} = -\varphi_{n\kappa}$ in the zone of reflection $z > 2\pi\delta$ and $\varphi^{\text{scat}}_{n\kappa} = \pi + \varphi_{n\kappa}$ and in the zone of transmission of the nonlinear layer $z < -2\pi\delta$, where all angles are measured counter-clockwise in the (y,z)-plane from the z-axis (cf. Fig. 1).

The conditions (C1), (C2) allow a further simplification of the system (9). Before we do so, we want to make a few comments on specific cases which have already been discussed in the literature. First we mention that the effect of a weak quasi-homogeneous electromagnetic field (C1) on the nonlinear dielectric structure such that harmonics at multiple frequencies are not generated, i.e. $E_1(\mathbf{r},2\kappa) = 0$ and $E_1(\mathbf{r},3\kappa) = 0$, reduces to find the electric field component $E_1(\mathbf{r},\kappa)$ determined by the first equation of the system (9). In this case, a diffraction problem for a plane wave on a nonlinear dielectric layer with a Kerr-type nonlinearity $\varepsilon_{n\kappa} = \varepsilon^{(L)}(z) + \alpha(z)|E_1(\mathbf{r},\kappa)|^2$ and a vanishing right-hand side is to be solved, see [1, 8, 11, 14, 19, 24, 25]. The generation process of a field at the triple frequency 3κ by the nonlinear dielectric structure is caused by a strong incident electromagnetic field at the frequency κ and can be described by the first and third equations of the system (9) only, i.e. $n = 1,3$. Since the right-hand side of the second equation in (9) is equal to zero, we may set $E_1(\mathbf{r},2\kappa) = 0$ corresponding to the homogeneous boundary condition w.r.t. $E_1(\mathbf{r},2\kappa)$. Therefore the second equation in (9) can be completely omitted, see [2] and the numerical results in [3], [4].

A further interesting problem consists in the investigation of the influence of a packet of waves on the generation of the third harmonic, if a strong incident field at the basic frequency κ and, in addition, weak incident quasi-homogeneous electromagnetic fields at the double and triple frequencies 2κ, 3κ (which alone do not generate harmonics at multiple frequencies) excite the nonlinear structure. Then we have to take into account all three equations of system (9). This is caused by the inhomogeneity of the corresponding problem, where a weak incident field at the double frequency 2κ (or two weak fields at the frequencies 2κ and 3κ) excites (resp. excite) the dielectric medium.

So we consider the problem of scattering and generation of waves on a nonlinear, layered, cubically polarisable structure, which is excited by a packet of plane waves consisting of a strong field at the frequency κ (which generates a field at the triple frequency 3κ) and of weak fields at the frequencies 2κ and 3κ (having an impact on the process of third harmonic generation due to the contribution of weak electromagnetic fields)

$$\left\{E_1^{\text{inc}}(\mathbf{r},n\kappa) := E_1^{\text{inc}}(n\kappa;y,z) := a^{\text{inc}}_{n\kappa} \exp\left(i(\phi_{n\kappa}y - \Gamma_{n\kappa}(z-2\pi\delta))\right)\right\}_{n=1}^{3}, \quad z > 2\pi\delta, \tag{12}$$

with amplitudes $a_{n\kappa}^{\text{inc}}$ and angles of incidence $\varphi_{n\kappa}$, $|\varphi| < \pi/2$ (cf. Fig. 1), where $\phi_{n\kappa} := n\kappa \sin \varphi_{n\kappa}$ are the longitudinal propagation constants (longitudinal wave-numbers) and $\Gamma_{n\kappa} := \sqrt{(n\kappa)^2 - \phi_{n\kappa}^2}$ are the transverse propagation constants (transverse wave-numbers).

In this setting, if a packet of plane waves excites a nonmagnetic, isotropic, linearly polarised (i.e.

$$\mathbf{E}(\mathbf{r}, n\kappa) = (E_1(n\kappa; y, z), 0, 0)^\top, \ \mathbf{H}(\mathbf{r}, n\kappa) = \left(0, \frac{1}{in\omega\mu_0} \frac{\partial E_1(n\kappa; y, z)}{\partial z}, -\frac{1}{in\omega\mu_0} \frac{\partial E_1(n\kappa; y, z)}{\partial y}\right)^\top$$

(E-polarisation)), transversely inhomogeneous $\varepsilon^{(L)} = \varepsilon^{(L)}(z) = 1 + 4\pi\chi_{11}^{(1)}(z)$ dielectric layer (see Fig. 1) with a cubic polarisability $\mathbf{P}^{(NL)}(\mathbf{r}, n\kappa) = (P_1^{(NL)}(n\kappa; y, z), 0, 0)^\top$ of the medium, the complex amplitudes of the total fields

$$E_1(\mathbf{r}, n\kappa) =: E_1(n\kappa; y, z) := U(n\kappa; z) \exp(i\phi_{n\kappa} y) := E_1^{\text{inc}}(n\kappa; y, z) + E_1^{\text{scat}}(n\kappa; y, z)$$

satisfy the system of equations (cf. (9) – (10))

$$\begin{cases} \nabla^2 E_1(\mathbf{r}, \kappa) + \kappa^2 \varepsilon_\kappa(z, \alpha(z), E_1(\mathbf{r}, \kappa), E_1(\mathbf{r}, 2\kappa), E_1(\mathbf{r}, 3\kappa)) E_1(\mathbf{r}, \kappa) \\ \qquad = -\alpha(z)\kappa^2 E_1^2(\mathbf{r}, 2\kappa) \overline{E}_1(\mathbf{r}, 3\kappa), \\ \nabla^2 E_1(\mathbf{r}, 2\kappa) + (2\kappa)^2 \varepsilon_{2\kappa}(z, \alpha(z), E_1(\mathbf{r}, \kappa), E_1(\mathbf{r}, 2\kappa), E_1(\mathbf{r}, 3\kappa)) E_1(\mathbf{r}, 2\kappa) = 0, \\ \nabla^2 E_1(\mathbf{r}, 3\kappa) + (3\kappa)^2 \varepsilon_{3\kappa}(z, \alpha(z), E_1(\mathbf{r}, \kappa), E_1(\mathbf{r}, 2\kappa), E_1(\mathbf{r}, 3\kappa)) E_1(\mathbf{r}, 3\kappa) \\ \qquad = -\alpha(z)(3\kappa)^2 \left\{ \frac{1}{3} E_1^3(\mathbf{r}, \kappa) + E_1^2(\mathbf{r}, 2\kappa) \overline{E}_1(\mathbf{r}, \kappa) \right\} \end{cases} \tag{13}$$

together with the following conditions, where $\mathbf{E}_{\text{tg}}(n\kappa; y, z)$ and $\mathbf{H}_{\text{tg}}(n\kappa; y, z)$ denote the tangential components of the intensity vectors of the full electromagnetic field $\{\mathbf{E}(n\kappa; y, z)\}_{n=1,2,3}$, $\{\mathbf{H}(n\kappa; y, z)\}_{n=1,2,3}$:

(C1) $E_1(n\kappa; y, z) = U(n\kappa; z) \exp(i\phi_{n\kappa} y), \quad n = 1, 2, 3$
(the quasi-homogeneity condition w.r.t. the spatial variable y introduced above),

(C2) $\phi_{n\kappa} = n\phi_\kappa, \quad n = 1, 2, 3,$
(the condition of phase synchronism of waves introduced above),

(C3) $\mathbf{E}_{\text{tg}}(n\kappa; y, z)$ and $\mathbf{H}_{\text{tg}}(n\kappa; y, z)$ (i.e. $E_1(n\kappa; y, z)$ and $H_2(n\kappa; y, z)$) are continuous at the boundary layers of the nonlinear structure,

(C4) $E_1^{\text{scat}}(n\kappa; y, z) = \begin{Bmatrix} a_{n\kappa}^{\text{scat}} \\ b_{n\kappa}^{\text{scat}} \end{Bmatrix} \exp\left(i\left(\phi_{n\kappa} y \pm \Gamma_{n\kappa}(z \mp 2\pi\delta)\right)\right), \quad z \gtrless \pm 2\pi\delta, \quad n = 1, 2, 3$
(the radiation condition w.r.t. the scattered and generated fields).

The condition (C4) provides a physically consistent behaviour of the energy characteristics of scattering and generation. It guarantees the absence of waves coming from infinity (i.e.

$z = \pm\infty$), see [12]. We study the scattering and generation properties of the nonlinear layer, where in (C4) we always have

$$\mathfrak{Im}\,\Gamma_{n\kappa} = 0, \quad \mathfrak{Re}\,\Gamma_{n\kappa} > 0. \tag{14}$$

Note that (C4) is also applicable for the analysis of the wave-guide properties of the layer, where $\mathfrak{Im}\,\Gamma_{n\kappa} > 0, \mathfrak{Re}\,\Gamma_{n\kappa} = 0$. The desired solution of the scattering and generation problem (13) under the conditions (C1) – (C4) can be represented as follows:

$$\begin{aligned} &E_1(n\kappa; y, z) = U(n\kappa; z)\exp(i\phi_{n\kappa}y) \\ &= \begin{cases} a^{\text{inc}}_{n\kappa}\exp(i(\phi_{n\kappa}y - \Gamma_{n\kappa}(z - 2\pi\delta))) + a^{\text{scat}}_{n\kappa}\exp(i(\phi_{n\kappa}y + \Gamma_{n\kappa}(z - 2\pi\delta))), & z > 2\pi\delta, \\ U(n\kappa; z)\exp(i\phi_{n\kappa}y), & |z| \le 2\pi\delta, \\ b^{\text{scat}}_{n\kappa}\exp(i(\phi_{n\kappa}y - \Gamma_{n\kappa}(z + 2\pi\delta))), & z < -2\pi\delta, \end{cases} \\ &n = 1, 2, 3. \end{aligned} \tag{15}$$

Substituting this representation into the system (13), the following system of nonlinear ordinary differential equations results, where "$'$" denotes the differentiation w.r.t. z:

$$\begin{aligned} &U''(n\kappa; z) + \{\Gamma^2_{n\kappa} - (n\kappa)^2\left[1 - \varepsilon_{n\kappa}(z, \alpha(z), U(\kappa; z), U(2\kappa; z), U(3\kappa; z))\right]\}\,U(n\kappa; z) \\ &= -(n\kappa)^2\alpha(z)\left(\delta_{n1}U^2(2\kappa; z)\overline{U}(3\kappa; z) + \delta_{n3}\left\{\frac{1}{3}U^3(\kappa; z) + U^2(2\kappa; z)\overline{U}(\kappa; z)\right\}\right), \\ &\qquad |z| \le 2\pi\delta, \quad n = 1, 2, 3. \end{aligned} \tag{16}$$

The boundary conditions follow from the continuity of the tangential components of the full fields of scattering and generation $\{\mathbf{E}_{\text{tg}}(n\kappa; y, z)\}_{n=1,2,3}$, $\{\mathbf{H}_{\text{tg}}(n\kappa; y, z)\}_{n=1,2,3}$ at the boundaries $z = \pm 2\pi\delta$ of the nonlinear layer (cf. (C3)). According to (C3) and the representation of the electrical components of the electromagnetic field (15), at the boundary of the nonlinear layer we obtain:

$$\begin{aligned} U(n\kappa; 2\pi\delta) &= a^{\text{scat}}_{n\kappa} + a^{\text{inc}}_{n\kappa}, & U'(n\kappa; 2\pi\delta) &= i\Gamma_{n\kappa}\left(a^{\text{scat}}_{n\kappa} - a^{\text{inc}}_{n\kappa}\right), \\ U(n\kappa; -2\pi\delta) &= b^{\text{scat}}_{n\kappa}, & U'(n\kappa; -2\pi\delta) &= -i\Gamma_{n\kappa}b^{\text{scat}}_{n\kappa}, \quad n = 1, 2, 3. \end{aligned} \tag{17}$$

Eliminating in (17) the unknown values of the complex amplitudes $\{a^{\text{scat}}_{n\kappa}\}_{n=1,2,3}$, $\{b^{\text{scat}}_{n\kappa}\}_{n=1,2,3}$ of the scattered and generated fields and taking into consideration that $a^{\text{inc}}_{n\kappa} = U^{\text{inc}}(n\kappa; 2\pi\delta)$, we arrive at the desired boundary conditions for the problem (16):

$$\begin{aligned} i\Gamma_{n\kappa}U(n\kappa; -2\pi\delta) + U'(n\kappa; -2\pi\delta) &= 0, \\ i\Gamma_{n\kappa}U(n\kappa; 2\pi\delta) - U'(n\kappa; 2\pi\delta) &= 2i\Gamma_{n\kappa}a^{\text{inc}}_{n\kappa}, \quad n = 1, 2, 3. \end{aligned} \tag{18}$$

The system of ordinary differential equations (16) and the boundary conditions (18) form a semi-linear boundary-value problem of Sturm-Liouville type, see also [2, 14, 15, 25]. The existence and uniqueness of a weak solution of the nonlinear boundary-value problem (16) and (18) have been demonstrated in [4, Sect. 4].

The problem (13), (C1) – (C4) can also be reduced to finding solutions of one-dimensional nonlinear integral equations w.r.t. the components $U(n\kappa; z)$, $n = 1,2,3$, $z \in [-2\pi\delta, 2\pi\delta]$, of the fields scattered and generated in the nonlinear layer (cf. [3], [2], [15], [25], [14], [8]):

$$\begin{aligned}
&U(n\kappa; z) + \frac{i(n\kappa)^2}{2\Gamma_{n\kappa}} \int_{-2\pi\delta}^{2\pi\delta} \exp(i\Gamma_{n\kappa}|z - z_0|) \times \\
&\quad \times \left[1 - \varepsilon_{n\kappa}\left(z_0,\ \alpha(z_0), U(\kappa; z_0)\ , U(2\kappa; z_0)\ , U(3\kappa; z_0)\right)\right] U(n\kappa; z_0) dz_0 \\
&= \delta_{n1} \frac{i(n\kappa)^2}{2\Gamma_{n\kappa}} \int_{-2\pi\delta}^{2\pi\delta} \exp(i\Gamma_{n\kappa}|z - z_0|) \alpha(z_0) U^2(2\kappa; z_0) \overline{U}(3\kappa; z_0) dz_0 \\
&+ \delta_{n3} \frac{i(n\kappa)^2}{2\Gamma_{n\kappa}} \int_{-2\pi\delta}^{2\pi\delta} \exp(i\Gamma_{n\kappa}|z - z_0|) \alpha(z_0) \left\{ \frac{1}{3} U^3(\kappa; z_0) + U^2(2\kappa; z_0) \overline{U}(\kappa; z_0) \right\} dz_0 \\
&+ U^{\text{inc}}(n\kappa; z), \quad |z| \le 2\pi\delta, \quad n = 1,2,3.
\end{aligned} \tag{19}$$

Here $U^{\text{inc}}(n\kappa; z) = a^{\text{inc}}_{n\kappa} \exp\left[-i\Gamma_{n\kappa}(z - 2\pi\delta)\right]$, $n = 1,2,3$.

The solution of the original problem (13), (C1) – (C4), represented as (15), can be obtained from (19) using the formulas

$$U(n\kappa; 2\pi\delta) = a^{\text{inc}}_{n\kappa} + a^{\text{scat}}_{n\kappa}, \quad U(n\kappa; -2\pi\delta) = b^{\text{scat}}_{n\kappa}, \quad n = 1,2,3, \tag{20}$$

(cf. (C3)). The system (19) can be regarded as an integral representation of the desired solution of (13), (C1) – (C4) (i.e. solutions of the form $E_1\left(n\kappa; y, z\right) = U(n\kappa; z)\ \exp\left(i\phi_{n\kappa} y\right), n = 1,2,3$, see (15)) for points located outside the nonlinear layer: $\{(y,z) :\ |y| < \infty,\ |z| > 2\pi\delta\}$. Indeed, given the solution of nonlinear integral equations (19) in the region $|z| \le 2\pi\delta$, the substitution into the integrals of (19) leads to explicit expressions of the desired solutions $U(n\kappa; z)$ for points $|z| > 2\pi\delta$ outside the nonlinear layer at each frequency $n\kappa$, $n = 1,2,3$.

In the case of a linear system (19), i.e. if $\alpha \equiv 0$, the problem of existence and uniqueness of solutions has been investigated in [16], [12]. Sufficient conditions for the existence and uniqueness of solutions $U(n\kappa; \cdot) \in L_2(-2\pi\delta, 2\pi\delta)$ to the nonlinear system (19) have been given in [4, Sect. 6].

3. A self-consistent approach to the numerical analysis of the nonlinear integral equations

According to [3], [2], the application of suitable quadrature rules to the system of nonlinear integral equations (19) leads to a system of complex-valued nonlinear algebraic equations of the second kind:

$$\begin{cases}
(\mathbf{I} - \mathbf{B}_\kappa(\mathbf{U}_\kappa, \mathbf{U}_{2\kappa}, \mathbf{U}_{3\kappa}))\mathbf{U}_\kappa = \mathbf{C}_\kappa(\mathbf{U}_{2\kappa}, \mathbf{U}_{3\kappa}) + \mathbf{U}^{\text{inc}}_\kappa, \\
(\mathbf{I} - \mathbf{B}_{2\kappa}(\mathbf{U}_\kappa, \mathbf{U}_{2\kappa}, \mathbf{U}_{3\kappa}))\mathbf{U}_{2\kappa} = \mathbf{U}^{\text{inc}}_{2\kappa}, \\
(\mathbf{I} - \mathbf{B}_{3\kappa}(\mathbf{U}_\kappa, \mathbf{U}_{2\kappa}, \mathbf{U}_{3\kappa}))\mathbf{U}_{3\kappa} = \mathbf{C}_{3\kappa}(\mathbf{U}_\kappa, \mathbf{U}_{2\kappa}) + \mathbf{U}^{\text{inc}}_{3\kappa},
\end{cases} \tag{21}$$

where $\{z_l\}_{l=1}^{N}$ is a discrete set of nodes such that $-2\pi\delta =: z_1 < z_2 < ... < z_l < ... < z_N =: 2\pi\delta$.

$\mathbf{U}_{n\kappa} := \{U_l(n\kappa)\}_{l=1}^N \approx \{U(n\kappa; z_l)\}_{l=1}^N$ denotes the vector of the unknown approximate solution values corresponding to the frequencies $n\kappa$, $n = 1, 2, 3$. The matrices are of the form

$$\mathbf{B}_{n\kappa}(\mathbf{U}_\kappa, \mathbf{U}_{2\kappa}, \mathbf{U}_{3\kappa}) = \{A_m K_{lm}(n\kappa, \mathbf{U}_\kappa, \mathbf{U}_{2\kappa}, \mathbf{U}_{3\kappa})\}_{l,m=1}^N$$

with entries

$$\begin{aligned} K_{lm}(n\kappa, \mathbf{U}_\kappa, \mathbf{U}_{2\kappa}, \mathbf{U}_{3\kappa}) &:= -\frac{i(n\kappa)^2}{2\Gamma_{n\kappa}} \exp\left(i\Gamma_{n\kappa}|z_l - z_m|\right) \Big[1 - \Big\{\varepsilon^{(L)}(z_m) \\ &+ \alpha(z_m)\Big(|U_m(\kappa)|^2 + |U_m(2\kappa)|^2 + |U_m(3\kappa)|^2 \\ &+ \delta_{n1}\, |U_m(\kappa)|\, |U_m(3\kappa)| \exp\{i[-3\arg U_m(\kappa) + \arg U_m(3\kappa)]\} \\ &+ \delta_{n2}\, |U_m(\kappa)|\, |U_m(3\kappa)| \exp\{i[-2\arg U_m(2\kappa) + \arg U_m(\kappa) + \arg U_m(3\kappa)]\}\Big)\Big\}\Big]. \end{aligned} \tag{22}$$

The numbers A_m are the coefficients determined by the quadrature rule, $\mathbf{I} := \{\delta_{lm}\}_{l,m=1}^N$ is the identity matrix, and δ_{lm} is Kronecker's symbol.

The right-hand side of (21) is defined by

$$\mathbf{U}_{n\kappa}^{\text{inc}} := \{a_{n\kappa}^{\text{inc}} \exp[-i\Gamma_{n\kappa}(z_l - 2\pi\delta)]\}_{l=1}^N,$$

$$\mathbf{C}_\kappa(\mathbf{U}_{2\kappa}, \mathbf{U}_{3\kappa}) := \left\{\frac{i\kappa^2}{2\Gamma_\kappa} \sum_{m=1}^N A_m \exp(i\Gamma_\kappa|z_l - z_m|)\alpha(z_m) U_m^2(2\kappa)\overline{U}_m(3\kappa)\right\}_{l=1}^N,$$

$$\mathbf{C}_{3\kappa}(\mathbf{U}_\kappa, \mathbf{U}_{2\kappa}) := \left\{\frac{i(3\kappa)^2}{2\Gamma_{3\kappa}} \sum_{m=1}^N A_m \exp(i\Gamma_{3\kappa}|z_l - z_m|)\alpha(z_m)\left[\frac{1}{3}U_m^3(\kappa) + U_m^2(2\kappa)\overline{U}_m(\kappa)\right]\right\}_{l=1}^N.$$

Given a relative error tolerance $\xi > 0$, the solution of (21) is approximated by means of the following iterative method:

$$\left\{\begin{array}{l} \Big\{\Big[\mathbf{I} - \mathbf{B}_\kappa\Big(\mathbf{U}_\kappa^{(s-1)}, \mathbf{U}_{2\kappa}^{(S_2(q))}, \mathbf{U}_{3\kappa}^{(S_3(q))}\Big)\Big]\mathbf{U}_\kappa^{(s)} \\ \qquad = \mathbf{C}_\kappa\Big(\mathbf{U}_{2\kappa}^{(S_2(q))}, \mathbf{U}_{3\kappa}^{(S_3(q))}\Big) + \mathbf{U}_\kappa^{\text{inc}}\Big\}_{s=1}^{S_1(q):\ \eta_1(S_1(q))<\xi} \\ \Big\{\Big[\mathbf{I} - \mathbf{B}_{2\kappa}\Big(\mathbf{U}_\kappa^{(S_1(q))}, \mathbf{U}_{2\kappa}^{(s-1)}, \mathbf{U}_{3\kappa}^{(S_3(q))}\Big)\Big]\mathbf{U}_{2\kappa}^{(s)} = \mathbf{U}_{2\kappa}^{\text{inc}}\Big\}_{s=1}^{S_2(q):\eta_2(S_2(q))<\xi} \\ \Big\{\Big[\mathbf{I} - \mathbf{B}_{3\kappa}\Big(\mathbf{U}_\kappa^{(S_1(q))}, \mathbf{U}_{2\kappa}^{(S_2(q))}, \mathbf{U}_{3\kappa}^{(s-1)}\Big)\Big]\mathbf{U}_{3\kappa}^{(s)} \\ \qquad = \mathbf{C}_{3\kappa}\Big(\mathbf{U}_\kappa^{(S_1(q))}, \mathbf{U}_{2\kappa}^{(S_2(q))}\Big) + \mathbf{U}_{3\kappa}^{\text{inc}}\Big\}_{s=1}^{S_3(q):\eta_3(S_3(q))<\xi} \end{array}\right\}_{q=1}^{Q} \tag{23}$$

with $\eta_n(s) := \|\mathbf{U}_{n\kappa}^{(s)} - \mathbf{U}_{n\kappa}^{(s-1)}\| / \|\mathbf{U}_{n\kappa}^{(s)}\|$, $n = 1, 2, 3$, where the terminating index $Q \in \mathbb{N}$ is defined by the requirement

$$\max\{\eta_1(Q), \eta_2(Q), \eta_3(Q)\} < \xi.$$

4. Eigen-modes of the linearised problems of scattering and generation of waves on the cubically polarisable layer

The solution of the system of nonlinear equations (19) is approximated by the solution of a linearised system, for given values of the induced dielectric permittivity and of the source functions at the right-hand side of the system. The solution can be found by the help of algorithm (23), where at each step a system of linearised nonlinear complex-valued algebraic equations of the second kind is solved iteratively. The analytic continuation of the linearised nonlinear problems into the region of complex values of the frequency parameter allows us to switch to the analysis of spectral problems. That is, the eigen-frequencies and the corresponding eigen-fields of the homogeneous linear problems with an induced nonlinear permittivity are to be determined. The results of the development of a spectral theory of linear problems for structures with noncompact boundaries can be found in [23], [13], [16], [12], [18], [17].

As mentioned above, the classical formulation of the problem of scattering and generation of waves, described by the *system* of boundary value problems (13), (C1) – (C4), can be reformulated as a *set* of independent spectral problems in the following way:

Find the eigen-frequencies κ_n and the corresponding eigen-functions $E_1(\mathbf{r},\kappa_n)$ (i.e. $\{\kappa_n \in \Omega_{n\kappa} \subset \mathsf{H}_{n\kappa}, E_1(\mathbf{r},\kappa_n)\}_{n=1}^{3}$, where $\Omega_{n\kappa}$ are the discrete sets of eigen-frequencies lying on the two-sheeted Riemann surfaces $\mathsf{H}_{n\kappa}$, see Fig. 2 and the more detailed explanations below) satisfying the equations

$$\nabla^2 E_1(\mathbf{r},\kappa_n) + \kappa_n^2 \varepsilon_{n\kappa}\left(z,\alpha(z),E_1(r,\kappa),E_1(\mathbf{r},2\kappa),E_1(\mathbf{r},3\kappa)\right) E_1(\mathbf{r},\kappa_n) = 0, \quad n = 1,2,3, \qquad (24)$$

together with the following conditions:

(CS1) $E_1(\kappa_n;y,z) = U(\kappa_n;z)\exp(i\phi_{n\kappa}y), \quad n = 1,2,3$
(the quasi-homogeneity condition w.r.t. the spatial variable y),

(CS2) $\phi_{n\kappa} = n\phi_\kappa, \quad n = 1,2,3$
(the condition of phase synchronism of waves),

(CS3) $\mathbf{E}_{\mathrm{tg}}(\kappa_n;y,z)$ and $\mathbf{H}_{\mathrm{tg}}(\kappa_n;y,z)$ (i.e. $E_1(\kappa_n;y,z)$ and $H_2(\kappa_n;y,z)$) are continuous at the boundary layers of the structure with the *induced* permittivity $\varepsilon_{n\kappa}$ for $\kappa := \kappa^{\mathrm{inc}}, \quad n = 1,2,3,$

(CS4) $E_1(\kappa_n;y,z) = \begin{Bmatrix} a_{\kappa_n} \\ b_{\kappa_n} \end{Bmatrix} \exp\left(i\left(\phi_{n\kappa}y \pm \Gamma_{\kappa_n}(\kappa_n,\phi_{n\kappa})(z \mp 2\pi\delta)\right)\right), \quad z \gtrless \pm 2\pi\delta, \quad n = 1,2,3$
(the radiation condition w.r.t. the eigen-field).

For real values of the parameters κ_n and $\phi_{n\kappa}$, the condition (CS4) meets the physically reasonable requirement of the absence of radiation fields of waves coming from infinity $z = \pm\infty$:

$$\mathfrak{Im}\,\Gamma_{\kappa_n}(\kappa_n,\phi_{n\kappa}) \geq 0, \; \mathfrak{Re}\,\Gamma_{\kappa_n}(\kappa_n,\phi_{n\kappa})\,\mathfrak{Re}\,\kappa_n \geq 0 \quad \text{for} \quad \mathfrak{Im}\,\phi_{n\kappa} = 0, \; \mathfrak{Im}\,\kappa_n = 0, \quad n = 1,2,3. \qquad (25)$$

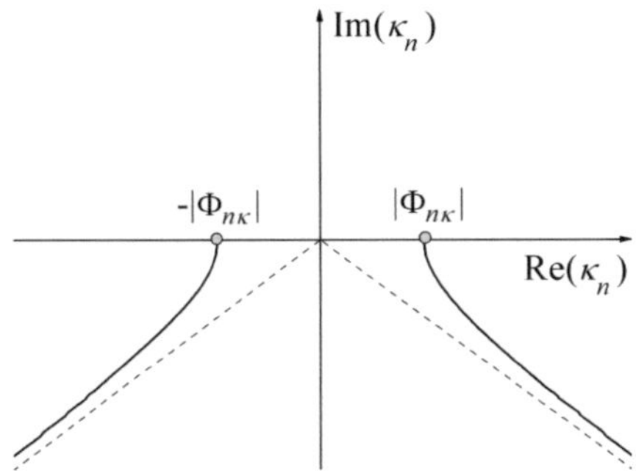

Figure 2. The geometry of the two-sheeted Riemann surfaces $\mathrm{H}_{n\kappa}$

The nontrivial solutions (eigen-fields) of problem (24), (CS1) – (CS4) can be represented as

$$E_1(\kappa_n; y, z) = U(\kappa_n; z)\exp(i\phi_{n\kappa}y) = \begin{cases} a_{\kappa_n}\exp(i(\phi_{n\kappa}y + \Gamma_{\kappa_n}(\kappa_n, \phi_{n\kappa})(z - 2\pi\delta))), & z > 2\pi\delta, \\ U(\kappa_n; z)\exp(i\phi_{n\kappa}y), & |z| \leq 2\pi\delta, \\ b_{\kappa_n}\exp(i(\phi_{n\kappa}y - \Gamma_{\kappa_n}(\kappa_n, \phi_{n\kappa})(z + 2\pi\delta))), & z < -2\pi\delta, \end{cases}$$
$$\kappa_n \in \Omega_{n\kappa} \subset \mathrm{H}_{n\kappa}, \quad n = 1, 2, 3, \qquad (26)$$

where $\kappa := \kappa^{\mathrm{inc}}$ is a given constant equal to the value of the excitation frequency of the nonlinear structure, $\Gamma_{\kappa_n}(\kappa_n, \phi_{n\kappa}) := (\kappa_n^2 - \phi_{n\kappa}^2)^{1/2}$ are the transverse propagation *functions* depending on the complex frequency spectral variables κ_n, $\phi_{n\kappa} := n\kappa\sin(\varphi_{n\kappa})$ denote the given real values of the longitudinal propagation constants, $\varepsilon_{n\kappa} = \varepsilon_{n\kappa}(z, \alpha(z), E_1(\mathbf{r}, \kappa), E_1(\mathbf{r}, 2\kappa), E_1(\mathbf{r}, 3\kappa))$ are the induced dielectric permittivities at the frequencies $n\kappa$ corresponding to the parameter $\kappa := \kappa^{\mathrm{inc}}$, $\Omega_{n\kappa}$ are the sets of eigen-frequencies and $\mathrm{H}_{n\kappa}$ are two-sheeted Riemann surfaces (cf. Fig. 2), $n = 1, 2, 3$. The range of the spectral parameters $\kappa_n \in \Omega_{n\kappa}$ is completely determined by the boundaries of those regions in which the analytic continuation (consistent with the condition (25)) of the canonical Green's functions

$$G_0(\kappa_n; q, q_0) = \frac{i}{4Y}\exp\{i[\phi_{n\kappa}(y - y_0) + \Gamma_{\kappa_n}(\kappa_n, \phi_{n\kappa})|z - z_0|]\}/\Gamma_{\kappa_n}(\kappa_n, \phi_{n\kappa}), \quad n = 1, 2, 3,$$

into the complex space of the spectral parameters κ_n of the unperturbed problems (24), (CS1) – (CS4) (i.e. for the special case $\varepsilon_{n\kappa} \equiv 1$, $n = 1, 2, 3$) is possible. These complex spaces are two-sheeted Riemann surfaces $\mathrm{H}_{n\kappa}$ (see Fig. 2) with real algebraic branch points of second order $\kappa_n^{\pm} : \Gamma_{\kappa_n}(\kappa_n^{\pm}, \phi_{n\kappa}) = 0$ (i.e. $\kappa_n^{\pm} = \pm|\phi_{n\kappa}|$, $n = 1, 2, 3$) and with cuts starting at these points and extending along the lines

$$(\mathfrak{Re}\,\kappa_n)^2 - (\mathfrak{Im}\,\kappa_n)^2 - \phi_{n\kappa}^2 = 0, \quad \mathfrak{Im}\,\kappa_n \leq 0, \; n = 1, 2, 3. \qquad (27)$$

The first, "physical" sheets (i.e. the pair of values $\{\kappa_n, \Gamma_{\kappa_n}(\kappa_n, \phi_{n\kappa})\}$) on each of the surfaces $\mathrm{H}_{n\kappa}$, $n = 1, 2, 3$, are completely determined by the condition (25) and the cuts (27). At the first sheets of $\mathrm{H}_{n\kappa}$ the signs of the pairs $\{\kappa_n, \mathfrak{Re}\,\Gamma_{\kappa_n}\}$ and $\{\kappa_n, \mathfrak{Im}\,\Gamma_{\kappa_n}\}$ are distributed as follows: $\mathfrak{Im}\,\Gamma_{\kappa_n} > 0$ for $0 < \arg\kappa_n < \pi$, $\mathfrak{Re}\,\Gamma_{\kappa_n} \geq 0$ for $0 < \arg\kappa_n < \pi/2$ and $\mathfrak{Re}\,\Gamma_{\kappa_n} \leq 0$ for

$\pi/2 \leq \arg\kappa_n < \pi$. For points κ_n with $3\pi/2 \leq \arg\kappa_n \leq 2\pi$, the function values (where $(\mathfrak{Re}\,\kappa_n)^2 - (\mathfrak{Im}\,\kappa_n)^2 - \phi_{n\kappa^2} > 0$) are determined by the condition $\mathfrak{Im}\,\Gamma_{\kappa_n} < 0$, $\mathfrak{Re}\,\Gamma_{\kappa_n} > 0$, for the remaining points κ_n the function $\Gamma_{\kappa_n}(\kappa_n, \phi_{n\kappa})$ is determined by the condition $\mathfrak{Im}\,\Gamma_{\kappa_n} > 0$, $\mathfrak{Re}\,\Gamma_{\kappa_n} \leq 0$. In the region $\pi < \arg\kappa_n < 3\pi/2$ the situation is similar to the previous one up to the change of the sign of $\mathfrak{Re}\,\Gamma_{\kappa_n}$. The second, "unphysical" sheets of the surfaces $\mathsf{H}_{n\kappa}$, $n = 1, 2, 3$ are different from the "physical" ones in that, for each κ_n, the signs of both $\mathfrak{Re}\,\Gamma_{\kappa_n}$ and $\mathfrak{Im}\,\Gamma_{\kappa_n}$ are reversed.

The qualitative analysis of the eigen-modes of the linearised problems (24), (CS1) – (CS4) is carried out using the equivalent formulation of spectral problems for the linearised nonlinear integral equations (19). It is based on the analytic continuation of (19) into the space of spectral values $\kappa_n \in \Omega_{n\kappa} \subset \mathsf{H}_{n\kappa}$, $n = 1, 2, 3$.

The spectral problem reduces to finding nontrivial solutions $U(\kappa_n; z)$ of a set of homogeneous (i.e. with vanishing right-hand sides), linear (i.e. linearised equations (19)) integral equations with the induced dielectric permittivity at the frequencies $n\kappa$ of excitation and generation:

$$\begin{array}{l} U(\kappa_n; z) + \dfrac{i\kappa_n^2}{2\Gamma_{\kappa_n}(\kappa_n, \phi_{n\kappa})} \displaystyle\int_{-2\pi\delta}^{2\pi\delta} \exp(i\Gamma_{\kappa_n}(\kappa_n, \phi_{n\kappa})|z - z_0|) \\ \times\ [1 - \varepsilon_{n\kappa}(z_0, \alpha(z_0), U(\kappa; z_0), U(2\kappa; z_0), U(3\kappa; z_0))]\, U(\kappa_n; z_0) dz_0 = 0; \\ \qquad |z| \leq 2\pi\delta, \quad \kappa := \kappa^{\text{inc}}, \quad \kappa_n \in \Omega_{n\kappa} \subset \mathsf{H}_{n\kappa}, \quad n = 1, 2, 3. \end{array} \tag{28}$$

The solution of the problem (24), (CS1) – (CS4) can be obtained from the solution of the equivalent problem (28), where – according to (CS3) – in the representation of the eigen-fields (26) the following formulas are used:

$$U(\kappa_n; 2\pi\delta) = a_{\kappa_n}, \quad U(\kappa_n; -2\pi\delta) = b_{\kappa_n}, \quad n = 1, 2, 3. \tag{29}$$

The qualitative analysis of the spectral characteristics allows to develop algorithms for solving the spectral problems (24), (CS1) – (CS4) by reducing them to the equivalent spectral problem of finding nontrivial solutions of the integral equations (28), see [13], [23]. The solvability of (28) follows from an analysis of the basic qualitative characteristics of the spectra. Applying to the integral equations (28) appropriate quadrature formulas, we obtain a set of independent systems of linear algebraic equations of second kind depending nonlinearly on the spectral parameter: $(\mathbf{I} - \mathbf{B}_{n\kappa}(\kappa_n))\mathbf{U}_{\kappa_n} = \mathbf{0}$, where $\kappa_n \in \mathsf{H}_{n\kappa}$, $\kappa := \kappa^{\text{inc}}$, $n = 1, 2, 3$. Consequently, the spectral problem of finding the eigen-frequencies $\kappa_n \in \Omega_{n\kappa} \subset \mathsf{H}_{n\kappa}$ and the corresponding eigen-fields (i.e. the nontrivial solutions of the integral equations (28)) reduces to the following algorithm:

$$\begin{cases} f_{n\kappa}(\kappa_n) := \det(\mathbf{I} - \mathbf{B}_{n\kappa}(\kappa_n)) = 0, \\ \quad (\mathbf{I} - \mathbf{B}_{n\kappa}(\kappa_n))\mathbf{U}_{\kappa_n} \qquad = \mathbf{0}, \\ \qquad\qquad \kappa := \kappa^{\text{inc}}, \ \kappa_n \in \Omega_{n\kappa} \subset \mathsf{H}_{n\kappa}, \quad n = 1, 2, 3. \end{cases} \tag{30}$$

Here we use a similar notation to that in Section 3. κ_n are the desired eigen-frequencies, and $\mathbf{U}_{\kappa_n} = \{U(\kappa_n; z_l)\}_{l=1}^{N} := \{U_l(\kappa_n)\}_{l=1}^{N}$ are the vectors of the unknown approximate solution values corresponding to the frequencies κ_n. The matrices are of the form

$$\mathbf{B}_{n\kappa}(\kappa_n) := \mathbf{B}_{n\kappa}(\kappa_n; \mathbf{U}_\kappa, \mathbf{U}_{2\kappa}, \mathbf{U}_{3\kappa}) = \{A_m K_{lm}(\kappa_n, \mathbf{U}_\kappa, \mathbf{U}_{2\kappa}, \mathbf{U}_{3\kappa})\}_{l,m=1}^{N} \tag{31}$$

with *given* values of the vectors of the scattered and generated fields $\mathbf{U}_{n\kappa} = \{U(n\kappa; z_l)\}_{l=1}^{N} := \{U_l(n\kappa)\}_{l=1}^{N}$, $n = 1,2,3$. The numbers A_m are the coefficients determined by the quadrature rule, and the entries $K_{lm}(\kappa_n, \mathbf{U}_\kappa, \mathbf{U}_{2\kappa}, \mathbf{U}_{3\kappa})$ are calculated by means of (22), where the first argument $n\kappa$ is replaced by κ_n. The eigen-frequencies $\kappa_n \in \Omega_{n\kappa} \subset \mathrm{H}_{n\kappa}$, $n = 1,2,3$, i.e. the characteristic numbers of the dispersion equations of the problem (30), are obtained by solving the corresponding dispersion equations $f_{n\kappa}(\kappa_n) := \det(\mathbf{I} - \mathbf{B}_{n\kappa}(\kappa_n)) = 0$ by the help of Newton's method or its modifications. The nontrivial solutions $\mathbf{U}_{\kappa_n}$ of the homogeneous systems of linear algebraic equations (30) corresponding to these characteristic numbers are the eigen-fields (26) of the linearised nonlinear layered structures with an induced dielectric constant. Since the solutions $\mathbf{U}_{\kappa_n}$ are unique up to multiplication by an arbitrary constant, we require $U(\kappa_n; 2\pi\delta) = a_{\kappa_n} := 1$ (cf. (26)). According to (31), the matrix entries in (30) depend on the dielectric permittivities. The latter are defined by the scattered and generated fields $\mathbf{U}_\kappa, \mathbf{U}_{2\kappa}, \mathbf{U}_{3\kappa}$ of the problem (19) by means of the algorithm (23). This defines the basic design of the implemented numerical algorithm. The investigation of the eigen-modes of the linearised nonlinear structures (30) should always precede the solution of the nonlinear scattering and generation problem in the self-consistent formulation (23). Note that, in the analysis of the linear structures, the problem of excitation (scattering) and the spectral problem can be solved independently.

5. Numerical examples

Consider the excitation of the nonlinear structure by a strong incident field at the basic frequency κ and, in addition, by weak incident quasi-homogeneous electromagnetic fields at the double and triple frequencies 2κ, 3κ (see (12)), i.e.

$$0 < \max\{|a_{2\kappa}^{\mathrm{inc}}|, |a_{3\kappa}^{\mathrm{inc}}|\} \ll |a_{1\kappa}^{\mathrm{inc}}|. \tag{32}$$

The desired solution of the scattering and generation problem (13), (C1) – (C4) (or of the equivalent problems (16) and (19)) are represented as in (15). According to (17) we determine the values of complex amplitudes $\{a_{n\kappa}^{\mathrm{scat}}, b_{n\kappa}^{\mathrm{scat}} : n = 1,2,3\}$ in (15) for the scattered and generated fields by means of the formulas (20). The solution of (21) is obtained by means of successive approximations using the self-consistent approach based on the iterative algorithm (23).

In order to describe the scattering and generation properties of the nonlinear structure in the zones of reflection $z > 2\pi\delta$ and transmission $z < -2\pi\delta$, we introduce the following notation:

$$R_{n\kappa} := |a_{n\kappa}^{\mathrm{scat}}|^2 \Big/ \sum_{n=1}^{3} |a_{n\kappa}^{\mathrm{inc}}|^2 \quad \text{and} \quad T_{n\kappa} := |b_{n\kappa}^{\mathrm{scat}}|^2 \Big/ \sum_{n=1}^{3} |a_{n\kappa}^{\mathrm{inc}}|^2, \quad n = 1,2,3.$$

The quantities $R_{n\kappa}$, $T_{n\kappa}$ are called *reflection, transmission* or *generation coefficients* of the waves w.r.t. the total intensity of the incident packet.

We note that, for nonabsorbing media with $\mathfrak{Im}\left[\varepsilon^{(L)}(z)\right]=0$, the energy balance equation

$$\sum_{n=1}^{3}\left[R_{n\kappa}+T_{n\kappa}\right]=1 \tag{33}$$

is satisfied. This equation generalises the law of conservation of energy which has been treated in [12], [21] for the case of a single incident field and a single equation. If we define by

$$W_{n\kappa}:=|a_{n\kappa}^{\text{scat}}|^2+|b_{n\kappa}^{\text{scat}}|^2 \tag{34}$$

the total energy of the scattered and generated fields at the frequencies $n\kappa$, $n=1,2,3$, then the energy balance equation (33) can be rewritten as

$$\sum_{n=1}^{3}W_{n\kappa}=\sum_{n=1}^{3}|a_{n\kappa}^{\text{inc}}|^2\,.$$

In the numerical experiments, the quantities $W_{3\kappa}/W_{\kappa}$ (which characterises the portion of energy generated in the third harmonic in comparison to the energy scattered in the nonlinear layer) and

$$W^{(\text{Error})}:=1-\sum_{n=1}^{3}\left[R_{n\kappa}+T_{n\kappa}\right] \tag{35}$$

(which characterises the numerical violation of the energy balance) are of particular interest. We emphasize that in the numerical simulation of scattering and generation processes without any weak fields, i.e. $a_{2\kappa}^{\text{inc}}=a_{3\kappa}^{\text{inc}}=0$, the residual of the energy balance equation (33) does not exceed the value $|W^{(\text{Error})}|<10^{-8}$. However, taking into consideration the impact of weak fields in the numerical simulation of the same scattering and generation processes, i.e. $a_{n\kappa}^{\text{inc}}\neq 0$, $n=2,3$, the error in the balance equation (33) can reach up to several percent. This indicates that the intensities of the exciting weak fields are sufficiently large such that these fields become also sources for the generation of oscillations. For such situations, the presented mathematical model (13), (C1) – (C4) and (24), (CS1) – (CS4) should take into account the complex Fourier amplitudes of the oscillations at the frequencies $n\kappa$ for numbers $n>3$. Furthermore we observe, on the one hand, situations in which the influence of a weak field $a_{2\kappa}^{\text{inc}}\neq 0$ on the scattering and generation process of oscillations leads to small errors in the energy balance equation (33) not exceeding 2% (that is $|W^{(\text{Error})}|<0.02$), and, on the other hand, situations in which the error can reach 10% (that is $|W^{(\text{Error})}|<0.1$). The scattering, generating, energetic, and dielectric properties of the nonlinear layer are illustrated by surfaces in dependence on the parameters of the particular problem. The bottom chart depicts the surface of the value of the residual $W^{(\text{Error})}$ of the energy balance equation (see (35)) and its projection onto the top horizontal plane of the figure. In particular, by the help of these graphs it is easy to localise that region of parameters of the problem, where the error of the energy balance does not exceed a given value, that is $|W^{(\text{Error})}|<\text{const}$.

The spectral characteristics of the linearised nonlinear problems (24), (CS1) – (CS4) with the induced dielectric permittivity (11) at the frequencies $n\kappa$, $n=1,2,3$, of excitation and generation were calculated by means of the algorithm (30). In the graphical illustration of the eigen-fields $\mathbf{U}_{\kappa_n}$ in the representation (26) we have set $a_{\kappa_n} := 1$ for $\kappa_n \in \Omega_{n\kappa} \subset \mathsf{H}_{n\kappa}$, $n = 1,2,3$. Finally we mention that the later-used classification of scattered, generated or eigen-fields of the dielectric layer by the $H_{m,l,p}$-type is identical to that given in [12], [13], [23]. In the case of E-polarisation, see (3), $H_{m,l,p}$ (or $TE_{m,l,p}$) denotes the type of polarisation of the wave field under investigation. The subscripts indicate the number of local maxima of $|E_1|$ (or $|U|$, as $|E_1| = |U|$, see (15), (26)) along the coordinate axes x, y and z (see Fig. 1). Since the considered waves are homogeneous along the x-axis and quasi-homogeneous along the y-axis, we study actually fields of the type $H_{0,0,p}$ (or $TE_{0,0,p}$), where the subscript p is equal to the number of local maxima of the function $|U|$ of the argument $z \in [-2\pi\delta, 2\pi\delta]$.

In what follows we present and discuss results of the numerical analysis of scattering and generation properties as well as the eigen-modes of the dielectric layer with a positive value of the cubic susceptibility of the medium. In more detail, we consider a single-layered structure with a dielectric permittivity $\varepsilon_{n\kappa}(z, \alpha(z), U(\kappa;z), U(2\kappa;z), U(3\kappa;z)) = \varepsilon^{(L)}(z) + \varepsilon^{(NL)}_{n\kappa}$, $n = 1,2,3$, where $\varepsilon^{(L)}(z) := 16$ and $\alpha(z) := 0.01$ for $z \in [-2\pi\delta, 2\pi\delta]$, $\varepsilon^{(NL)}_{n\kappa}$ is given by (11), $\delta := 0.5$, $\kappa^{\text{inc}} := \kappa := 0.375$, and $\varphi_\kappa \in [0°, 90°)$. The Figs. 3 – Fig. 13 illustrate the following cases of the incident fields:

$$a^{\text{inc}}_{2\kappa} = \tfrac{1}{3}a^{\text{inc}}_{\kappa},\ a^{\text{inc}}_{3\kappa} = 0 \quad \ldots \quad \text{graphs labeled by “1/3”},$$
$$a^{\text{inc}}_{2\kappa} = \tfrac{2}{3}a^{\text{inc}}_{\kappa},\ a^{\text{inc}}_{3\kappa} = 0 \quad \ldots \quad \text{graphs labeled by “2/3”},$$
$$a^{\text{inc}}_{2\kappa} = a^{\text{inc}}_{3\kappa} = 0 \quad \ldots \quad \text{graphs labeled by “0”}.$$

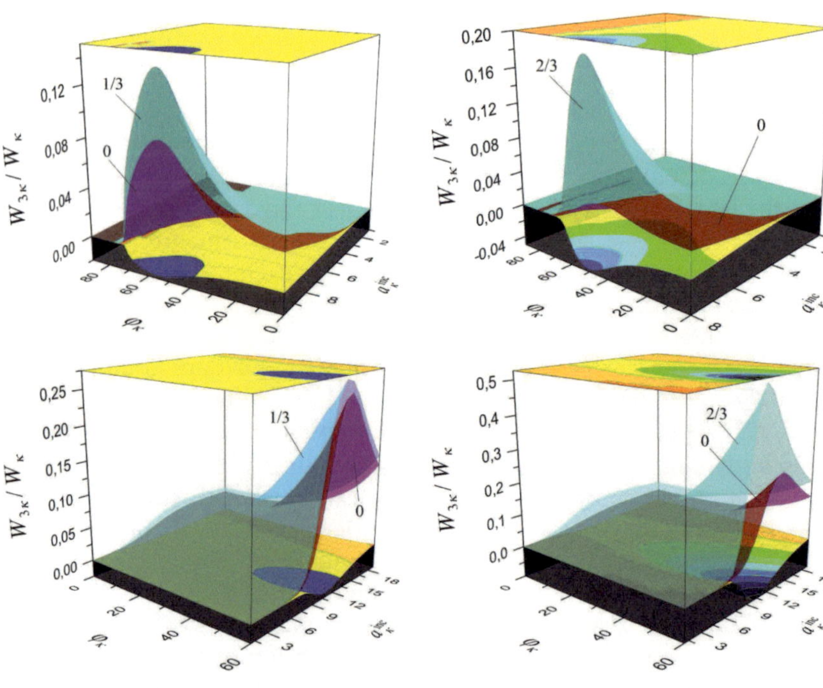

Figure 3. The portion of energy generated in the third harmonic: $a^{\text{inc}}_{2\kappa} = \frac{1}{3}a^{\text{inc}}_{\kappa}$ (left), $a^{\text{inc}}_{2\kappa} = \frac{2}{3}a^{\text{inc}}_{\kappa}$ (right)

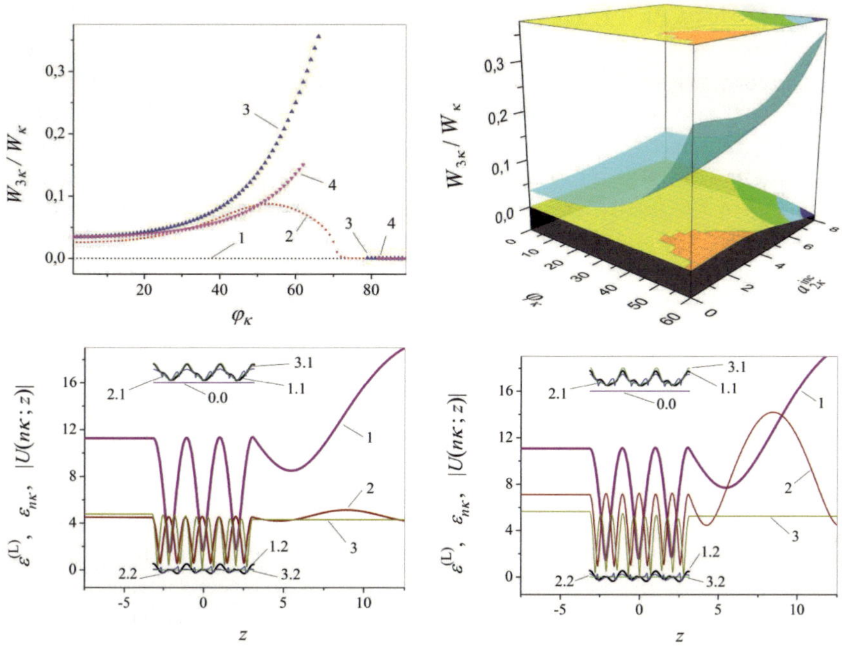

Figure 4. The portion of energy generated in the third harmonic: #1 ... $a_\kappa^{\text{inc}} = 1$, #2 ... $a_\kappa^{\text{inc}} = 9.93$, #3 ... $a_\kappa^{\text{inc}} = 14$, #4 ... $a_\kappa^{\text{inc}} = 19$ for $a_{2\kappa}^{\text{inc}} = 0$ (top left), the dependence of $W_{3\kappa}/W_\kappa$ on φ_κ, $a_{2\kappa}^{\text{inc}}$ for $a_\kappa^{\text{inc}} = 14$ (top right), some graphs describing the properties of the non-linear layer for $\varphi_\kappa = 60°$, $a_\kappa^{\text{inc}} = 14$ and $a_{2\kappa}^{\text{inc}} = \frac{1}{3}a_\kappa^{\text{inc}}$ (bottom left), $a_{2\kappa}^{\text{inc}} = \frac{2}{3}a_\kappa^{\text{inc}}$ (bottom right): #0.0 ... $\varepsilon^{(L)}$, #1 ... $|U(\kappa;z)|$, #2 ... $|U(2\kappa;z)|$, #3 ... $|U(3\kappa;z)|$, # n.1 ... $\Re e(\varepsilon_{n\kappa})$, # n.2 ... $\Im m(\varepsilon_{n\kappa})$

The results shown in Fig. 3 allow us to track the dynamic behaviour of the quantity $W_{3\kappa}/W_\kappa$ characterising the ratio of the generated and scattered energies. Fig. 3 shows the dependence of $W_{3\kappa}/W_\kappa$ on the angle of incidence φ_κ and on the amplitude a_κ^{inc} of the incident field for different relations between $a_{2\kappa}^{\text{inc}}$ and a_κ^{inc}. It describes the portion of energy generated in the third harmonic by the nonlinear layer when a plane wave at the excitation frequency κ and with the amplitude a_κ^{inc} is passing the layer under the angle of incidence φ_κ. It can be seen that the weaker incident field at the frequency 2κ leads to an increase of $W_{3\kappa}/W_\kappa$ in comparison with the situation where the structure is excited only by a single field at the basic frequency κ. For example, in Fig. 3 the maximum value of $W_{3\kappa}/W_\kappa$ and the value $W^{(\text{Error})}$ are reached at the following parameters $[a_\kappa^{\text{inc}}, a_{2\kappa}^{\text{inc}}, \varphi_\kappa]$: $W_{3\kappa}/W_\kappa = 0.08753$, $W^{(\text{Error})} = -1.98292 \cdot 10^{-9}$, $[a_\kappa^{\text{inc}} = 9.93, a_{2\kappa}^{\text{inc}} = 0, \varphi_\kappa = 53°]$ ... graph #0 and, taking into consideration the weak field at the double frequency, $W_{3\kappa}/W_\kappa = 0.13903$, $W^{(\text{Error})} = -0.01692$, $[a_\kappa^{\text{inc}} = 9.93, a_{2\kappa}^{\text{inc}} = \frac{1}{3}a_\kappa^{\text{inc}}, \varphi_\kappa = 53°]$ ... graph #1/3 (top left); $W_{3\kappa}/W_\kappa = 0.03265$, $W^{(\text{Error})} = -8.53239 \cdot 10^{-9}$, $[a_\kappa^{\text{inc}} = 8, a_{2\kappa}^{\text{inc}} = 0, \varphi_\kappa = 42°]$ ... graph #0 and, taking into consideration the weak field at the double frequency, $W_{3\kappa}/W_\kappa = 0.1864$, $W^{(\text{Error})} = -0.04625$, $[a_\kappa^{\text{inc}} = 8, a_{2\kappa}^{\text{inc}} = \frac{2}{3}a_\kappa^{\text{inc}}, \varphi_\kappa = 50°]$ ... graph #2/3 (top right); $W_{3\kappa}/W_\kappa = 0.25054$, $W^{(\text{Error})} = -9.29243 \cdot 10^{-10}$, $[a_\kappa^{\text{inc}} = 14, a_{2\kappa}^{\text{inc}} = 0, \varphi_\kappa = 60°]$ ... graph #0 (bottom left and right) and, taking into consideration the weak field at the double frequency, $W_{3\kappa}/W_\kappa = 0.26589$, $W^{(\text{Error})} = -0.00578$, $[a_\kappa^{\text{inc}} = 14, a_{2\kappa}^{\text{inc}} = \frac{1}{3}a_\kappa^{\text{inc}}, \varphi_\kappa = 60°]$ ... graph #1/3 (bottom left); $W_{3\kappa}/W_\kappa = 0.42662$, $W^{(\text{Error})} = -0.06037$ $[a_\kappa^{\text{inc}} = 13, a_{2\kappa}^{\text{inc}} = \frac{2}{3}a_\kappa^{\text{inc}}, \varphi_\kappa = 60°]$ ... graph #2/3 (bottom right).

The numerical analysis of the processes displayed in Fig. 4 (top left) by the curves #3 in the range of angles $\varphi_\kappa \in (66^\circ, 79^\circ)$ and #4 in the range of angles $\varphi_\kappa \in (62^\circ, 82^\circ)$ did not lead to the convergence of the computational algorithm. The value $W_{3\kappa}/W_\kappa = 0.3558$ for $a_\kappa^{\text{inc}} = 14$, $a_{2\kappa}^{\text{inc}} = 0$ and $\varphi_\kappa = 66^\circ$ (see the graph #3 in Fig. 4 (top left)) indicates that $W_{3\kappa}$ is 35.58% of W_κ. This is the maximal value of $W_{3\kappa}/W_\kappa$ that has been achieved in the case of a single incident field at the basic frequency κ. Fig. 4 (top right) illustrates the portion of energy generated in the third harmonic in dependence on the angle of incidence φ_κ and on the amplitude $a_{2\kappa}^{\text{inc}}$ of the incident field at the double frequency. Here the maximum value of $W_{3\kappa}/W_\kappa$ is reached at $[a_\kappa^{\text{inc}} = 14, a_{2\kappa}^{\text{inc}} = 8, \varphi_\kappa = 60^\circ]$. If the structure is excited by a single field at the basic frequency κ only, then the portion of energy generated in the third harmonic is $\approx$ 25.05%, i.e. for $[a_\kappa^{\text{inc}} = 14, a_{2\kappa}^{\text{inc}} = 0, \varphi_\kappa = 60^\circ]$ we have $W_{3\kappa}/W_\kappa = 0.25054$ and $W^{(\text{Error})} = -9.29243 \cdot 10^{-10}$. If we take into consideration the weak exciting field at the double frequency, then we get for $[a_\kappa^{\text{inc}} = 14, a_{2\kappa}^{\text{inc}} = 8, \varphi_\kappa = 60^\circ]$ the following results: $W_{3\kappa}/W_\kappa = 0.35084$ and $W^{(\text{Error})} = -0.0377$ (which corresponds to a relative error of 3.78% in the energy balance).

The bottom diagrams in Fig. 4 display some graphs characterising the scattering and generation properties of the nonlinear structure. Graph #0.0 illustrates the value of the linear part $\varepsilon^{(L)} = 16$ of the permittivity of the nonlinear layered structure. Graphs #n.1 and #n.2 show the real and imaginary parts of the permittivities at the frequencies $n\kappa$, $n = 1, 2, 3$. The figure also shows the absolute values $|U(\kappa; z)|$, $|U(2\kappa; z)|$ of the amplitudes of the full scattered fields at the frequencies of excitation κ, 2κ (graphs #1, #2) and $|U(3\kappa; z)|$ of the generated field at the frequency 3κ (graph #3). The values $|U(n\kappa; z)|$ are given in the nonlinear layered structure ($|z| \leq 2\pi\delta$) and outside it (i.e. in the zones of reflection $z > 2\pi\delta$ and transmission $z < -2\pi\delta$). Here $W^{(\text{Error})} = -5.782328 \cdot 10^{-3}$, i.e. the error in the energy balance is less than 0.58% (bottom left) and $W^{(\text{Error})} = -4.567534 \cdot 10^{-2}$, i.e. the error in the energy balance is less than 4.57% (bottom right).

Figs. 5 and 6 show the numerical results obtained for the scattered and the generated fields in the nonlinear structure and for the residual $W^{(\text{Error})}$ of the energy balance equation (33) for an incident angle $\varphi_\kappa = 60^\circ$ in dependence on the amplitudes a_κ^{inc} and $a_{2\kappa}^{\text{inc}}$ of the plane incident waves at the basic frequency κ and at the double frequency 2κ, resp. The figures show the graphs of $\left|U_{n\kappa}\left[a_\kappa^{\text{inc}}, a_{2\kappa}^{\text{inc}}, z\right]\right|$, $n = 1, 2, 3$, demonstrating the dynamic behaviour of the scattered and the generated fields $|U(n\kappa; z)|$ in the nonlinear layered structure in dependence on increasing amplitudes a_κ^{inc} and $a_{2\kappa}^{\text{inc}}$ for an incident angle $\varphi_\kappa = 60^\circ$ of the plane waves. We mention that, in the range $a_\kappa^{\text{inc}} \in (0, 22]$ and $a_{2\kappa}^{\text{inc}} = \frac{1}{3} a_\kappa^{\text{inc}}$ (see Fig. 5) or $a_{2\kappa}^{\text{inc}} = \frac{2}{3} a_\kappa^{\text{inc}}$ (see Fig. 6) of the amplitudes of the incident fields and for an incident angle $\varphi_\kappa = 60^\circ$ of the plane waves, the scattered field has the type $H_{0,0,4}$ at the frequency κ and $H_{0,0,7}$ at the frequency 2κ. The generated field, observed in the range $a_\kappa^{\text{inc}} \in [5, 22]$, is of the type $H_{0,0,10}$ at the frequency 3κ, see Figs. 5, 6 (bottom left).

The nonlinear parts $\varepsilon_{n\kappa}^{(NL)}$ of the dielectric permittivity at each frequency $n\kappa$ depend on the values $U_{n\kappa} := U(n\kappa; z)$, $n = 1, 2, 3$, of the fields. The variation of the nonlinear parts $\varepsilon_{n\kappa}^{(NL)}$ of the dielectric permittivity for increasing amplitudes a_κ^{inc} and $a_{2\kappa}^{\text{inc}}$ of the incident fields are illustrated by the behaviour of $\mathfrak{Re}\left(\varepsilon_{n\kappa}\left[a_\kappa^{\text{inc}}, a_{2\kappa}^{\text{inc}}, z\right]\right)$ and $\mathfrak{Im}\left(\varepsilon_{n\kappa}\left[a_\kappa^{\text{inc}}, a_{2\kappa}^{\text{inc}}, z\right]\right)$ at the frequencies $n\kappa$ in Fig. 7 (case $a_{2\kappa}^{\text{inc}} = \frac{1}{3} a_\kappa^{\text{inc}}$) and Fig. 8 (case $a_{2\kappa}^{\text{inc}} = \frac{2}{3} a_\kappa^{\text{inc}}$). The quantities $\mathfrak{Im}\left(\varepsilon_{n\kappa}\right)$ take both positive and negative values along the height of the nonlinear layer (i.e.

http://dx.doi.org/10.5772/ 50679

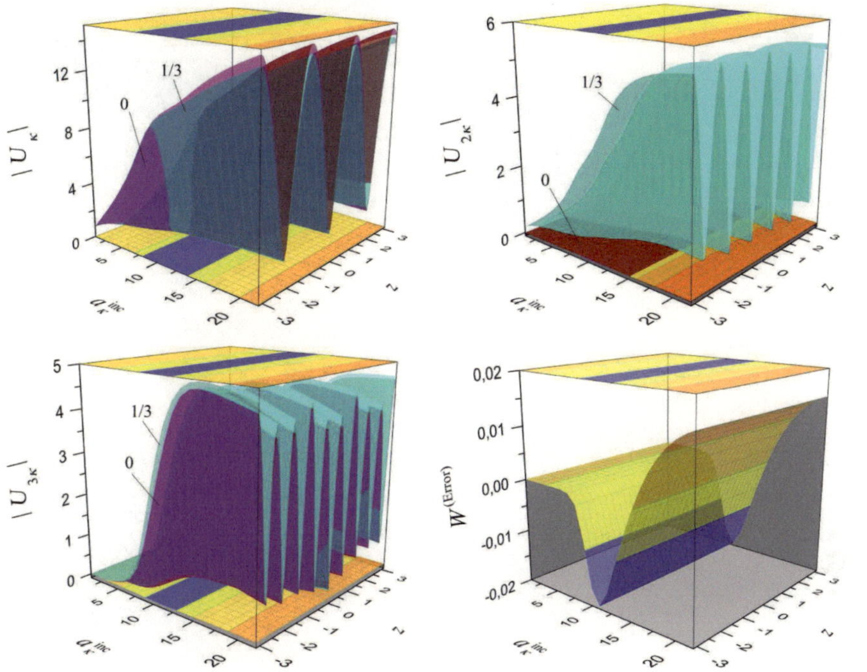

Figure 5. Graphs of the scattered and generated fields in the nonlinear layered structure in dependence on $\left[a_\kappa^{\text{inc}}, a_{2\kappa}^{\text{inc}}, z\right]$ for $\varphi_\kappa = 60^\circ$ and $a_{2\kappa}^{\text{inc}} = \frac{1}{3} a_\kappa^{\text{inc}}$: $|U_\kappa|$ (top left), $|U_{2\kappa}|$ (top right) $|U_{3\kappa}|$ (bottom left), and the residual $W^{(\text{Error})}$ of the energy balance equation (bottom right)

in the interval $z \in [-2\pi\delta, 2\pi\delta]$), see Figs. 7, 8 (right). For given amplitudes a_κ^{inc} and $a_{2\kappa}^{\text{inc}}$, the graph of $\mathfrak{Im}\left(\varepsilon_{n\kappa}\left[a_\kappa^{\text{inc}}, a_{2\kappa}^{\text{inc}}, z\right]\right)$ characterises the *loss of energy* in the nonlinear layer at the excitation frequencies $n\kappa$, $n = 1, 2$, caused by the *generation* of the electromagnetic field of the third harmonic. Such a situation arises because of the right-hand side of (13) at the triple frequency and the generation which is evoked by the right-hand side of (13) at the basic frequency. In our case $\mathfrak{Im}\left[\varepsilon^{(L)}(z)\right] = 0$ and $\mathfrak{Im}\left[\alpha(z)\right] = 0$, therefore, according to (11),

$$
\begin{aligned}
&\mathfrak{Im}(\varepsilon_{n\kappa}(z, \alpha(z), U(\kappa; z), U(2\kappa; z), U(3\kappa; z))) \\
&= \alpha(z)\left[\delta_{n1}|U(\kappa; z)||U(3\kappa; z)|\mathfrak{Im}\left(\exp\left\{i\left[-3\arg(U(\kappa; z)) + \arg(U(3\kappa; z))\right]\right\}\right)\right. \\
&\quad \left. + \delta_{n2}|U(\kappa; z)||U(3\kappa; z)|\mathfrak{Im}\left(\exp\left\{i\left[-2\arg(U(2\kappa; z)) + \arg(U(\kappa; z)) + \arg(U(3\kappa; z))\right]\right\}\right)\right], \\
&\qquad n = 1, 2, 3.
\end{aligned}
\tag{36}
$$

From Figs. 7, 8 (right) we see that small values of a_κ^{inc} and $a_{2\kappa}^{\text{inc}}$ induce a small amplitude of the function $\mathfrak{Im}(\varepsilon_{n\kappa})$, i.e. $|\mathfrak{Im}(\varepsilon_{n\kappa})| \approx 0$. The increase of a_κ^{inc} corresponds to a strong incident field and leads to the generation of a third harmonic field $U(3\kappa; z)$, and the increase of $a_{2\kappa}^{\text{inc}}$ changes the behaviour of $\varepsilon_{n\kappa}$ (compare the surface #0 with the surfaces #1/3 and #2/3 in Figs. 7, 8). Figs. 7, 8 (right) show the dynamic behaviour of $\mathfrak{Im}(\varepsilon_{n\kappa})$. It can be seen that $\mathfrak{Im}(\varepsilon_{3\kappa}) = 0$, whereas at the same time the values of $\mathfrak{Im}(\varepsilon_{n\kappa})$, $n = 1, 2$, may be *positive or negative* along the height of the nonlinear layer, i.e. in the interval $z \in [-2\pi\delta, 2\pi\delta]$, see (36).

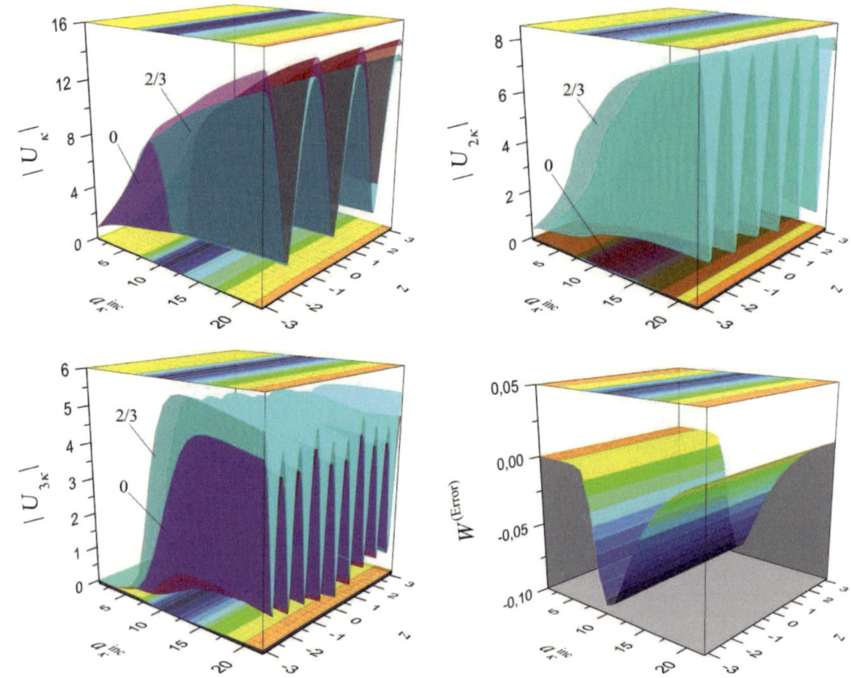

Figure 6. Graphs of the scattered and generated fields in the nonlinear layered structure in dependence on $\left[a_{\kappa}^{\rm inc}, a_{2\kappa}^{\rm inc}, z\right]$ for $\varphi_\kappa = 60^\circ$ and $a_{2\kappa}^{\rm inc} = \frac{2}{3} a_{\kappa}^{\rm inc}$: $|U_\kappa|$ (top left), $|U_{2\kappa}|$ (top right) $|U_{3\kappa}|$ (bottom left), and the residual $W^{(\rm Error)}$ (bottom right)

The zero values of $\mathfrak{Im}(\varepsilon_{n\kappa})$, $n = 1, 2$, are determined by the phase relations between the scattered and the generated fields in the nonlinear layer, namely at the basic frequency κ by the phase relation between $U(\kappa; z)$, $U(3\kappa; z)$, and at the double frequency 2κ by the phases of $\{U(n\kappa; z)\}_{n=1,2,3}$, see (36):

$$\begin{aligned}&\delta_{n1}\left[-3\arg(U(\kappa;z)) + \arg(U(3\kappa;z))\right]\\&+\delta_{n2}\left[-2\arg(U(2\kappa;z)) + \arg(U(\kappa;z)) + \arg(U(3\kappa;z))\right] = p\pi, \quad p = 0, \pm 1, \ldots, \; n = 1, 2.\end{aligned}$$

We mention that the behaviour of both the quantities $\mathfrak{Im}(\varepsilon_{n\kappa})$ and

$$\begin{aligned}&\mathfrak{Re}(\varepsilon_{n\kappa}(z,\alpha(z),U(\kappa;z),U(2\kappa;z),U(3\kappa;z)) - \varepsilon_{3\kappa}(z,\alpha(z),U(\kappa;z),U(2\kappa;z),U(3\kappa;z)))\\&= \alpha(z)\left[\delta_{n1}|U(\kappa;z)||U(3\kappa;z)|\mathfrak{Re}\left(\exp\left\{i\left[-3\arg(U(\kappa;z)) + \arg(U(3\kappa;z))\right]\right\}\right)\right.\\&\left.\quad+\delta_{n2}|U(\kappa;z)||U(3\kappa;z)|\mathfrak{Re}\left(\exp\left\{i\left[-2\arg(U(2\kappa;z)) + \arg(U(\kappa;z)) + \arg(U(3\kappa;z))\right]\right\}\right)\right],\\&\qquad n = 1, 2,\end{aligned} \tag{37}$$

plays an essential role in the process of third harmonic generation. Figs. 7, 8 (bottom) show the graphs describing the behaviour of $\mathfrak{Re}\left(\varepsilon_\kappa\left[a_\kappa^{\rm inc}, a_{2\kappa}^{\rm inc}, z\right] - \varepsilon_{3\kappa}\left[a_\kappa^{\rm inc}, a_{2\kappa}^{\rm inc}, z\right]\right)$ and $\mathfrak{Re}\left(\varepsilon_{2\kappa}\left[a_\kappa^{\rm inc}, a_{2\kappa}^{\rm inc}, z\right] - \varepsilon_{3\kappa}\left[a_\kappa^{\rm inc}, a_{2\kappa}^{\rm inc}, z\right]\right)$.

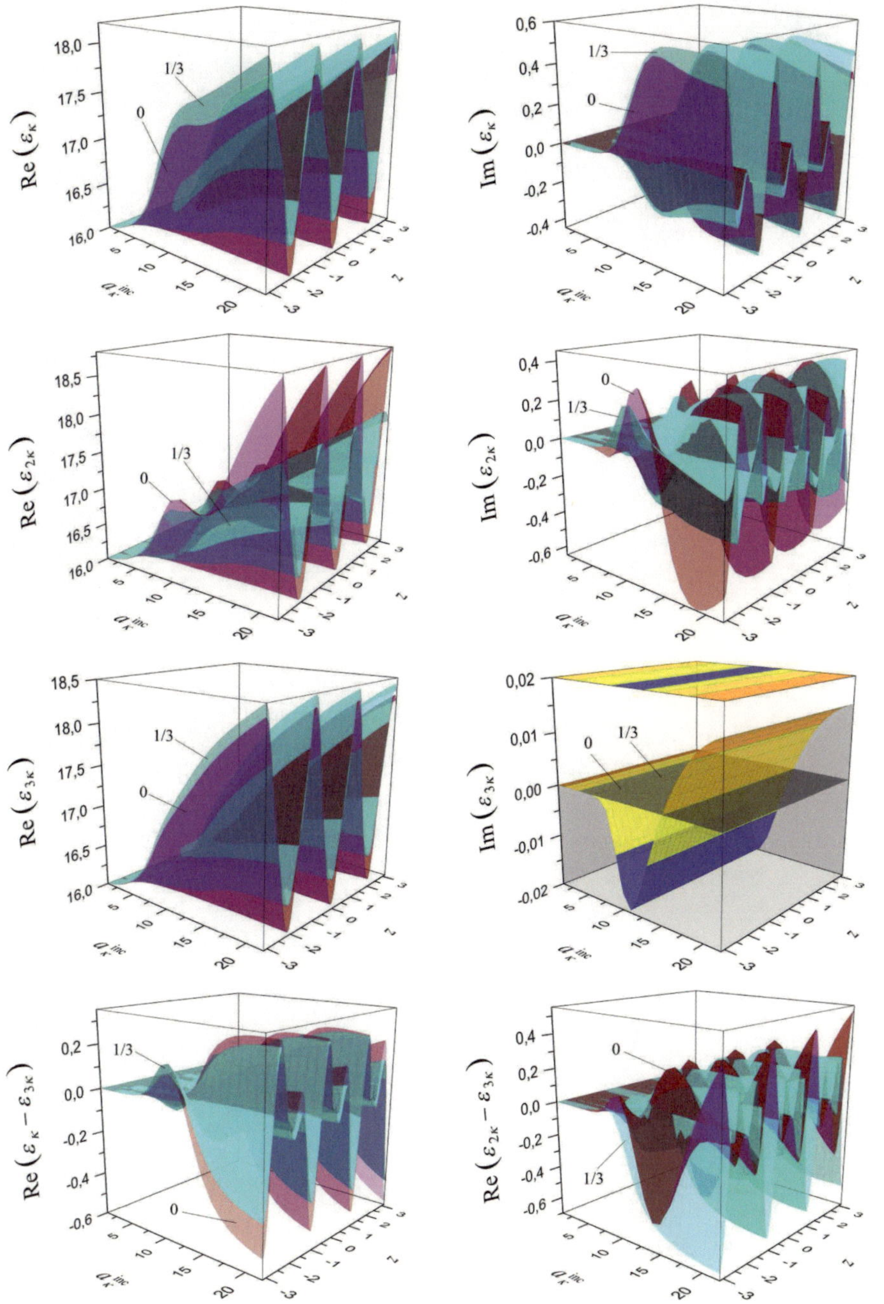

Figure 7. Graphs characterising the nonlinear dielectric permittivity in dependence on $\left[a_{\kappa}^{\text{inc}}, a_{2\kappa}^{\text{inc}}, z\right]$ for $\varphi_{\kappa} = 60^{\circ}$ and $a_{2\kappa}^{\text{inc}} = \frac{1}{3} a_{\kappa}^{\text{inc}}$: $\mathfrak{Re}\left(\varepsilon_{\kappa}\right)$ (top left), $\mathfrak{Im}\left(\varepsilon_{\kappa}\right)$ (top right), $\mathfrak{Re}\left(\varepsilon_{2\kappa}\right)$ (second from top left), $\mathfrak{Im}\left(\varepsilon_{2\kappa}\right)$ (second from top right), $\mathfrak{Re}\left(\varepsilon_{3\kappa}\right)$ (second to the last left), $\mathfrak{Im}\left(\varepsilon_{3\kappa}\right)$ (second to the last right), $\mathfrak{Re}\left(\varepsilon_{\kappa} - \varepsilon_{3\kappa}\right)$ (bottom left), $\mathfrak{Re}\left(\varepsilon_{2\kappa} - \varepsilon_{3\kappa}\right)$ (bottom right)

We mention that the impact of a strong electromagnetic field with an amplitude a_{κ}^{inc} even in the absence of a weak field $a_{2\kappa}^{\text{inc}} = 0$ (where $U(2\kappa; z) = 0$, see (11) and the surface #0 in Fig. 5

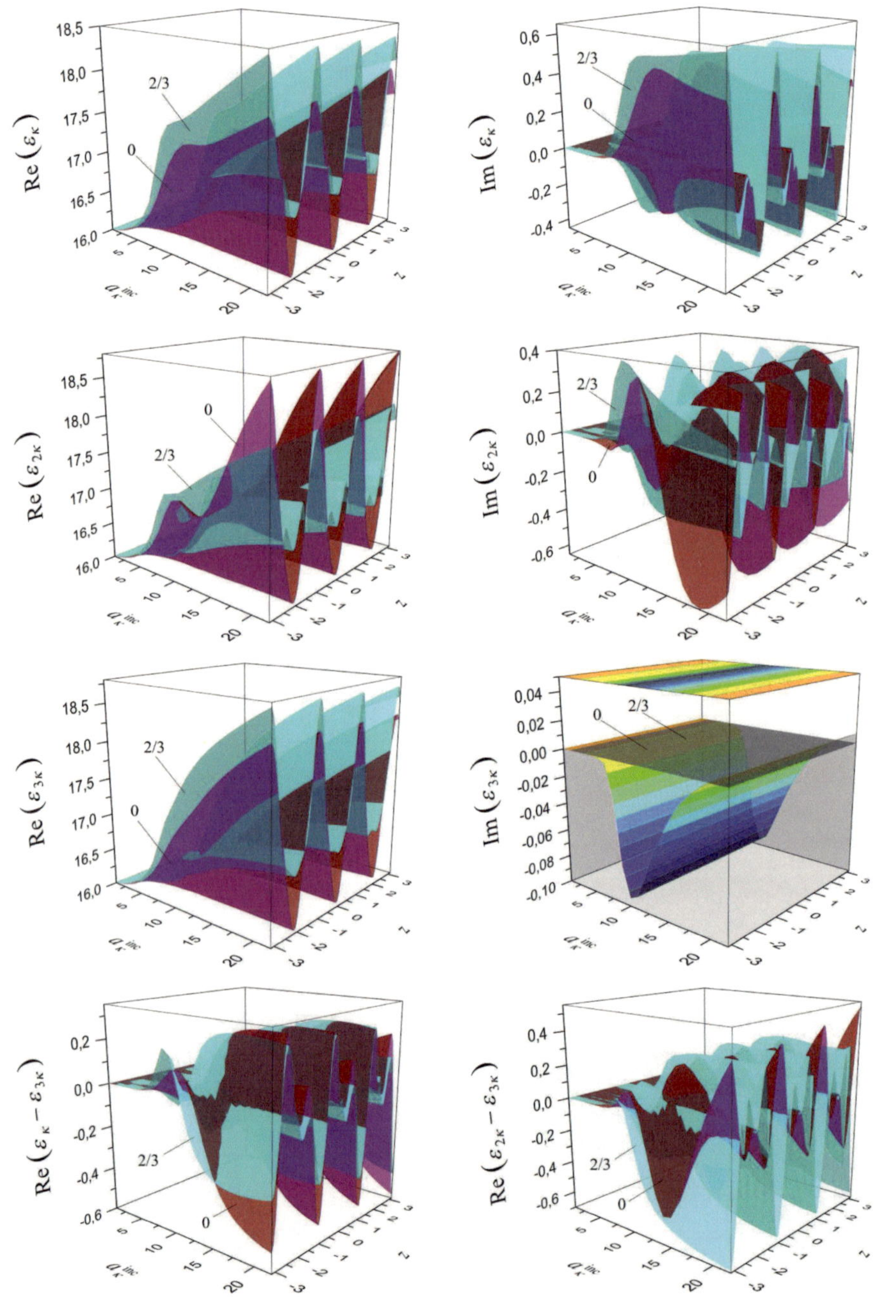

Figure 8. Graphs characterising the nonlinear dielectric permittivity in dependence on $\left[a_\kappa^{\text{inc}}, a_{2\kappa}^{\text{inc}}, z\right]$ for $\varphi_\kappa = 60^\circ$ and $a_{2\kappa}^{\text{inc}} = \frac{2}{3}a_\kappa^{\text{inc}}$: $\mathfrak{Re}\left(\varepsilon_\kappa\right)$ (top left), $\mathfrak{Im}\left(\varepsilon_\kappa\right)$ (top right), $\mathfrak{Re}\left(\varepsilon_{2\kappa}\right)$ (second from top left), $\mathfrak{Im}\left(\varepsilon_{2\kappa}\right)$ (second from top right), $\mathfrak{Re}\left(\varepsilon_{3\kappa}\right)$ (second to the last left), $\mathfrak{Im}\left(\varepsilon_{3\kappa}\right)$ (second to the last right), $\mathfrak{Re}\left(\varepsilon_\kappa - \varepsilon_{3\kappa}\right)$ (bottom left), $\mathfrak{Re}\left(\varepsilon_{2\kappa} - \varepsilon_{3\kappa}\right)$ (bottom right)

and Fig. 6 (top right)) induces a nontrivial component of the nonlinear dielectric permittivity at the frequency 2κ. Figs. 7, 8 (second from top) show that the existence of nontrivial values

$\mathfrak{Re}\left(\varepsilon_{2\kappa}\right) \neq \mathfrak{Re}\left(\varepsilon^{(L)}\right)$ and $\mathfrak{Im}\left(\varepsilon_{2\kappa}\right) \neq 0$ is caused by the amplitude and phase characteristics of the fields $U(\kappa;z)$ and $U(3\kappa;z)$, see (11) taking into account $U(2\kappa;z) = 0$. Moreover, the nonlinear component of the dielectric permittivity, which is responsible for the variation of $\mathfrak{Re}\left(\varepsilon_{n\kappa} - \varepsilon_{3\kappa}\right)$ and $\mathfrak{Im}\left(\varepsilon_{n\kappa}\right)$, does not depend on the absolute value of the amplitude of the field at the double frequency $|U(2\kappa;z)|$, see (37) and (36). Thus, even a weak field (see (11), #1/3 in Fig. 5 and #2/3 in Fig. 6 (top right)) includes a mechanism for the redistribution of the energy of the incident wave packet which is consumed for the scattering process and the generation of waves, cf. the dynamics of the surfaces #0 with #1/3 and #2/3 in Figs. 5 – 8.

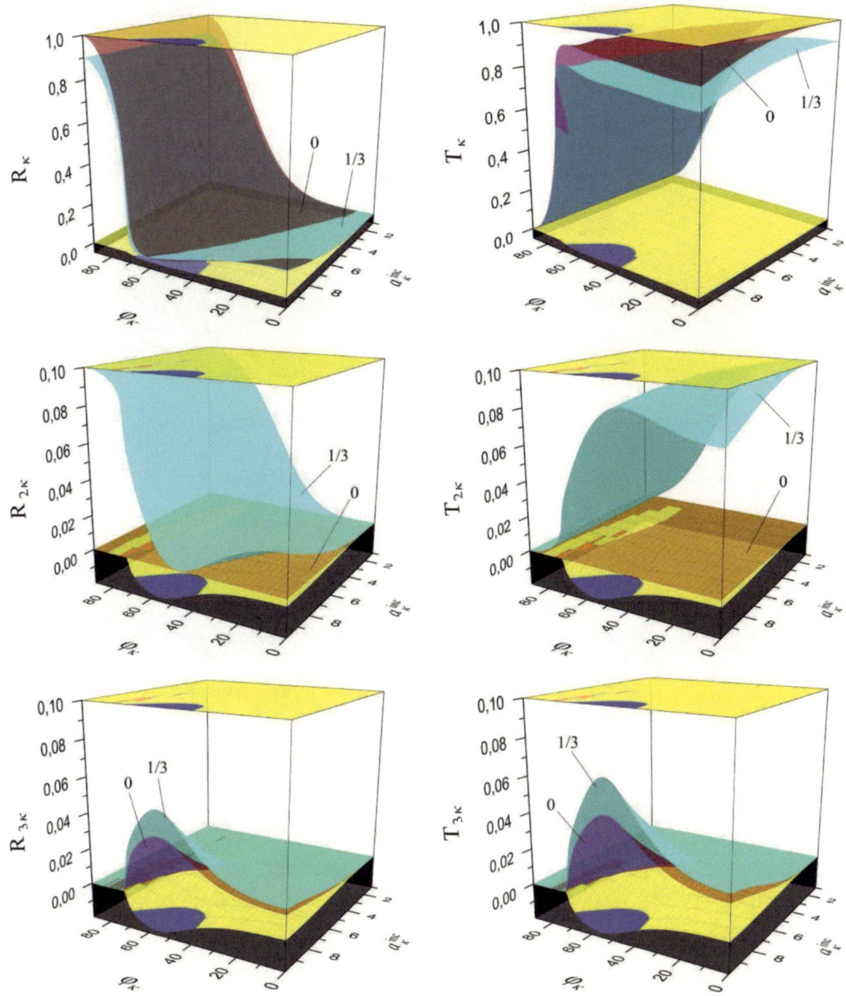

Figure 9. The scattering and generation properties of the nonlinear structure in dependence on $\left[\varphi_\kappa, a_\kappa^{\text{inc}}, a_{2\kappa}^{\text{inc}}\right]$ for $a_{2\kappa}^{\text{inc}} = \frac{1}{3}a_\kappa^{\text{inc}}$: R_κ, T_κ (top), $R_{2\kappa}$, $T_{2\kappa}$ (second from top), $R_{3\kappa}$, $T_{3\kappa}$ (bottom)

The scattering and generation properties of the nonlinear structure in the ranges $\varphi_\kappa \in [0°, 90°)$, $a_\kappa^{\text{inc}} \in [1, 9.93]$, $a_{2\kappa}^{\text{inc}} = \frac{1}{3}a_\kappa^{\text{inc}}$ and $\varphi_\kappa \in [0°, 90°)$, $a_\kappa^{\text{inc}} \in [1, 8]$ of the parameters of the incident field are presented in Figs. 9 – 11. The graphs show the dynamics of the scattering ($R_\kappa\left[\varphi_\kappa, a_\kappa^{\text{inc}}, a_{2\kappa}^{\text{inc}}\right]$, $T_\kappa\left[\varphi_\kappa, a_\kappa^{\text{inc}}, a_{2\kappa}^{\text{inc}}\right]$, $R_{2\kappa}\left[\varphi_\kappa, a_\kappa^{\text{inc}}, a_{2\kappa}^{\text{inc}}\right]$, $T_{2\kappa}\left[\varphi_\kappa, a_\kappa^{\text{inc}}, a_{2\kappa}^{\text{inc}}\right]$, see Figs.

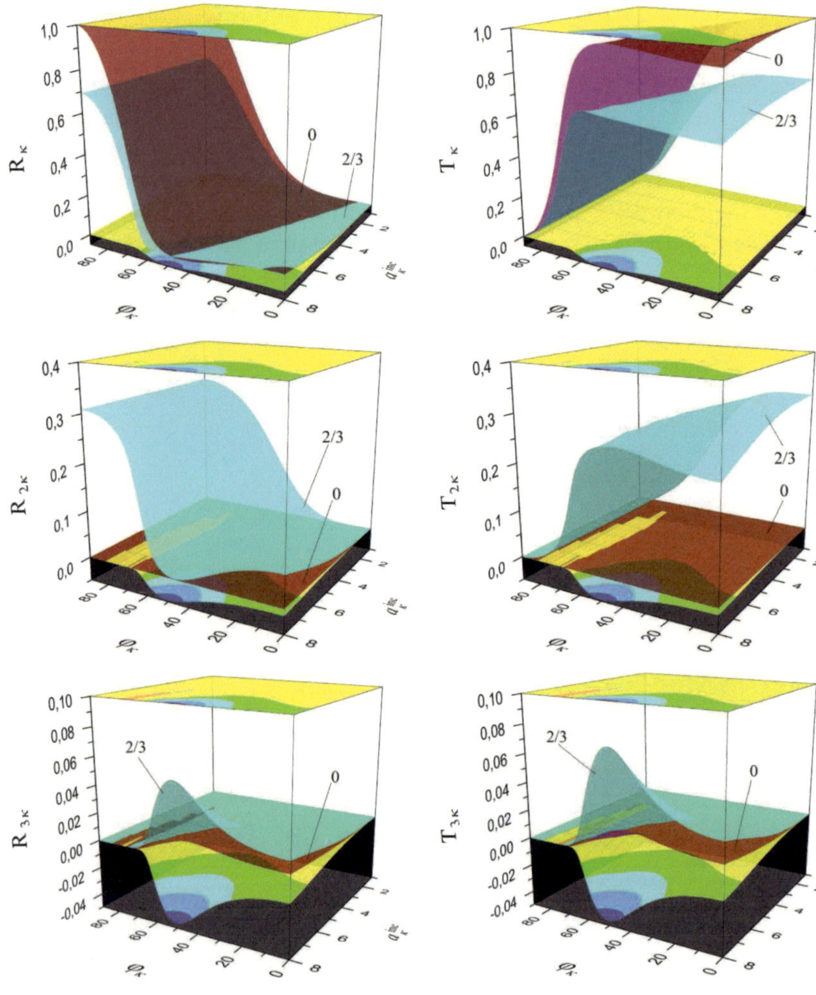

Figure 10. The scattering and generation properties of the nonlinear structure in dependence on $\left[\varphi_\kappa, a_\kappa^{\text{inc}}, a_{2\kappa}^{\text{inc}}\right]$ for $a_{2\kappa}^{\text{inc}} = \frac{2}{3} a_\kappa^{\text{inc}}$: R_κ, T_κ (top), $R_{2\kappa}$, $T_{2\kappa}$ (second from top), $R_{3\kappa}$, $T_{3\kappa}$ (bottom)

9, 10 (top 2)) and generation ($R_{3\kappa}\left[\varphi_\kappa, a_\kappa^{\text{inc}}, a_{2\kappa}^{\text{inc}}\right]$, $T_{3\kappa}\left[\varphi_\kappa, a_\kappa^{\text{inc}}, a_{2\kappa}^{\text{inc}}\right]$, see Figs. 9, 10 (bottom)) properties of the structure. Fig. 11 (top) shows cross sections of the surfaces depicted in Fig. 9 and of the graph of $W_{3\kappa}\left[\varphi_\kappa, a_\kappa^{\text{inc}}, a_{2\kappa}^{\text{inc}}\right] / W_\kappa\left[\varphi_\kappa, a_\kappa^{\text{inc}}, a_{2\kappa}^{\text{inc}}\right]$ (see Fig. 3 (left)) by the planes $\varphi_\kappa = 60^\circ$ and $a_\kappa^{\text{inc}} = 9.93$.

In Figs. 11 – 13, a slightly more detailed illustration for the situation of a single incident field (i.e. $a_{2\kappa}^{\text{inc}} = 0$) is given, cf. also the graphs labeled by "0" in Fig. 9. In the resonant range of wave scattering and generation frequencies, i.e. $\kappa^{\text{scat}} := \kappa^{\text{inc}} = \kappa$ and $\kappa^{\text{gen}} = 3\kappa$, resp., the dynamic behaviour of the characteristic quantities depicted in Figs. 11 – 13 has the following causes. The scattering and generation frequencies are close to the corresponding eigen-frequencies of the linear ($\alpha = 0$) and linearised nonlinear ($\alpha \neq 0$) spectral problems (24), (CS1) – (CS4). Furthermore, the distance between the corresponding eigen-frequencies of the spectral problems with $\alpha = 0$ and $\alpha \neq 0$ is small. Thus, the graphs in Fig. 11 (top)

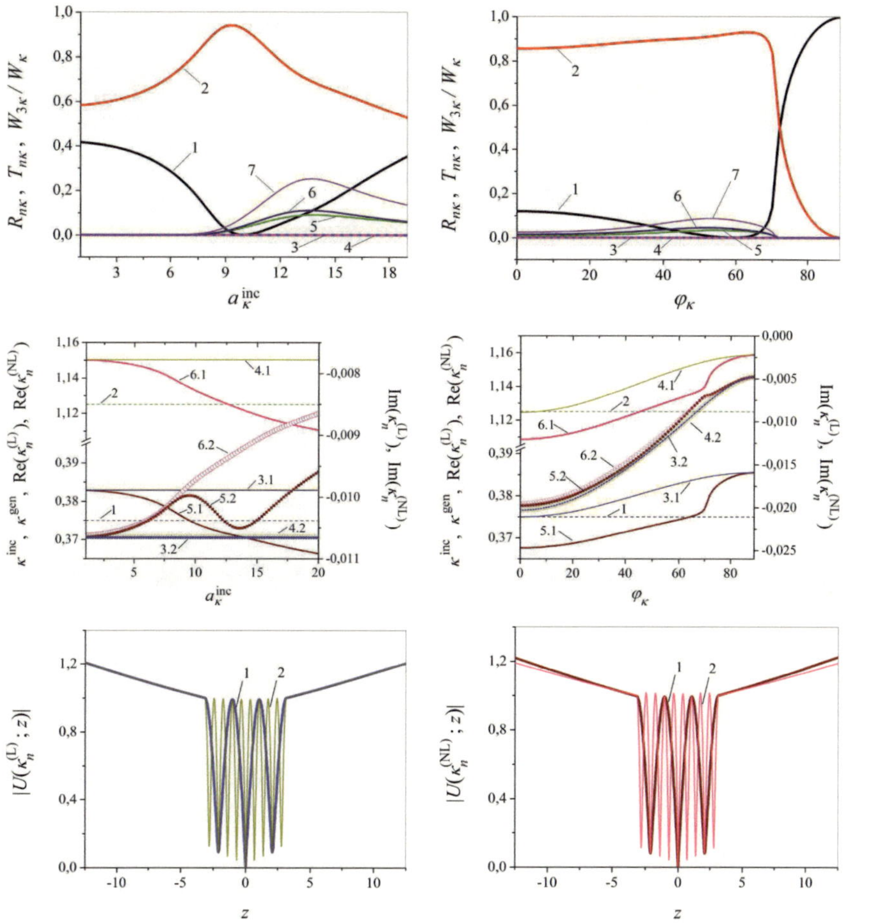

Figure 11. The curves R_κ (#1), T_κ (#2), $R_{2\kappa}$ (#3), $T_{2\kappa}$ (#4), $R_{3\kappa}$ (#5), $T_{3\kappa}$ (#6), $W_{3\kappa}/W_\kappa$ (#7) for $\varphi_\kappa = 60°$ (top left) and $a_\kappa^{inc} = 9.93$ (top right), the curves $\kappa := \kappa^{inc} := 0.375$ (#1), $3\kappa = \kappa^{gen} = 3\kappa^{inc} = 1.125$ (#2), the complex eigen-frequencies $\mathfrak{Re}(\kappa_1^{(L)})$ (#3.1), $\mathfrak{Im}(\kappa_1^{(L)})$ (#3.2), $\mathfrak{Re}(\kappa_3^{(L)})$ (#4.1), $\mathfrak{Im}(\kappa_3^{(L)})$ (#4.2) of the linear problem ($\alpha = 0$) and $\mathfrak{Re}(\kappa_1^{(NL)})$ (#5.1), $\mathfrak{Im}(\kappa_1^{(NL)})$ (#5.2), $\mathfrak{Re}(\kappa_3^{(NL)})$ (#6.1), $\mathfrak{Im}(\kappa_3^{(NL)})$ (#6.2) of the linearised nonlinear problem ($\alpha = +0.01$) for $\varphi_\kappa = 60°$ (second from top left) and $a_\kappa^{inc} = 9.93$ (second from top right), and the graphs of the eigen-fields of the layer for $\varphi_\kappa = 60°$, $a_\kappa^{inc} = 14$. The linear problem ($\alpha = 0$, bottom left): $|U(\kappa_1^{(L)};z)|$ with $\kappa_1^{(L)} = 0.3829155 - i\,0.01066148$ (#1), $|U(\kappa_3^{(L)};z)|$ with $\kappa_3^{(L)} = 1.150293 - i\,0.01062912$ (#2), the linearised nonlinear problem ($\alpha = +0.01$, bottom right): $|U(\kappa_1^{(NL)};z)|$ with $\kappa_1^{(NL)} = 0.3705110 - i\,0.01049613$ (#1), $|U(\kappa_3^{(NL)};z)|$ with $\kappa_3^{(NL)} = 1.121473 - i\,0.009194824$ (#2)

can be compared with the dynamic behaviour of the branches of the eigen-frequencies of the spectral problems presented in Fig. 11 (second from top). The graphs of the eigen-fields corresponding to the branches of the considered eigen-frequencies are shown in Fig. 11 (bottom).

Fig. 11 (second from top) illustrates the dispersion characteristics of the linear ($\alpha = 0$) and the linearised nonlinear ($\alpha = +0.01$) layer $\varepsilon_{n\kappa} = \varepsilon^{(L)} + \varepsilon_{n\kappa}^{(NL}$, $n = 1,2,3$. The nonlinear components of the permittivity at the scattering (excitation) frequencies $\kappa^{scat} := \kappa^{inc} = \kappa$ and the generation frequencies $\kappa^{gen} := 3\kappa$ depend on the amplitude a_κ^{inc} and the angle of

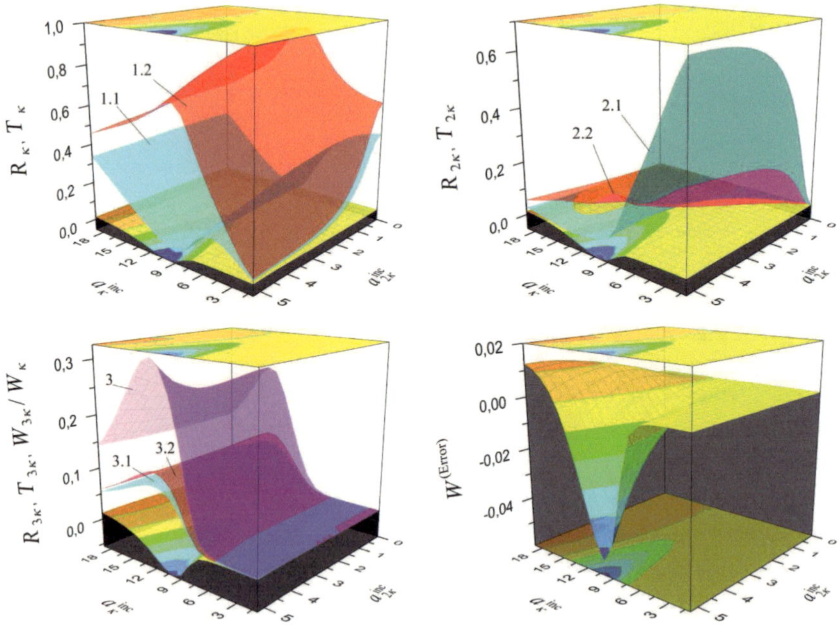

Figure 12. The scattering and generation properties of the nonlinear structure in dependence on $\left[\varphi_\kappa, a_\kappa^{\text{inc}}, a_{2\kappa}^{\text{inc}}\right]$ for $\varphi_\kappa = 60°$: R_κ, T_κ (#1.1, #1.2 top left), $R_{2\kappa}$, $T_{2\kappa}$ (#2.1, #2.2 top right), $W_{3\kappa}/W_\kappa$, $R_{3\kappa}$, $T_{3\kappa}$ (#3, #3.1, #3.2 bottom left), $W^{(\text{Error})}$ (bottom right)

incidence φ_κ of the incident field. This is reflected in the dynamics of the behaviour of the complex-valued eigen-frequencies of the linear and the linearised nonlinear layer.

We start the analysis of the results of our calculations with the comparison of the dispersion relations given by the branches of the eigen-frequencies (curves #3.1, #3.2 and #5.1, #5.2) near the scattering frequency (curve #1, corresponding to the excitation frequency) and (curves #4.1, #4.2, #6.1, #6.2) near the oscillation frequency (line #2) in the situations presented in Fig. 11 (second from top). The graph #5.1 lies below the graph #3.1 and the graph #6.1 below the graph #4.1. That is, canalising properties (properties of transparency) of the nonlinear layer occur if $\alpha > 0$. This case is characterised by the increase of the angle of transparency of the nonlinear structure at the excitation frequency with an increasing amplitude of the incident field (see Fig. 9 (top left), there where the reflection coefficient is close to zero). The analysis of the eigen-modes of Fig. 11 (second from top) allows us to explain the mechanisms of the canalisation phenomena (transparency) (see Fig. 9 (top left), Fig. 11 (top, graph #1)) and wave generation (see Fig. 9 (bottom), Fig. 11 (top, graphs #5, #6)).

Comparing the results shown in Fig. 11 (top) and Fig. 11 (second from top) we note the following. The intersection of the curves #1 and #5.1 in Fig. 11 (second from top) defines certain parameters, in the neighborhood of which the canalisation effect (transparency) of the nonlinear structure can be observed in Fig. 11 (top). For example, in Fig. 11 (second from top left) the curves #1 and #5.1 intersect at $a_\kappa^{\text{inc}} = 9.5$, also here the curve #5.2 achieves a local maximum. Near this value, we see the phenomenon of canalisation (transparency) of the layer in Fig. 11 (top left). If we compare the Figs. 11 (top right) and 11 (second from top right), we detect a similar situation. The intersection of the curves #1 and #5.1 defines the parameter

$\varphi_\kappa = 64°$, near which we observe the canalisation effect in Fig. 11 (second from top right). The same is true – to some extent – for the description of the wave generation processes. For example, for similar values of the imaginary parts of the branches of the eigen-frequencies #5.2 and #6.2 in Fig. 11 (second from top right), the intersection of the curves #2 and #6.1 defines the parameter $\varphi_\kappa = 45°$. Near this value, stronger generation properties of the layer can be observed, see Fig. 11 (top) and Fig. 9 (bottom), at $\varphi_\kappa = 45°$. Let us also consider the situation in Fig. 11 (second from top left). Here, at the point of intersection of the curves #2 and #6.1, the graph #5.2 starts to decrease monotonically in some interval. The intersection of the curves #2 and #6.1 defines the parameter $a_\kappa^{\text{inc}} = 12.6$, which falls into the range $[9.5, 13.6]$ of values of the amplitudes at which the curve #5.2 is monotonically decreasing. This leads to a shift in the imaginary part of the eigen-frequency of the scattering structure (graph #5.2) with respect to the eigen-frequency of the generating structure (graph #6.2). The magnitude of the shift depends on the distance between the curves of #6.2 and #5.2 at the given value a_κ^{inc}. The maximal distance between the graphs #6.2 and #5.2 is achieved at the local minimum of the graph #5.2 at $a_\kappa^{\text{inc}} = 13.6$. Right from this point, i.e. with an increasing amplitude a_κ^{inc}, the distance between the graphs #6.2 and #5.2 shows no significant change. The maximum value of the generation is achieved at an amplitude close to the intersection of curves #2 and #6.1, but shifted to the point of the local minimum of the curve #5.2, see $R_{3\kappa}$, $T_{3\kappa}$, $W_{3\kappa}/W_\kappa$ in Fig. 11 (top left), Fig. 9 (bottom) and Fig. 3 (top right).

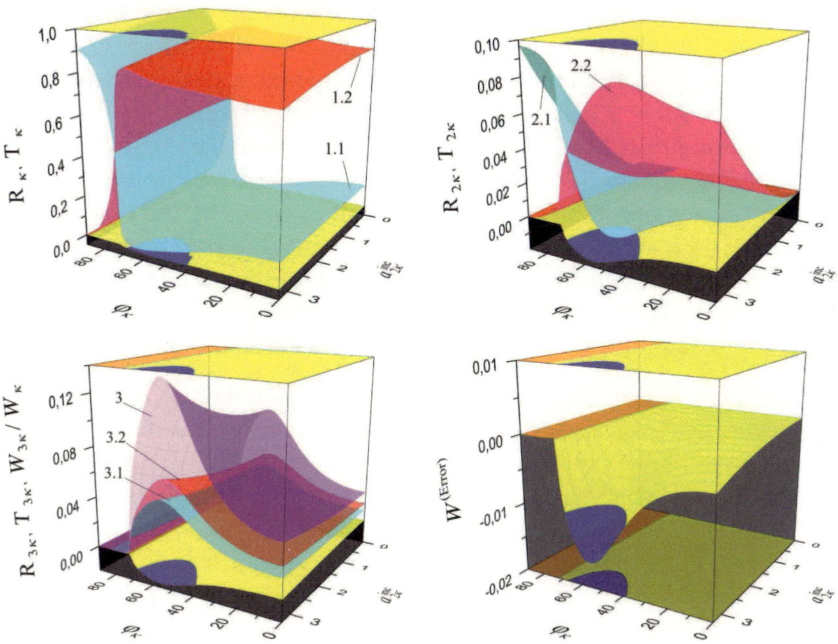

Figure 13. The scattering and generation properties of the nonlinear structure in dependence on $[\varphi_\kappa, a_\kappa^{\text{inc}}, a_{2\kappa}^{\text{inc}}]$ for $a_\kappa^{\text{inc}} = 9.93$: R_κ, T_κ (#1.1, #1.2 top left), $R_{2\kappa}$, $T_{2\kappa}$ (#2.1, #2.2 top right), $W_{3\kappa}/W_\kappa$, $R_{3\kappa}$, $T_{3\kappa}$ (#3, #3.1, #3.2 bottom left), $W^{(\text{Error})}$ (bottom right)

Fig. 11 (bottom) presents the characteristic distribution of the eigen-fields corresponding to the branches of the eigen-frequencies under consideration. The graphs of the eigen-fields of type $H_{0,0,4}$ are labeled by #1, the graphs of the eigen-fields of type $H_{0,0,10}$ by #2.

Figs. 12, 13 show the same dependencies as in Fig. 11 (top) but with the additional parameter $a_{2\kappa}^{\text{inc}}$. Here we can track the dynamics of the scattering, generation and energy characteristics of the nonlinear layer under the influence of the wave package. The incident package consists of a strong and a weak magnetic field with amplitudes a_{κ}^{inc} and $a_{2\kappa}^{\text{inc}}$, resp.

The numerical results presented in this chapter were obtained using an approach based on the description of the wave scattering and generation processes in a nonlinear, cubically polarisable layer by a system of nonlinear integral equations (19), and of the corresponding spectral problems by the nontrivial solutions of the integral equations (28). We have considered an excitation of the nonlinear layer defined by the condition (32). For this case we passed from (19) to (21) and from (28) to (30) by the help of Simpson's quadrature rule. The numerical solution of (21) was obtained using the self-consistent iterative algorithm (23). The problem (30) was solved by means of Newton's method. In the investigated range of parameters, the dimension of the resulting systems of algebraic equations was $N = 301$, and the relative error of calculations did not exceed $\xi = 10^{-7}$.

6. Conclusion

We presented a mathematical model of resonance scattering and generation of waves on an isotropic nonmagnetic nonlinear layered dielectric structure excited by a packet of plane waves in a self-consistent formulation, where the analysis is performed in the domain of resonance frequencies [3, 4, 15]. Here, both the radio [6] and optical [9] frequency ranges are of interest. The wave packets consist of both strong electromagnetic fields at the excitation frequency of the non-linear structure (leading to the generation of waves) and of weak fields at the multiple frequencies (which do not lead to the generation of harmonics but influence on the process of scattering and generation of waves by the non-linear structure). The model reduces to a system of nonlinear boundary-value problems of Sturm-Liouville type or, equivalently, to a system of nonlinear Fredholm integral equations. The solution of these nonlinear problems was obtained rigorously in a self-consistent formulation and without using approximations of the given field, slowly varying amplitudes etc.

The approximate solution of the nonlinear problems was obtained by means of solutions of linear problems with an induced nonlinear dielectric permeability. The analytical continuation of these linear problems into the region of complex values of the frequency parameter allowed us to switch to the analysis of spectral problems. In the frequency domain, the resonant scattering and generation properties of nonlinear structures are determined by the proximity of the excitation frequencies of the nonlinear structures to the complex eigen-frequencies of the corresponding homogeneous linear spectral problems with the induced nonlinear dielectric permeability of the medium.

We presented a collection of numerical results that describe interesting properties of the nonlinear permittivities of the layers as well as their scattering and generation characteristics. The results demonstrate the possibility to control the scattering and generating properties of a nonlinear structure via the intensities of its excitation fields. They also indicate a possibility of designing a frequency multiplier and other electrodynamic devices containing nonlinear dielectrics with controllable permittivity.

Author details

Lutz Angermann[1] and Vasyl V. Yatsyk[2]

1 Institut für Mathematik, Technische Universität Clausthal, Clausthal-Zellerfeld, Germany
2 Usikov Institute of Radiophysics and Electronics, National Academy of Sciences of Ukraine, Kharkov, Ukraine

References

[1] Angermann, L. & Yatsyk, V. [2008]. Numerical simulation of the diffraction of weak electromagnetic waves by a Kerr-type nonlinear dielectric layer, *Int. J. Electromagnetic Waves and Electronic Systems* 13(12): 15–30.

[2] Angermann, L. & Yatsyk, V. [2010]. Mathematical models of the analysis of processes of resonance scattering and generation of the third harmonic by the diffraction of a plane wave through a layered, cubically polarisable structure, *Int. J. Electromagnetic Waves and Electronic Systems* 15(1): 36–49. In Russian.

[3] Angermann, L. & Yatsyk, V. [2011a]. Generation and resonance scattering of waves on cubically polarisable layered structures, *in* L. Angermann (ed.), *Numerical Simulations – Applications, Examples and Theory*, InTech, Rijeka/Vienna, Croatia/Austria, pp. 175–212.

[4] Angermann, L. & Yatsyk, V. [2011b]. Resonance properties of scattering and generation of waves on cubically polarisable dielectric layers, *in* V. Zhurbenko (ed.), *Electromagnetic Waves*, InTech, Rijeka/Vienna, Croatia/Austria, pp. 299–340.

[5] Butcher, P. [1965]. Nonlinear optical phenomena, *Bulletin 200*, Ohio State University, Columbus.

[6] Chernogor, L. [2004]. *Nonlinear Radiophysics*, V.N. Karazin Kharkov National University, Kharkov.

[7] Kleinman, D. [1962]. Nonlinear dielectric polarization in optical media, *Phys. Rev.* 126(6): 1977–1979.

[8] Kravchenko, V. & Yatsyk, V. [2007]. Effects of resonant scattering of waves by layered dielectric structure with Kerr-type nonlinearity, *Int. J. Electromagnetic Waves and Electronic Systems* 12(12): 17–40.

[9] Miloslavsky, V. [2008]. *Nonlinear Optics*, V.N. Karazin Kharkov National University, Kharkov.

[10] Schürmann, H. W., Serov, V. & Shestopalov, Y. [2001]. Reflection and transmission of a TE-plane wave at a lossless nonlinear dielectric film, *Physica D* 158: 197–215.

[11] Serov, V., Schürmann, H. & Svetogorova, E. [2004]. Integral equation approach to reflection and transmission of a plane te-wave at a (linear/nonlinear) dielectric film with spatially varying permittivities, *J. Phys. A: Math. Gen.* 37: 3489–3500.

[12] Shestopalov, V. & Sirenko, Y. [1989]. *Dynamical Theory of Gratings*, Naukova, Dumka, Kiev.

[13] Shestopalov, V. & Yatsyk, V. [1997]. Spectral theory of a dielectric layer and the Morse critical points of dispersion equations, *Ukrainian J. of Physics* 42(7): 861–869.

[14] Shestopalov, Y. & Yatsyk, V. [2007]. Resonance scattering of electromagnetic waves by a Kerr nonlinear dielectric layer, *Radiotekhnika i Elektronika (J. of Communications Technology and Electronics)* 52(11): 1285–1300.

[15] Shestopalov, Y. & Yatsyk, V. [2010]. Diffraction of electromagnetic waves by a layer filled with a Kerr-type nonlinear medium, *J. of Nonlinear Math. Physics* 17(3): 311–335.

[16] Sirenko, Y., Shestopalov, V. & Yatsyk, V. [1985]. Elements of the spectral theory of gratings, *Preprint 266*, IRE NAS Ukraine, Kharkov.

[17] Sirenko, Y. & Ström, S. (eds) [2010]. *Modern Theory of Gratings. Resonant Scattering: Analysis Techniques and Phenomena*, Springer-Verlag, New York. Springer Series in Optical Sciences, Vol. 153.

[18] Sirenko, Y., Ström, S. & Yashina, N. [2007]. *Modeling and Analysis of Transient Processes in Open Resonant Structures. New Methods and Techniques.*, Springer-Verlag, New York.

[19] Smirnov, Y., Schürmann, H. & Shestopalov, Y. [2005]. Propagation of TE-waves in cylindrical nonlinear dielectric waveguides, *Physical Review E* 71: 0166141–10.

[20] Sukhorukov, A. P. [1988]. *Nonlinear wave interactions in optics and radio physics*, Nauka, Moskva. (In Russian).

[21] Vainstein, L. [1988]. *Electromagnetic Waves*, Radio i Svyas, Moscow. In Russian.

[22] Vinogradova, M., Rudenko, O. & Sukhorukov, A. [1990]. *Wave Theory*, Nauka, Moscow.

[23] Yatsyk, V. [2000]. A constructive approach to construction of local equations of irregular dispersion and evolution of fields in a quasi-homogeneous electrodynamic structure, *Usp. Sovr. Radioelektroniki* 10: 27–44. Translated in: *Telecommunications and Radio Engineering*, 56(8&9): 89-113, 2001.

[24] Yatsyk, V. [2006]. Diffraction by a layer and layered structure with positive and negative susceptibilities of Kerr-nonlinear media, *Usp. Sovr. Radioelektroniki* 8: 68–80.

[25] Yatsyk, V. [2007]. About a problem of diffraction on transverse non-homogeneous dielectric layer of Kerr-like nonlinearity, *Int. J. Electromagnetic Waves and Electronic Systems* 12(1): 59–69.

Chapter 12

Matter Wave Interferometry, the Gouy Phase and Complementarity Principle

Luis A. Cabral, Irismar G. da Paz,
José G. G. de Oliveira Júnior, Saulo Moreira,
Marcos D. R. Sampaio and Maria C. Nemes

Additional information is available at the end of the chapter

http://dx.doi.org/10.5772/53651

1. Introduction

Over a hundred years ago L. G. Gouy described an anomalous behavior of the phase of a converging diffracted spherical wave as it passes through a focus [1, 2]. He also performed the experiment which confirmed his theoretical analysis. He considered a point source and let it impinge onto two mirrors with different curvature radius, one concave of other plane, parallel to each other. The observational plane was transverse to the so generated beams. Their interference yielded circular interference fringes. Since then many observations of such phase anomaly have been reported, some of them in [3–12].

It turns out that the Gouy phase has found several important applications. As one example we quote the influence of this anomalous phase as being of critical importance for any application as ultrashort laser pulses, including high-harmonics and attosecond pulse generation [13]. It also has important consequences in the optical range of the electromagnetic spectrum such as phase matching in high order harmonic generation governed by atomic response and the Gouy phase of the fundamental radiation [14–16]. On the theoretical side numerous authors have treated the problem an explanation for such phase. One of the first authors to consider the question was P. Debye in 1909, who found an exact solution to the wave equation for all space treating the boundary condition in an elegant way, avoiding its mathematical complication [17]. The anomalous propagation through the focus has been treated by several authors since then. In refs [18, 19] one also finds early rather involved theories explaining the phase anomaly.

Interesting enough even nowadays several authors still consider the Gouy phase as "the most poorly understood subjects in physical optics" [20], although several simple explanation have been offered in terms of Berry geometrical phase [4], the uncertainty principle [10] and so one [5, 6, 21–25].

A successful description of optical waves crossing an aperture is the so called paraxial approximation to the Helmholtz equation (explained below). A physically appealing aspect

INTECH
open science | open minds

© 2012 A. Cabral et al.; licensee InTech. This is an open access article distributed under the terms of the Creative Commons Attribution License (http://creativecommons.org/licenses/by/3.0), which permits unrestricted use, distribution, and reproduction in any medium, provided the original work is properly cited.

of this equation is its complete correspondence to Schrödinger equation for matter waves. However in the latter case the corresponding phase is a time dependent overall phase and therefore omitted in quantum mechanics text books. In spite of that it is manifest and in fact the agent responsible for the growth of the width gaussian wave packet under a free evolution. One of the objectives of the present contribution is to unveil the similarities between these to apparently independent subjects, which, however bear a strong conceptual basis, which has been recently show [9, 10]. Also an experimental proposal was put forth for the observation of the Gouy phase of matter waves in the optical regime using Ramsey interferometry with Rydberg atoms [12].

In fact the basis for the interpretation of the anomalous phase of matter waves is the Schrödinger Robertson uncertainty relation which can be immediately carried over for laser light. As will be shown here the Gouy phase in both cases is related to the $x - p$ cross correlations, i.e., the expectation value of the position-symmetrized-momentum. It becomes thus clear that the phase is intimately connected to the correlations generated by the uncertainty relation.

Our second purpose, which is a rather novel investigation relates to the role of the Gouy phase in interference phenomena. Here in particular we restrict ourselves to two slit experiments which, again, are equally both for light or matter.

The first theoretical to be faced is the bridge between the complementarity principle as usually constructed using q-bits and their equivalent for continuous systems. We define and explore the constructed definition for visibility and predictability, showing, in particular that the Gouy phase will be manifest in the number of fringes of a given arrangement. Thus by changing the slit aperture of the grating and measuring the corresponding change in the number of fringes on the screen gives an indirect evidence of this phase. Again the idea that the Gouy phase is at the root of this phenomena is not exclusive of matter or light waves, since it has only to do with wave nature.

In section 2 we make the analogy between the paraxial wave equation for light and the Schrödinger's equation for a free particle. In section 3 we discuss the generalized complementarity principle within a double slit experiment. We quantify the wave-particle duality through the definition of quantities called visibility and predictability which characterize the wave and particle behavior respectively [26, 32]. Section 4 is reserved to a wave quantitative analogies of interferometry of matter waves in double slit experiment showing that the Gouy phase is related to number of fringes. Moreover we show that the Gouy phase can be indirectly extracted by analyzing the change in the interference fringes with the variation of the slit width or the distance of the detector. The results that we will obtain here for matter waves can be equally applied for classical and quantum electromagnetic waves in the paraxial regime. In section 5 we make our conclusions.

2. From Maxwell to matter: Waves the Gouy phase

We proceed to exploring the analogies between properties of electromagnetic waves in the paraxial approximation and matter waves. In Fig. 1 we compare plane, spherical and Gaussian wavefronts. The latter is obtained when a converging spherical wave is obstructed by a circular aperture. Notice that a Gaussian wavefront acquires an anomalous phase shift as it passes through the focal region, i.e., their wavefront are slighted advanced in relation to the others. This phase shift is the Gouy phase.

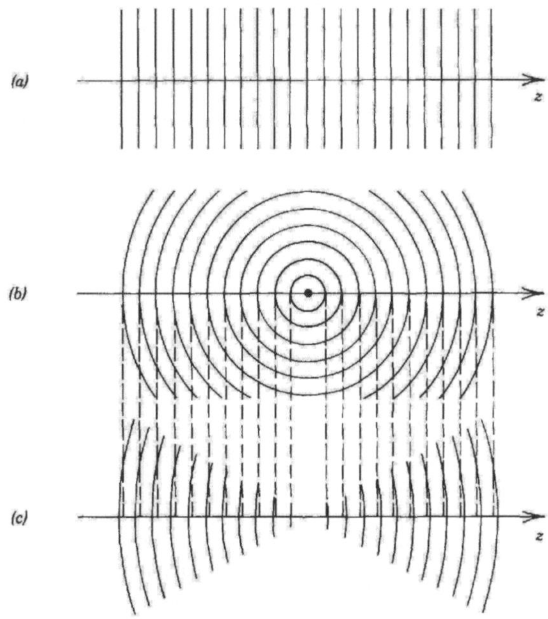

Figure 1. Wavefronts: a) plane wave, b) spherical wave, c) Gaussian beam. Extracted from [ref. B. E. A. Saleh and M. Teich, Fundamentals of Photonics (John Wiley Sons, New York, 1991).]

Consider a stationary electric field in vacuum

$$E(\vec{r}) = A(\vec{r})\exp(ikz), \tag{1}$$

where k is the wave number. In the paraxial approximation we assume that the modulus of the wave vector $|\vec{k}| \approx k_z$, i.e., $k_z \gg k_x, k_y$. This amounts to saying that the function $A(\vec{r})$ which modulates the field varies slowly compared with the wavelength $\lambda = 2\pi/k$. Under this condition the Helmholtz equation for the electric field is [25]

$$\left(\frac{\partial^2}{\partial x^2} + \frac{\partial^2}{\partial y^2} + i4\pi\frac{1}{\lambda_L}\frac{\partial}{\partial z}\right) A\left(x,y,z\right) = 0, \tag{2}$$

where λ_L is the light wavelength.

On the other hand the two-dimensional Schrödinger equation for a free particle of mass m reads

$$\left(\frac{\partial^2}{\partial x^2} + \frac{\partial^2}{\partial y^2} + 2i\frac{m}{\hbar}\frac{\partial}{\partial t}\right)\psi(x,y,t) = 0. \tag{3}$$

Notice that snapshots of $\psi(x,y,t_i)$ is identical to slices for fixed z_i of $A(x,y,z_i)$. Therefore we can interpret the time evolution of $\psi(x,y,t)$ in terms of parameter z related to time as $t = z/v_z$, where v_z can be interpreted in terms of the de Broglie wavelength $\lambda_P = h/mv_z$ as we assume $p \sim p_z$.

Based on this discussion Equation (3) can formally written as [7–9]

$$\left(\frac{\partial^2}{\partial x^2}+\frac{\partial^2}{\partial y^2}+i4\pi\frac{1}{\lambda_P}\frac{\partial}{\partial z}\right)\psi\left(x,y,t=z/v_z\right)=0. \tag{4}$$

Next we show that the Gouy phase is intimately connected to the generalized Schrödinger-Robertson uncertainty relation [7–10] which provides for a physically appealing interpretation of this phase which can be carried over to light waves in paraxial approximation.

The solution of Equation (2) for a transverse Gaussian beam collimated in the z direction is

$$A_G(x,z)=A_0\frac{w_0}{w(z)}\exp\left[-\frac{x^2}{w(z)^2}+i\frac{kx^2}{2R(z)}+i\xi(z)\right] \tag{5}$$

where we dropped the y component for brevity. In the equation above, $w(z)$ is the beam width, $R(z)$ is the curvature radius, $\xi(z)$ is the Gouy phase and z_0 is the Rayleigh range given by

$$w(z)=w_0\left[1+\left(\frac{z}{z_0}\right)^2\right]^{\frac{1}{2}},\quad R(z)=z\left[1+\left(\frac{z_0}{z}\right)^2\right], \tag{6}$$

$$\xi(z)=-\frac{1}{2}\arctan\left(\frac{z}{z_0}\right),\quad z_0=\frac{kw_0^2}{2}. \tag{7}$$

The total variation of the Gouy phase as we go from $z=-\infty$ to $z=+\infty$ is $\pi/2$ as we can immediately verify from (7). This change is abrupt and effectively takes place in the so called Rayleigh range z_0. On the other hand, solution of Equation (4) for an initial condition given by a Gaussian wave packet is given by [7–9]

$$\psi\left(x,t\right)=\frac{1}{\sqrt{\sqrt{\pi}B\left(t\right)}}\exp\left[-\frac{x^2}{2B(t)^2}+i\frac{mx^2}{2\hbar R(t)}+i\mu(t)\right]. \tag{8}$$

From the solutions (5) and (8) we can make the following identifications

$$w\left(z\right)\longrightarrow B\left(t\right)=b_0\left[1+\left(\frac{t}{\tau_0}\right)^2\right]^{\frac{1}{2}}, \tag{9}$$

$$R\left(z\right)\longrightarrow R\left(t\right)=t\left[1+\left(\frac{\tau_0}{t}\right)^2\right], \tag{10}$$

$$\xi(z) \longrightarrow \mu(t) = -\frac{1}{2}\arctan\left(\frac{t}{\tau_0}\right), \tag{11}$$

and

$$z_0 \longrightarrow \tau_0 = \frac{mb_0^2}{\hbar}. \tag{12}$$

To get a better insight inter the role played by τ_0 (or equivalently the Rayleigh range z_0) is convenient to rewrite (8) as

$$\psi(x,t) = \tilde{\varphi}(x,t)\exp[iS(x,t)], \tag{13}$$

where

$$\tilde{\varphi}(x,t) = \frac{1}{\sqrt{\sqrt{\pi}B(t)}}\exp\left[-\frac{x^2}{2B(t)^2}\right], \tag{14}$$

and

$$S(x,t) = \frac{x^2}{2B(t)^2}\frac{t}{\tau_0} + \mu(t). \tag{15}$$

Notice that the position dependence phase S contains the time scale τ_0. the ratio t/τ_0 will determine the importance of this x-dependent phase to the interference pattern. In the experimental setups using fullerene molecules [36] $t/\tau_0 \approx 10^4$ which is also the condition for Frauhoffer diffraction (see ref. [27]). The time scale τ_0 is fundamentally determined by Heisenberg's uncertainty relation, given the initial position dispersion $\Delta x(0) = \sigma_0/\sqrt{2}$. In fact, the corresponding momentum dispersion is $\Delta p = \hbar/(\sigma_0\sqrt{2})$. Because the momentum is a constant of motion this momentum spread will be preserved in time. Both Δx and Δp constitute *intrinsic* properties of the initial wave packet, in terms of which the time scale τ_0 is expressed as

$$\tau_0 = \frac{\Delta x(0)}{(\Delta p)/m}. \tag{16}$$

The numerator in the above relation represents the spatial dimensions of the initial wave packet, whilst the denominator stands for the scale of *velocity difference* enforced by the uncertainty principle. Therefore the time scale τ_0 corresponds essentially to the time during which a distance of the order of the wave packet extension is traversed with a speed corresponding to the dispersion in velocity. It can therefore be viewed as a characteristic time for the "aging" of the initial state, which consists in components with larger velocities (relatively to the group velocity of the wave packet) concentrating at the frontal region of the packet. This can be seen explicitly by deriving the velocity field associated with the phase S in Equation (15), which reads

$$v(x,t) = \frac{\hbar}{m}\frac{\partial S}{\partial x} = \frac{tx}{\tau_0^2 + t^2}. \tag{17}$$

This expression shows that for $t > 0$ the initial velocity field $v(x,0) = 0$ varies linearly with respect to the distance from center of the wave packet ($x = 0$).

Next we relate quantitatively this "aging" effect to position-momentum correlations. This is readily achieved using the generalized uncertainty relation devised by Schrödinger [28], which is expressed in this case in terms of the determinant of the covariance matrix

$$det \sum \equiv \det \begin{pmatrix} \sigma_{xx} & \sigma_{xp} \\ \sigma_{xp} & \sigma_{pp} \end{pmatrix} \geq \frac{\hbar^2}{4}. \tag{18}$$

For the minimum uncertainty wave packet of Equation (8) we obtain, *at all times*,

$$\det \begin{pmatrix} \sigma_{xx} & \sigma_{xp} \\ \sigma_{xp} & \sigma_{pp} \end{pmatrix} = \frac{\hbar^2}{4}. \tag{19}$$

Therefore we establish a direct relation between the Gouy phase and the position-momentum correlation for matter waves,

$$\sigma_{xp} = \frac{\hbar t}{2\tau_0} = -\frac{\hbar}{2}\tan 2\mu(t). \tag{20}$$

For light waves a similar interpretation applies,

$$\sigma_{xk_x} = \frac{z}{2z_0} = -\frac{1}{2}\tan 2\zeta(z). \tag{21}$$

The formulation of the generalized uncertainty relation for light waves in terms of operators can be found in [29]

3. The complementarity principle in double slit experiment

Wave-particle duality as first conceived by [30] stated that in a given experimental set up one observes either wave or particle properties. Using his words: "we are presented with a choice of either tracing the path of a particle or observing interference effects". In double slit experiments the wave nature of the object is reflected by the interference pattern exhibited on a screen. We know however that once the knowledge of the object's path (i.e., through which slit it crossed) is obtained, the interference pattern is completely destroyed. A natural question that can arise is how partial knowledge of path affects such pattern. This discussion was first addressed by [31] and later by [32]. The quantification of which-way information

http://dx.doi.org/10.5772/53651

(predictability) and visibility was established by [33] and the relation between both given by [34].

We proceed to define visibility and predictability in two slit experiments. Consider a double slit experiment as depicted in the Fig. 2. B and C denote the slits, A the position from which the particles are emitted and D the detector which covers the whole extension of the screen and counts the numbers of particles. Notice that A need not be placed on the symmetry axis Ox.

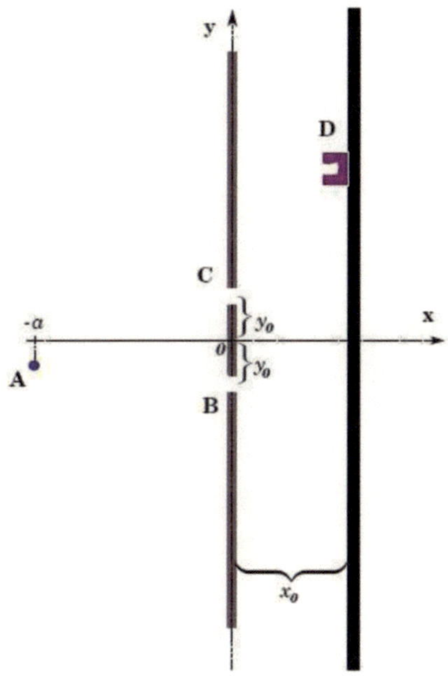

Figure 2. Illustration of the double slit with a displaced source of particles.

Suppose N particles are sent to the grating. Consider that (N_B) N_C particles cross slit (B) C leaving slit (C) B closed with relative probability (P_B) P_C.

The probability amplitude at the screen is

$$\psi(x_0, y) = \sqrt{P_B}\psi_B(x_0, y) + \sqrt{P_C}\psi_C(x_0, y). \tag{22}$$

The probability density of finding a particle at the point (x_0, y) on the screen knowing that it crossed slit B is obviously

$$\mathsf{P}_B(y) = P_B|\psi_B(x_0, y)|^2. \tag{23}$$

Analogously

$$\mathsf{P}_C(y) = P_C|\psi_C(x_0, y)|^2, \tag{24}$$

for slit C.

Now the quantification of which way information, also known as predictability is

$$\mathcal{P}(y) = \left|\frac{\mathrm{P}_B(y) - \mathrm{P}_C(y)}{\mathrm{P}_B(y) + \mathrm{P}_C(y)}\right|. \tag{25}$$

The visibility $\mathcal{V}$ is defined through the expression for the intensity $|\psi(x_0, y)|^2$ on the screen

$$I(y) = P_B|\psi_B(x_0, y)|^2 + P_C|\psi_C(x_0, y)|^2 + 2|\sqrt{P_B P_C}\psi_B(x_0, y)\psi_C(x_0, y)|\cos\phi, \tag{26}$$

with $\phi = \arg\psi_B(x_0, y) - \arg\psi_C(x_0, y)$.

We can rewrite $I(y)$ as

$$I(y) = [\mathrm{P}_B(y) + \mathrm{P}_C(y)]\left[1 + 2\left|\frac{\sqrt{P_B P_C}\psi_B(x_0, y)\psi_C(x_0, y)}{\mathrm{P}_B(y) + \mathrm{P}_C(y)}\right|\cos\phi\right], \tag{27}$$

from which the visibility [37]

$$\mathcal{V} = \frac{I_{max} - I_{min}}{I_{max} + I_{min}}, \tag{28}$$

can be read off as

$$\mathcal{V} = \frac{2|\sqrt{P_B P_C}\psi_B(x_0, y)\psi_C(x_0, y)|}{\mathrm{P}_B(y) + \mathrm{P}_C(y)}. \tag{29}$$

We can immediately verify that [32, 34]

$$\mathcal{P}^2 + \mathcal{V}^2 = 1, \tag{30}$$

for all y. The relation above is known as the complementarity relation.

4. Double slit interference: Complementarity principle and the Gouy phase

In this section we discuss double-slit interference of matter waves, for instance neutrons or fullerene molecules [35], to illustrate the ideas that we discussed so far. Consider the experimental set up illustrated in the Fig. 3. b_0 is the characteristic width of a wavepacket, assumed to be coherent, i.e., not affected by indeterminacies inherent to the production process, σ_0 is the slit widths, x_0 the distance between the slit grating and screen and d is the distance between the slits. The particles travel in the x direction with speed v, assuming that the center of mass moves classically, that is $\Delta p_x \ll p_x$.

The intensity on the screen reads

http://dx.doi.org/10.5772/53651

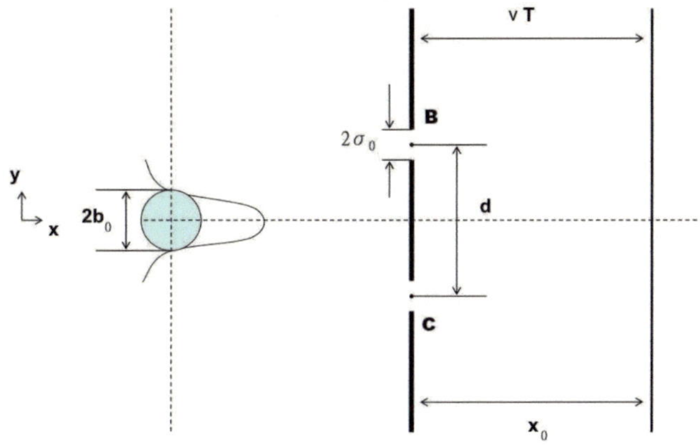

Figure 3. Experimental set up.

$$I(y, T = x_0/v) = \left|\sqrt{P_B}\psi_B(y, T = x_0/v) + \sqrt{P_C}\psi_C(y, T = x_0/v)\right|^2 \tag{31}$$

where

$$\psi_B(y, T) = \frac{1}{\sqrt{B(T)\sqrt{\pi}}} \exp\left[-\frac{(y + \frac{d}{2})^2(1 - i\frac{T}{\tau_0})}{2B(T)^2} + i\mu(T)\right], \tag{32}$$

$$\psi_C(y, T) = \frac{1}{\sqrt{B(T)\sqrt{\pi}}} \exp\left[-\frac{(y - \frac{d}{2})^2(1 - i\frac{T}{\tau_0})}{2B(T)^2} + i\mu(T)\right], \tag{33}$$

and $\tau_0 = \frac{m\sigma_0^2}{\hbar}$. An analytical expression for P_B and P_C can in principle be derived in terms of the parameters as illustrated in Fig. 3 as performed in [35] but it is not essential to the purposes of our discussion.

It is easy to show that $I(y, T)$ can be written as

$$I(y) = F(y)\left[1 + \frac{2\sqrt{P_B P_C}}{P_B e^{\alpha y} + P_C e^{-\alpha y}}\cos(\beta y)\right], \tag{34}$$

where

$$F(y) = I_0 e^{-\gamma(y)} \frac{P_B e^{\alpha y} + P_C e^{-\alpha y}}{2\sqrt{P_B P_C}}, \tag{35}$$

$$\gamma(y) = \frac{y^2 + (\frac{d}{2})^2}{B(T)^2}, \tag{36}$$

$$\alpha = \frac{d}{B(T)^2} = \frac{1}{\sigma_{xx}} \frac{d}{2}, \tag{37}$$

and

$$\beta = \frac{T}{\tau_0} \frac{d}{B(T)^2} = \frac{\sigma_{xp}}{\sigma_{xx}} \frac{d}{\hbar}, \tag{38}$$

where hereafter we write $I(y, T) = I(y)$ for sake of brevity.

Recalling the expression for predictability and visibility (25), (29) we get

$$\mathcal{V}(y) = \frac{2\sqrt{P_B P_C}}{P_B e^{\alpha x} + P_C e^{-\alpha y}}, \tag{39}$$

$$\mathcal{P}(y) = \left| \frac{P_B e^{\alpha y} - P_C e^{-\alpha y}}{P_B e^{\alpha y} + P_C e^{-\alpha y}} \right|, \tag{40}$$

and

$$\mathcal{P}^2(y) + \mathcal{V}^2(y) = 1. \tag{41}$$

Let $p = \frac{P_C}{P_B}$. From expression (39) we can find the point on the screen where the visibility reaches its maximal value,

$$y_{max} = \frac{\ln \sqrt{p}}{\alpha}. \tag{42}$$

At this point the which way information is *minimum*, that is to say $\mathcal{P}(y_{max}) = 0$ and $\mathcal{V}(y_{max}) = 1$. Notice that it. In Fig. 4 we illustrate this with $p = 1,\ 0.15$ and 0.015.

For $p = 1$, illustrated in Fig. 4, we have

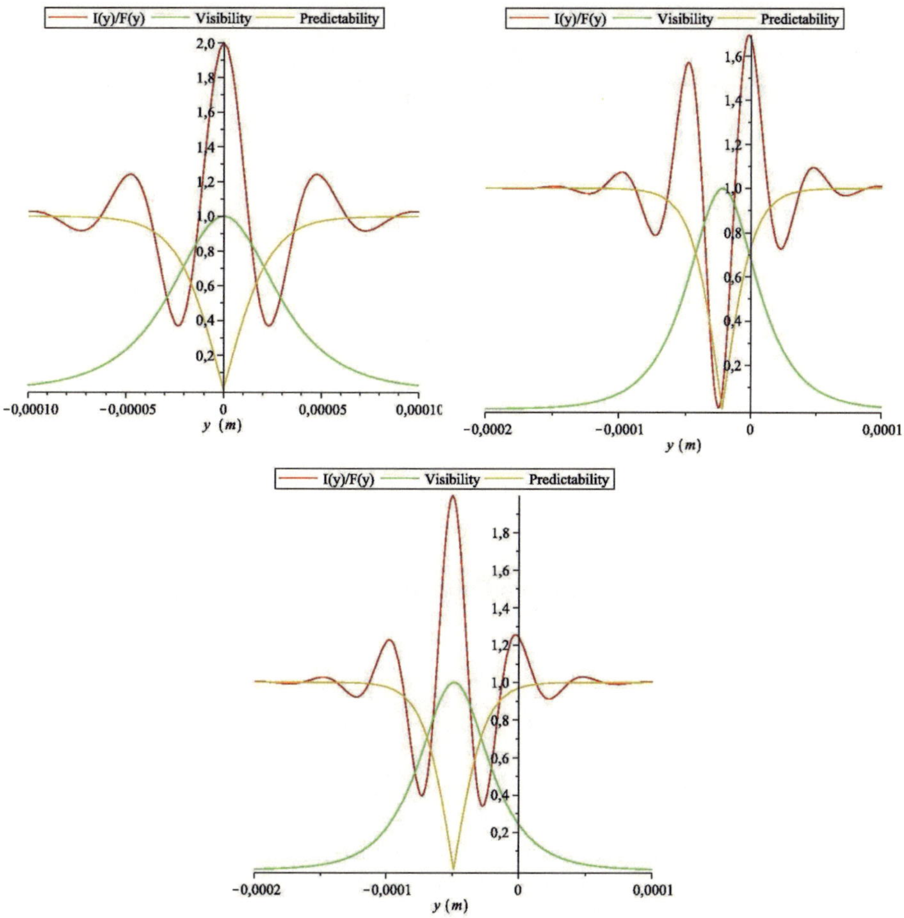

Figure 4. Intensity, visibility and predictability for $p = 1$, $p = 0.15$ and $p = 0.015$, respectively.

$$I(y) = F(y)\left[1 + \frac{\cos(\beta y)}{\cosh(\alpha y)}\right], \tag{43}$$

$$\mathcal{V}(y) = \frac{1}{\cosh(\alpha y)}, \tag{44}$$

and

$$\mathcal{P}(y) = \left|\tanh(\alpha y)\right|. \tag{45}$$

The peaks of intensity and visibility occurs at $y = 0$ where predictability is minimum. In terms of the complementarity principle we have the minimum of knowledge about from which slit the particle cames from. Note also that we have intermediate situations where we have interference fringes with $0 < \mathcal{P} < 1$ and $0 < \mathcal{V} < 1$, signalizing the existence of situations where the wave like property does not preclude the particle like property.

An important quantity is the ratio R between α and β,

$$R = \frac{\alpha}{\beta} = \frac{\hbar}{2\sigma_{xp}}. \quad (46)$$

It can be used to estimate the effective number of fringes (contrast) ν before the visibility decreases by a factor of $1/e$ [26]

$$\nu = \frac{0.264}{R}, \quad (47)$$

for instance, $R = 1$ for $\nu = 0.264$.

It is also related to the Gouy phase. From Equations (20) and (46) we get

$$\mu(T) = -\frac{1}{2}\arctan\left(\frac{1}{R}\right). \quad (48)$$

In Fig. 5 we plot μ against the slit width σ_0. We immediately verify that for σ_0 between 0 and $0.02\mu m$ the phase varies abruptly from -0.05 to $\frac{\pi}{4}$ rad. For values of the Gouy phase within this range we associated a noticeable contrast in the intensity curves as shown in Fig. 5. When the Gouy phase tends to zero the interference fringes tends to disappear because the geometrical optical regime is attained.

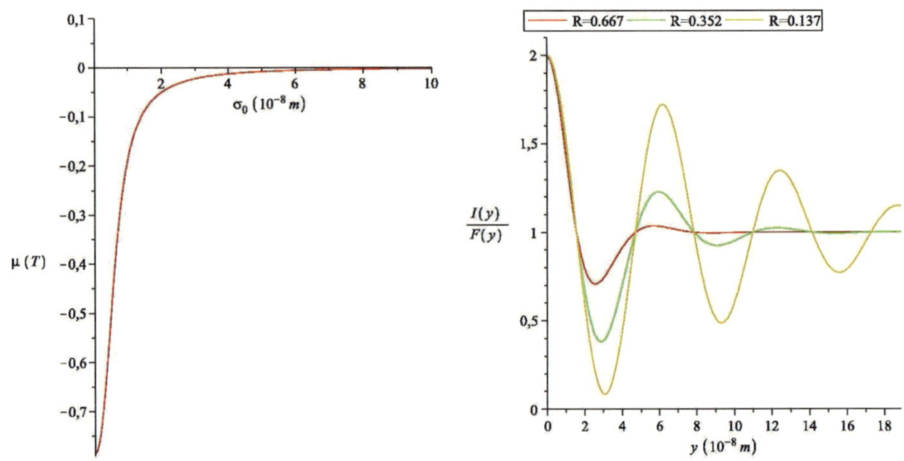

Figure 5. Gouy phase and intensity contrast according Gouy phase.

http://dx.doi.org/10.5772/53651

5. Conclusion

Our results show how partial knowledge of path affects the interference pattern. The position of the peaks of intensity and visibility are affected by the probabilities of the particles cross the slit B (P_B) or C (P_C). When $P_B = P_C$ these peaks are centered in the position $y_{max} = 0$ and when $P_B \neq P_C$ they are displaced from this position. In positions $y \neq y_{max}$ we have the situations where $\mathcal{V}$ and $\mathcal{P}$ are both different of zero, i.e., the wave like property not exclude the particle like property. The crucial point here is the connection between the complementarity principle and the Gouy phase of matter waves. We see that this phase appears in the number of interference fringes exhibiting the wave behavior of the matter. We see that the number of fringes tends to disappear when the Gouy phase tends to zero.

Acknowledgments

This work was in part supported by the Brazilian agencies CNPq, Capes, Fapepi and Fapemig.

Author details

Luis A. Cabral[1], Irismar G. da Paz[2],
José G. G. de Oliveira Júnior[3], Saulo Moreira[4],
Marcos D. R. Sampaio[4] and Maria C. Nemes[4]

1 Departamento de Física, Universidade Federal do Tocantins, Brazil
2 Departamento de Física, Universidade Federal do Piauí, Brazil
3 Departamento de Ciências Exatas e Tecnológicas, Universidade Estadual de Santa Cruz, Brazil
4 Departamento de Física, Universidade Federal de Minas Gerais, Brazil

References

[1] Gouy, L. G. (1890). Sur une propriete nouvelle des ondes lumineuses, *Acad. Sci. Paris* 110, 1251.

[2] Gouy, L. G. (1891). Sur la propagation anomale des ondes, *Ann. Chim. Phip. Ser. 6* 24, 145.

[3] Berry, M. V. (1984). Quantal phase factors accompanying adiabatic changes, *Proc. R. Soc. London A* 392, 45.

[4] Simon, R., and Mukunda, N. (1993). Bargman Invariant and the Geometry of the Gouy Effect, *Phys. Rev. Lett.* 70, 880.

[5] Feng, S., and Winful, H. G. (2001). Physical origin of the Gouy phase shift, *Opt. Lett.* 26, 485.

[6] Yang, J., and Winful, H. G. (2006) Generalized eikonal treatment of the Gouy phase shift, *Opt. Lett.* 31, 104.

[7] da Paz, I. G. (2006). *Matter Waves and Paraxial Propagation of Light*, Department of Physics, UFMG, MG, Brazil.

[8] da Paz, I. G. (2011). *Gouy phase in matter waves: from pure and mixed Gaussian states*, Department of Physics, UFMG, MG, Brazil.

[9] da Paz, I. G., Nemes, M. C., and Peixoto de Faria, J. G. (2011). *Electromagnetic Waves - Chapter four - Edited by Vitaliy Zhurbenko*, InTech, Janeza Trdine 9, 51000 Rijeka, Croatia.

[10] da Paz, I. G., Monken, C. H., Padua, S., Nemes, M. C., and Peixoto de Faria, J. G. (2010). Indirect evidence for the Gouy phase for matter waves, *Phys. Lett. A* 374, 1660.

[11] da Paz, I. G., Nemes, M. C., and Peixoto de Faria, J. G. (2007). Gouy phase and matter waves, *J. Phys.: Conference Series* 84, 012016.

[12] da Paz, I. G., Saldanha, P. L., Nemes, M. C., and Peixoto de Faria, J. G. (2011). Experimental proposal for measuring the Gouy phase of matter waves, *New Journal of Physics* 13, 125005.

[13] Lindner, F., Paulus, G. G., Walther, H., Batuska, A., Goulielmakis, E., Lezius, M., and Krausz, F. (2004). Gouy Phase Shift for Few-Cycle Laser Pulses, *Phys. Rev. Lett.* 92, 113001.

[14] Balcou, P., and LŠHuillier, A. (1993). Phase-matching effects in strong-field harmonic generation, *Phys. Rev. A* 47, 1447.

[15] Lewenstein, M., Salières, P., and LŠHuillier, A. (1995). Phase of the atomic polarization in high-order harmonic generation, *Phys. Rev. A* 52, 4747.

[16] Lindner, F., Stremme, W., Schätzel, M. G., Grasbon, F., Paulus, G. G., Walther, H., Hartmann, R.,and Strüder, L. (2003). High-order harmonic generation at a repetition rate of 100 kHz, *Phys. Rev. A* 68, 013814.

[17] Debye, P. (1905). *Ann. Phys.* 30, 755.

[18] Rubinowicz, A. (1938). On the Anomalous Propagation of Phase in the Focus, *Phys. Rev.* 54, 931.

[19] Linfoot, E. H. (1956). *Proc. Phys. Soc. London* 69, 823.

[20] Visser, T. D., and Wolf, E. (2010). The origin of the Gouy phase anomaly and its generalization to astigmatic wavefields, *Opt. Comm.* 283, 3371.

[21] Boyd, R. W. (1980). Intuitive explanation of the phase anomaly of focused light beams, *J. Opt. Soc. Am.* 70, 877.

[22] Hariharan, P., and Robinson, P. A. (1996). The Gouy phase shift as a geometrical quantum effect, *J. Mod. Opt.* 43, 219.

[23] Born, M., and Wolf, E. (1999). *Principles of Optics*, Cambridge University Press, Cambridge.

[24] Mandel, L., and Wolf, E. (1995). *Optical Coherence and Quantum Optics*, Cambridge, New York.

[25] Saleh, B. E. A., and Teich, M. C. (1991). *Fundamentals of Photonics*, John Wiley et Sons, New York.

[26] Bramon, A., Garbarino.,and Hiesmayr, B. C. (2004). Quantitative complementarity in two-path interferometry, *Phys. Rev. A* 69, 022112.

[27] Viale, A., Vicari, M., and Zanghì, N. (2003). Analysis of the loss of coherence in interferometry with macromolecules, *Phys. Rev. A* 68, 063610.

[28] Schrödinger, E. (1930). Zum Heisenbergschen Unsch"arfeprinzip,*Proceedings of the Prussian Academy of Sciences XIX* pp. 296Ű303.

[29] Stoler, D. (1981). Operator methods in physical optics, *J. Opt. Soc. Am.* 71, 334.

[30] Bohr, N. (1928). The quantum postulate and the recent development of atomic theory, *Nature* 121, 580.

[31] Wootters, W. K., and Zurek, W. K. (1979). Complementarity in the double-slit experiment: Quantum nonseparability and a quantitative statement of Bohr's principle, *Phys. Rev. D* 19, 473.

[32] Greenberger, D. M., and Yasin, A. (1988). Simultaneous wave and particle knowledge in a neutron interferometer, *Phys. Lett. A* 128, 391.

[33] Mandel, L. (1991). Coherence and indistinguishability, *Opt. Lett.* 16, 1882.

[34] Englert, B. G. (1996). Fringe Visibility and Which-Way Information: An Inequality, *Phys. Rev. Lett.* 77, 2154.

[35] Glionna, G., Blin, A. H., Nemes, M. C., Sampaio, M., and Toledo Piza, A. F. R. (2008). Dispersion and uncertainty in multislit matter wave diffraction, *Physica A* 387, 1485.

[36] Nairz, O., Arndt, M., and Zeilinger, A. (2002). Experimental verification of the Heisenberg uncertainty principle for fullerene molecules, *Phys. Rev. A* 65, 032109.

[37] Scully, M. O., and Zubairy, M. S. (1997). *Quantum Optics*, Cambridge University Press, Cambridge.